Research Notes in Mathematics

ubmission of proposals for consideration
uggestions for publication, in the form of outlines and representative
amples, are invited by the editorial board for assessment. Intending
uthors should contact either the main editor or another member of the
ditorial board, citing the relevant AMS subject classifications. Refereeing
by members of the board and other mathematical authorities in the
opic concerned, located throughout the world.

reparation of accepted manuscripts
n acceptance of a proposal, the publisher will supply full instructions
r the preparation of manuscripts in a form suitable for direct photo-
thographic reproduction. Specially printed grid sheets are provided
nd a contribution is offered by the publisher towards the cost of typing.

lustrations should be prepared by the authors, ready for direct
production without further improvement. The use of hand-drawn
mbols should be avoided wherever possible, in order to maintain
aximum clarity of the text.

he publisher will be pleased to give any guidance necessary during the
eparation of a typescript, and will be happy to answer any queries.

nportant note
order to avoid later retyping, intending authors are strongly urged
ot to begin final preparation of a typescript before receiving the
ublisher's guidelines and special paper. In this way it is hoped to
eserve the uniform appearance of the series.

Titles in this series

1 Improperly posed boundary value problems
A Carasso and A P Stone

2 Lie algebras generated by finite dimensional ideals
I N Stewart

3 Bifurcation problems in nonlinear elasticity
R W Dickey

4 Partial differential equations in the complex domain
D L Colton

5 Quasilinear hyperbolic systems and waves
A Jeffrey

6 Solution of boundary value problems by the method of integral operators
D L Colton

7 Taylor expansions and catastrophes
T Poston and I N Stewart

8 Function theoretic methods in differential equations
R P Gilbert and R J Weinacht

9 Differential topology with a view to applications
D R J Chillingworth

10 Characteristic classes of foliations
H V Pittie

11 Stochastic integration and generalized martingales
A U Kussmaul

12 Zeta-functions: An introduction to algebraic geometry
A D Thomas

13 Explicit *a priori* inequalities with applications to boundary value problems
V G Sigillito

14 Nonlinear diffusion
W E Fitzgibbon III and H F Walker

15 Unsolved problems concerning lattice points
J Hammer

16 Edge-colourings of graphs
S Fiorini and R J Wilson

17 Nonlinear analysis and mechanics: Heriot-Watt Symposium Volume I
R J Knops

18 Actions of fine abelian groups
C Kosniowski

19 Closed graph theorems and webbed spaces
M De Wilde

20 Singular perturbation techniques applied to integro-differential equations
H Grabmüller

21 Retarded functional differential equations: A global point of view
S E A Mohammed

22 Multiparameter spectral theory in Hilbert space
B D Sleeman

24 Mathematical modelling techniques
R Aris

25 Singular points of smooth mappings
C G Gibson

26 Nonlinear evolution equations solvable by the spectral transform
F Calogero

27 Nonlinear analysis and mechanics: Heriot-Watt Symposium Volume II
R J Knops

28 Constructive functional analysis
D S Bridges

29 Elongational flows: Aspects of the behaviour of model elasticoviscous fluids
C J S Petrie

30 Nonlinear analysis and mechanics: Heriot-Watt Symposium Volume III
R J Knops

31 Fractional calculus and integral transforms of generalized functions
A C McBride

32 Complex manifold techniques in theoretical physics
D E Lerner and P D Sommers

33 Hilbert's third problem: scissors congruence
C-H Sah

34 Graph theory and combinatorics
R J Wilson

35 The Tricomi equation with applications to the theory of plane transonic flow
A R Manwell

36 Abstract differential equations
S D Zaidman

37 Advances in twistor theory
L P Hughston and R S Ward

38 Operator theory and functional analysis
I Erdelyi

39 Nonlinear analysis and mechanics: Heriot-Watt Symposium Volume IV
R J Knops

40 Singular systems of differential equations
S L Campbell

41 N-dimensional crystallography
R L E Schwarzenberger

42 Nonlinear partial differential equations in physical problems
D Graffi

43 Shifts and periodicity for right invertible operators
D Przeworska-Rolewicz

44 Rings with chain conditions
A W Chatters and C R Hajarnavis

45 Moduli, deformations and classifications of compact complex manifolds
D Sundararaman

46 Nonlinear problems of analysis in geometry and mechanics
M Atteia, D Bancel and I Gumowski

47 Algorithmic methods in optimal control
W A Gruver and E Sachs

48 Abstract Cauchy problems and functional differential equations
F Kappel and W Schappacher

49 Sequence spaces
W H Ruckle

50 Recent contributions to nonlinear partial differential equations
H Berestycki and H Brezis

51 Subnormal operators
J B Conway

52 Wave propagation in viscoelastic media
F Mainardi

53 Nonlinear partial differential equations and their applications: Collège de France Seminar. Volume I
H Brezis and J L Lions

54 Geometry of Coxeter groups
H Hiller

55 Cusps of Gauss mappings
T Banchoff, T Gaffney and C McCrory

56 An approach to algebraic K-theory
A J Berrick
57 Convex analysis and optimization
J-P Aubin and R B Vintner
58 Convex analysis with applications in the differentiation of convex functions
J R Giles
59 Weak and variational methods for moving boundary problems
C M Elliott and J R Ockendon
60 Nonlinear partial differential equations and their applications: Collège de France Seminar. Volume II
H Brezis and J L Lions
61 Singular systems of differential equations II
S L Campbell
62 Rates of convergence in the central limit theorem
Peter Hall
63 Solution of differential equations by means of one-parameter groups
J M Hill
64 Hankel operators on Hilbert space
S C Power
65 Schrödinger-type operators with continuous spectra
M S P Eastham and H Kalf
66 Recent applications of generalized inverses
S L Campbell
67 Riesz and Fredholm theory in Banach algebra
B A Barnes, G J Murphy, M R F Smyth and T T West
68 Evolution equations and their applications
F Kappel and W Schappacher
69 Generalized solutions of Hamilton-Jacobi equations
P L Lions
70 Nonlinear partial differential equations and their applications: Collège de France Seminar. Volume III
H Brezis and J L Lions
71 Spectral theory and wave operators for the Schrödinger equation
A M Berthier
72 Approximation of Hilbert space operators I
D A Herrero
73 Vector valued Nevanlinna Theory
H J W Ziegler
74 Instability, nonexistence and weighted energy methods in fluid dynamics and related theories
B Straughan
75 Local bifurcation and symmetry
A Vanderbauwhede
76 Clifford analysis
F Brackx, R Delanghe and F Sommen
77 Nonlinear equivalence, reduction of PDEs to ODEs and fast convergent numerical methods
E E Rosinger
78 Free boundary problems, theory and applications. Volume I
A Fasano and M Primicerio
79 Free boundary problems, theory and applications. Volume II
A Fasano and M Primicerio
80 Symplectic geometry
A Crumeyrolle and J Grifone
81 An algorithmic analysis of a communication model with retransmission of flawed messages
D M Lucantoni
82 Geometric games and their applications
W H Ruckle
83 Additive groups of rings
S Feigelstock
84 Nonlinear partial differential equations and their applications: Collège de France Seminar. Volume IV
H Brezis and J L Lions
85 Multiplicative functionals on topological algebras
T Husain
86 Hamilton-Jacobi equations in Hilbert spaces
V Barbu and G Da Prato
87 Harmonic maps with symmetry, harmonic morphisms and deformations of metrics
P Baird
88 Similarity solutions of nonlinear partial differential equations
L Dresner
89 Contributions to nonlinear partial differential equations
C Bardos, A Damlamian, J I Díaz and J Hernández
90 Banach and Hilbert spaces of vector-valued functions
J Burbea and P Masani
91 Control and observation of neutral systems
D Salamon
92 Banach bundles, Banach modules and automorphisms of C*-algebras
M J Dupré and R M Gillette
93 Nonlinear partial differential equations and their applications: Collège de France Seminar. Volume V
H Brezis and J L Lions
94 Computer algebra in applied mathematics: an introduction to MACSYMA
R H Rand
95 Advances in nonlinear waves. Volume I
L Debnath
96 FC-groups
M J Tomkinson
97 Topics in relaxation and ellipsoidal methods
M Akgül
98 Analogue of the group algebra for topological semigroups
H Dzinotyiweyi
99 Stochastic functional differential equations
S E A Mohammed
100 Optimal control of variational inequalities
V Barbu
101 Partial differential equations and dynamical systems
W E Fitzgibbon III
102 Approximation of Hilbert space operators. Volume II
C Apostol, L A Fialkow, D A Herrero and D Voiculescu
103 Nondiscrete induction and iterative processes
V Ptak and F-A Potra
104 Analytic functions – growth aspects
O P Juneja and G P Kapoor
105 Theory of Tikhonov regularization for Fredholm equations of the first kind
C W Groetsch

106 Nonlinear partial differential equations and free boundaries. Volume I
J I Díaz

107 Tight and taut immersions of manifolds
T E Cecil and P J Ryan

108 A layering method for viscous, incompressible L_p flows occupying R^n
A Douglis and E B Fabes

109 Nonlinear partial differential equations and their applications: Collège de France Seminar. Volume VI
H Brezis and J L Lions

110 Finite generalized quadrangles
S E Payne and J A Thas

111 Advances in nonlinear waves. Volume II
L Debnath

112 Topics in several complex variables
E Ramírez de Arellano and D Sundararaman

113 Differential equations, flow invariance and applications
N H Pavel

114 Geometrical combinatorics
F C Holroyd and R J Wilson

115 Generators of strongly continuous semigroups
J A van Casteren

116 Growth of algebras and Gelfand–Kirillov dimension
G R Krause and T H Lenagan

117 Theory of bases and cones
P K Kamthan and M Gupta

118 Linear groups and permutations
A R Camina and E A Whelan

119 General Wiener–Hopf factorization methods
F-O Speck

120 Free boundary problems: applications and theory, Volume III
A Bossavit, A Damlamian and M Fremond

121 Free boundary problems: applications and theory, Volume IV
A Bossavit, A Damlamian and M Fremond

122 Nonlinear partial differential equations and their applications: Collège de France Seminar. Volume VII
H Brezis and J. L. Lions

123 Geometric methods in operator algebras
H Araki and E G Effros

124 Infinite dimensional analysis–stochastic processes
S Albeverio

Tight and taut immersions of manifolds

T E Cecil & P J Ryan

College of the Holy Cross, Worcester, Massachusetts/
McMaster University

Tight and taut immersions of manifolds

Pitman Advanced Publishing Program
BOSTON · LONDON · MELBOURNE

PITMAN PUBLISHING INC
1020 Plain Street, Marshfield, Massachusetts 02050

PITMAN PUBLISHING LIMITED
128 Long Acre, London WC2E 9AN

Associated Companies
Pitman Publishing Pty Ltd, Melbourne
Pitman Publishing New Zealand Ltd, Wellington
Copp Clark Pitman, Toronto

First published 1985

AMS Subject Classifications: (main) 53C42, 53C40
(subsidiary) 53B25, 57R42, 52A20

ISSN 0743-0337

Library of Congress Cataloging in Publication Data

Cecil, T. E. (Thomas E.), 1945–
Tight and taut immersions of manifolds.

Bibliography: p.
Includes index.
1. Immersions (Mathematics) 2. Submanifolds.
I. Ryan, P. J. II. Title.
QA649.C43 1985 516.3'62 84-25378
ISBN 0-273-08631-6

British Library Cataloguing in Publication Data

Cecil, T. E.
Tight and taut immersions of manifolds.—
(Research Notes in mathematics, ISSN 0743-0337; 107)
1. Differentiable manifolds.
I. Title II. Ryan, P. J. III. Series
516.3'6 QA614.3

ISBN 0-273-08631-6

Reproduced and printed by photolithography
in Great Britain by Biddles Ltd, Guildford

To my wife,
Patsy
and my parents,
Thomas and Marian Cecil

Thomas E. Cecil

To my wife,
Ellen
and in memory of my parents,
Cecil and Kathleen Ryan

Patrick J. Ryan

Contents

Preface

This book is intended to provide a relatively complete and self-contained survey of the work done in the past twenty-five years in the areas of tight (minimal total absolute curvature) and taut immersions of manifolds into Euclidean space, and the related field of isoparametric hypersurfaces in spheres. It could be used for a second course in differential geometry, for a faculty seminar or as a reference. The presentation is aimed at a reader who has completed a one-year graduate course in differential geometry, and who knows the elements of homology theory and critical point theory.

There are three chapters covering, respectively, tight immersions, taut immersions, and isoparametric hypersurfaces. Although the chapters are closely related, we have tried to make each as self-contained as possible. The introduction to each chapter is a survey of the known results in that field, and the three introductions taken together provide a fairly complete overview of the general area. The definitions and theorems are numbered within each chapter. If a theorem from a different chapter is cited, then the chapter is listed along with the number of the theorem.

Chapter 1 on tight immersions begins with a recollection of the facts from critical point theory and the theory of convex sets required to formulate the definition of a tight immersion and prove several basic results. In Section 5, we introduce the two-piece property (TPP), relate it to tightness and establish Kuiper's bound on the codimension of a substantial TPP immersion. Of particular importance is Section 7, which contains a thorough survey of the known results for tight immersions of surfaces into Euclidean m-dimensional space E^m. The proof of the non-existence of tight immersions of the projective plane and Klein bottle into E^3 is an excellent application of the method of analysis of top-sets. Section 8 contains Kuiper's proof of the Chern-Lashof theorem for topological immersions, i.e., a tight topological immersion of the n-sphere S^n into E^m must actually embed S^n as a convex hypersurface in a subspace E^{n+1} of E^m. The proof also involves top-set analysis and differs significantly from the original proof of Chern and

Lashof for smooth immersions. The chapter concludes with a presentation of the Veronese embeddings of projective spaces and the classification of tight immersions of maximal codimension due to Kuiper, Little and Pohl.

Chapter 2 treats taut immersions of manifolds into Euclidean space and related topics such as focal sets, shape operators of tubes, principal foliations, totally focal embeddings and Dupin hypersurfaces, particularly the cyclides of Dupin. Section 5 contains a detailed discussion of the classical cyclides in E^3, including fourteen computer graphics illustrations produced by Thomas F. Banchoff. The next section provides a thorough introduction to the study of Dupin hypersurfaces, extending to the current frontiers. The chapter also includes the classification of taut surfaces in Euclidean space, the known results for taut hypersurfaces in E^m, and a survey of tight and taut immersions into hyperbolic space. Section 9 contains a complete presentation of the characterization of isoparametric hypersurfaces in E^m and S^m as totally focal hypersurfaces, due to Carter and West.

Chapter 3 contains an in-depth treatment of the fundamental results of H.F. Münzner [1] on isoparametric hypersurfaces in spheres. Isoparametric hypersurfaces are shown to be taut and totally focal, and several key examples are presented in detail and from different points of view. With this background, the reader can begin to study the many important recent papers in the subject which are noted in the introduction to the chapter.

We wish to acknowledge the published work of the many mathematicians who have labored in this field. In particular, we are indebted to the two survey articles on tight immersions of N.H. Kuiper [8] and [13] which guided our presentation of Chapter 1 and the paper of H.F. Münzner mentioned above which forms the basis for Chapter 3. We have also drawn extensively from the papers of S.S. Tai [1] for Section 9 of Chapter 1, J. Little and W. Pohl [1] for Section 10 of Chapter 1, S. Carter and A. West [2]-[4] for Section 9 of Chapter 2, and K. Nomizu [2] for various parts of Chapter 3.

We also want to thank several mathematicians for their personal contributions. Katsumi Nomizu, who served as our doctoral thesis advisor, was a formative influence in our development as mathematicians, and has remained a constant source of inspiration and encouragement throughout the years. Thomas Banchoff introduced us to tight and taut immersions and has given us several key insights. Nicolaas Kuiper has been very helpful to us in recent years, answering many questions both in person and in correspondence.

Lawrence Conlon, Paul Schweitzer, S.J. and Patrick Shanahan have clarified several topological arguments for us while Sheila Carter, Martin Magid, Alan West, Andrew Whitman, S.J. and Thomas Willmore have all contributed to our understanding of the subject through lectures and conversations.

This book grew out of lectures given to the Clavius group and others in the summers of 1978-82 at the Institut des Hautes Études Scientifiques, the Institute for Advanced Study, the University of Notre Dame, the College of the Holy Cross and the I.H.E.S. again. We wish to thank our fellow members of Clavius for their enthusiasm and support of these lectures. We also acknowledge with gratitude the hospitality and stimulating atmosphere which we have experienced at the institutions mentioned above.

We would like to acknowledge the support of fellowships and grants from the College of the Holy Cross, Indiana University, McMaster University and the National Science and Engineering Research Council of Canada.

We thank Joy Bousquet, Rosemary Comerford, and the McMaster University Faculty of Science word-processing group for their assistance in preparation of the manuscript, and Bridget Buckley and William Roberts of Pitman Publishing for their support and cooperation.

Finally, we are most deeply grateful to our wives, Patsy and Ellen, and our families for their patience, generosity and constant support of this project.

Thomas E. Cecil

Department of Mathematics
College of the Holy Cross
Worcester, Ma. 01610
U.S.A.

Patrick J. Ryan

Department of Mathematical Sciences
McMaster University
Hamilton, Ontario L8S 4K1
Canada

Tight immersions

0. INTRODUCTION.

We begin with some examples to motivate the subject, and follow with a survey of the major results, particularly those which will be covered in these notes.

Let $\gamma(s)$, $a \leq s \leq b$ be a closed plane curve parametrized by arc-length. Let

$$t(s) = \cos\theta(s)\, e_1 + \sin\theta(s)\, e_2$$

be the unit tangent vector to $\gamma(s)$, where e_1, e_2 are the standard orthonormal basis for E^2, and $\theta(s)$ is a well-defined differentiable function once $\theta(a)$ is chosen. The curvature of γ is defined by $\kappa(s) = \theta'(s)$. The <u>total curvature of γ</u> is

$$T(\gamma) = \int_a^b \kappa(s)ds = \theta(b) - \theta(a).$$

Since γ is closed, $T(\gamma) = 2\pi n$ for some $n \in Z$. One can prove that if γ is a simple closed curve, $T(\gamma) = \pm 2\pi$, with the sign depending on the orientation of the curve.

A second measure of the curvature of γ is the <u>total absolute curvature</u> of γ,

$$\tau(\gamma) = \int_a^b |\kappa(s)|\, ds.$$

This measures the total arc-length of the curve $t(s)$ on the unit circle, not merely the difference between $\theta(a)$ and $\theta(b)$. The minimum value for $\tau(\gamma)$ is 2π, and this minimum is attained precisely when γ is a convex curve. The curve γ is called <u>tight</u> if $\tau(\gamma) = 2\pi$, i.e., $\tau(\gamma)$ is the minimum value possible.

Similarly, suppose $f{:}M \to E^3$ is a smooth immersion of a smooth orientable compact surface. Let ξ be a global field of unit normal vectors to $f(M)$. The Gauss map $\nu: M \to S^2$ is defined by letting $\nu(x)$ be the vector at the origin in E^3 which is parallel to $\xi(x)$. The Gaussian curvature $K(x)$ is the determinant of the Jacobian ν_* at x. The <u>total curvature</u> of the immersion is

$$T(M,f) = \int_M K \, dA,$$

where dA is the volume element on M. The celebrated Gauss-Bonnet theorem states that $T(M,f)$ does not depend on the immersion f at all, but only on the topology of M, i.e., for any f,

$$T(M,f) = 2\pi\chi(M),$$

where $\chi(M)$ is the Euler characteristic of M. Thus, if one distorts a given immersion of M introducing more positive curvature in a certain neighborhood, this must be offset by the introduction of more negative curvature in some other region of M.

As in the case of curves, one can consider also the <u>total absolute curvature</u> of the immersion.

$$\tau(M,f) = \int_M |K| \, dA.$$

$\tau(M,f)$ is the actual total area covered by the image of the Gauss map. Using Morse theory, it is not hard to prove that

$$\tau(M,f) \geq 2\pi(4 - \chi(M)).$$

Thus, given the topology of M, one can ask if a certain immersion f attains the minimum possible value for $\tau(M,f)$. If so, the immersion is called <u>tight</u>.

Corresponding to the result for curves, a special case of a theorem of Chern and Lashof states that a tight immersion of S^2 into E^3 must be an embedding of S^2 as a convex surface.

The problem can actually be posed in the general setting of an immersion $f: M^n \to E^m$ for arbitrary n,m. Generalizing the Gaussian curvature, the

Lipschitz-Killing curvature is defined on the bundle of unit normal vectors (see Chern-Lashof [1]). The absolute value of this curvature is then integrated over the bundle of unit normals to obtain the total absolute curvature $\tau(M,f)$. As in the case of surfaces in E^3, $\tau(M,f)$ represents the total area covered by the Gauss map from the bundle of unit normals to the unit sphere S^{m-1} in E^m.

The simplest case of this more general definition is that of a space curve $f:S^1 \to E^3$. In this case,

$$\tau(S^1,f) = \int_{S^1} |\kappa(s)|\, ds,$$

where $\kappa(s)$ is the ordinary curvature of a space curve. In 1929, Fenchel [1] proved that the minimum value of 2π is attained for $\tau(S^1,f)$ if and only if f embeds S^1 as a convex plane curve. (See Chern [2] for a proof of Fenchel's theorem). The same conclusion holds for curves in E^m (Borsuk [1]). When f is knotted, $\tau(f)$ must be greater than 4π (Fary [1], Milnor [1,2]). Analogous results for knotted surfaces have recently been obtained by Kuiper and Meeks [1] (see Remark 7.30).

It is often convenient to normalize the volume of the unit sphere in E^m to be 1, as we shall do. The first tightness result in higher dimension was that of Chern and Lashof [1] who proved that if $f:S^n \to E^m$ has the minimum total absolute curvature possible, namely 2, then f embeds S^n as a convex hypersurface in a Euclidean space $E^{n+1} \subset E^m$. This is known as the Chern-Lashof theorem (see Section 8).

An important step in the development of the theory was the reformulation of the problem in terms of critical point theory by Kuiper [1]. He showed that for a given compact manifold M, the infimum of $\tau(M,f)$ over all immersions of M into all Euclidean spaces is the Morse number γ of M, i.e., γ is the minimum number of critical points which any non-degenerate real-valued function on M can possess. Moreover, this lower bound is attained if and only if every non-degenerate linear height function in E^m has γ critical points on M. Such an immersion f is said to have <u>minimum total absolute curvature</u>.

For a manifold M such that γ is equal to the sum of the Betti numbers of M with respect to some field F, Kuiper [6,13] showed that minimal total absolute curvature is equivalent to requiring that for every closed half-space h in E^m, the induced homomorphism

$$H_*(f^{-1}h) \to H_*(M) \text{ (coefficients in F)}$$

is injective. This definition has the advantage that it makes sense even if f is only assumed to be a continuous map of a compact topological space into E^m. The most powerful approach to classifying smooth tight immersions has been Kuiper's method of analyzing top-sets, i.e. sets on which height functions, particularly degenerate ones, attain their absolute maxima. The formulation of tightness given above makes it possible to state and prove the key result that top-sets of tight immersions must be tight sets, even though they are not necessarily smooth manifolds (see Theorem 7.11).

Kuiper [5] produced tight smooth embeddings of all orientable surfaces into E^3 and tight smooth immersions into E^3 of all non-orientable surfaces with Euler characteristic $\chi < -1$. At the same time, he showed [4] that there do not exist tight smooth immersions of the projective plane and the Klein bottle into E^3. Essentially the same proof shows that there does not exist a tight continuous immersion of the projective plane into E^3, but the non-existence of a tight continuous immersion of the Klein bottle is more delicate and has only recently been established by Kuiper [16]. The existence of a tight (continuous or smooth) immersion in the case $\chi = -1$ is still a major open question. These results are proven in Section 7, which is meant to serve as an introduction to Kuiper's method of top-set analysis. In Section 8, we give his proof of the Chern-Lashof theorem which only assumes that f is a tight continuous immersion of S^n into E^m.

In the same spirit as the top-set approach, Banchoff [4] introduced the two-piece property (TPP) for subsets of E^m, i.e. a subset M has the TPP if every hyperplane π in E^m separates M into at most two pieces. For a smooth immersion of a compact manifold M, this is equivalent to requiring that every non-degenerate linear height function have exactly one local maximum and one local minimum on M. Hence, it is weaker than tightness, except when M is two-dimensional in which case the two concepts are equivalent. In addition, Banchoff [1,2,6] and Kühnel [2]-[5] have proven several interesting results for piece-wise linear tight immersions (see Example 5.21).

As the Chern-Lashof theorem shows, tightness can force a submanifold to be situated "economically" in the ambient space. In particular, a tight sphere S^n must lie in an (n+1)-plane. Although tight submanifolds with more complicated topology need not satisfy such a severe restriction, Kuiper [1] showed that a TPP immersion $f:M^n \to E^m$ must lie in a hyperplane if $m > n(n+3)/2$ (see Theorem 5.18). Moreover, if $m = n(n+3)/2$, and f(M) does not lie in any hyperplane, then f must be a Veronese embedding of the real projective space P^n (see Kuiper [6] and Little-Pohl [1]). These results are treated in Sections 9 and 10. The Veronese embeddings belong to an extensive class of submanifolds, the standard embeddings of symmetric R-spaces, which were shown to be tight by Takeuchi and Kobayashi [1] (see Section 9 for more discussion of this.)

In Chapter 2, we deal with taut immersions. A smooth immersion $f:M^n \to E^m$ is called taut if every non-degenerate Euclidean distance function L_p has the minimum number of critical points. For n = 2, this is equivalent to the spherical two-piece property of Banchoff [3] who showed that a taut compact surface in E^3 must be a round sphere or a cyclide of Dupin (image under stereographic projection of a product torus in the 3-sphere). Tautness is a stronger condition than tightness. For example, a taut n-sphere must be a round sphere, not merely convex (Nomizu-Rodriguez [1] and Carter-West [1]). On the other hand, the tautness condition can be extended to the non-compact case and a taut non-compact surface in E^3 must be a plane, circular cylinder or parabolic cyclide (see Cecil [3]). The results of Banchoff and Cecil, together with Kuiper's work on tight surfaces of maximal codimension, enable one to give a complete list of all taut surfaces in Euclidean spaces. Cecil and Ryan [2] generalized Banchoff's result to show that a taut immersion of $S^k \times S^{n-k}$ into E^{n+1} must be an n-dimensional cyclide. The proof involves a careful analysis of the behavior of the focal set. In particular, the cyclides were characterized as the only complete hypersurfaces in Euclidean space whose focal set consists of two submanifolds of codimension greater than one (Cecil-Ryan [2], [5]). From this point of view, the cyclides are examples of so-called Dupin hypersurfaces, which have recently been studied by Pinkall [1]-[3] and Thorbergsson [1] (see Chapter 2 for more detail).

A product of spheres $S^k \times S^{n-k}$ is well-known to be an isoparametric hypersurface in S^{n+1}, i.e., its principal curvatures are constant. In 1979, it was shown that all isoparametric hypersurfaces in the sphere are taut, Cecil-Ryan [6]. The proof is primarily an application of the results of H.F. Münzner [1] on isoparametric hypersurfaces and their focal sets. This is a rich theory which was first studied by E. Cartan and further developed by Nomizu, Takagi and Takahashi, Ozeki and Takeuchi, and most extensively by Münzner alone and later with his collaborators Ferus and Karcher. In Chapter 3, we will develop the major results of Münzner, and prove that all isoparametric hypersurfaces in spheres are taut. From Münzner's work, it is easy to see that isoparametric hypersurfaces satisfy the condition of being totally focal, Carter-West [2], i.e., every distance function is either non-degenerate or has only degenerate critical points. Carter and West [4] proved conversely that every totally focal hypersurface in the sphere is isoparametric.

The reader is referred to the introductions of Chapters 2 and 3 for surveys of those fields and more detailed description of the contents of those chapters. For other surveys of the field of tight immersions, see Kuiper [8, 13] and Willmore [1, 2].

All manifolds and maps are assumed to be smooth unless explicitly stated otherwise. In particular, all immersions are smooth until Section 7 when topological immersions are introduced. Notation generally follows Kobayashi-Nomizu [1]. See also the summary of notation at the end of the book.

1. REVIEW OF CRITICAL POINT THEORY.

We begin by recalling some basic terminology and results from critical point theory (see, for example, Milnor [3, pp. 4-6]). Let M_1, M_2 be smooth manifolds and $f:M_1 \to M_2$ a smooth map. A point $x \in M_1$ is called a critical point of f if the Jacobian f_* at x is singular. If $y \in M_2$ is the image under f of a critical point, then y is called a critical value of f. All other points in M_2 are called regular values of f.

In the special case of a real-valued function ϕ on a manifold M, ϕ has a critical point at $x \in M$ if and only if $\phi_* = 0$ at x. If $(x^1,\ldots,x^n)$ are local coordinates in a neighborhood of x, this means

$$\frac{\partial\phi(x)}{\partial x^1} = \cdots = \frac{\partial\phi(x)}{\partial x^n} = 0.$$

If x is a critical point of ϕ, one examines the Hessian of ϕ which is represented in local coordinates by the symmetric matrix

$$H_x = [\frac{\partial^2 \phi}{\partial x^i \partial x^j}].$$

Definition 1.1: Let x be a critical point of the function $\phi: M \to R$.

(a) x is called a degenerate critical point if rank $H_x < n$, where n = dimension M.

(b) If rank $H_x = n$, x is called a non-degenerate critical point. The index of ϕ at x is equal to the number of negative eigenvalues of H_x.

The behavior of ϕ in a neighborhood of a non-degenerate critical point is determined by the index, as follows.

Lemma 1.2 (Lemma of Morse): Let x be a non-degenerate critical point of $\phi: M \to R$ of index k. Then there is a local coordinate system $(x^1,\ldots,x^n)$ in a neighborhood U with origin at x such that the identity

$$\phi = \phi(x) - (x^1)^2 - \ldots - (x^k)^2 + (x^{k+1})^2 + \ldots + (x^n)^2$$

holds throughout U.

Thus a critical point of index n is a local maximum of ϕ, one of index 0 is a local minimum, and all others are various types of saddle points. Note also that non-degenerate critical points are isolated.

A function ϕ which has only non-degenerate critical point is called a non-degenerate function or a Morse function. If M is compact, then a Morse function ϕ on M can have only a finite number of critical points.

2. THE MORSE INEQUALITIES.

Suppose that ϕ is a Morse function on M such that the set

$$M_r(\phi) = \{x \in M \mid \phi(x) < r\}$$

is compact for all $r \in R$. This is always true, of course, if M is compact. Let $\mu_k(\phi,r)$ be the number of critical points of ϕ of index k on $M_r(\phi)$. For compact M, let $\mu_k(\phi)$ be the number of critical points of ϕ with index k on M, and $\mu(\phi)$ the number of critical points of ϕ, For a field F, let

$$\beta_k(\phi,r,F) = \dim_F(H_k(M_r(\phi));F),$$

the kth F-Betti number of $M_r(\phi)$, and let $\beta_k(M,F)$ be the kth F-Betti number of M. The Morse inequalities (see, for example, Morse-Cairns [1, p. 270]) state that

$$\mu_k(\phi,r) \geq \beta_k(\phi,r,F)$$

for all F, r, k. Let γ be the <u>Morse number of M</u>, for M compact, i.e.,

$$\gamma = \min\{\mu(\phi) \mid \phi \text{ is a Morse function on M}\}.$$

The Morse inequalities imply that

$$\gamma \leq \beta(M;F) = \sum_{k=0}^{n} \beta_k(M;F)$$

for any F. The following reformulation of the condition that the Morse inequalities are actually equalities is quite important in the theory of tight immersions.

<u>Theorem 2.1</u>: <u>Let ϕ be a Morse function on a compact manifold M. For a given field F, $\mu_k(\phi,r) = \beta_k(\phi,r,F)$ for all k, r if and only if the map on homology</u>

$$H_*(M_r(\phi);F) \to H_*(M;F)$$

<u>induced by the inclusion of $M_r(\phi)$ in M is injective for all r.</u>

This result follows immediately from Theorem 2.2 of Morse-Cairns [1, p. 260], and it was first pointed out in the context of tight immersions by Kuiper [6]. A good deal of critical point theory is involved in the proof and we will not give it here, but will briefly mention a few points.

Suppose p is a critical point of ϕ and $\phi(p) = r$. Assume for the sake of simplicity that p is the only critical point at critical level r. A fundamental result of critical point theory (see, for example, Milnor [3, pp. 12-24] or Morse-Cairns [1, pp. 184-202]) states that $M_r(\phi)$ has the homotopy type of $M_r^-(\phi)$ with a k-cell attached, where $M_r^-(\phi)$ consists of all points for which $\phi < r$. Morse and Cairns characterize the effect of attaching this k-cell as follows. Let

$$\Delta\beta_i(r) = \beta_i(M_r(\phi)) - \beta_i(M_r^-(\phi)).$$

Then the $\Delta\beta_i(r)$ are all 0, except that

$\Delta\beta_k(r) = 1$, when the critical point p is of "linking" type,

$\Delta\beta_{k-1}(r) = -1$, when the critical point p is of "non-linking" type.

From this, it is clear that the two conditions in Theorem 2.1 are equivalent and that they hold precisely when every critical point is of linking type. See Morse-Cairns [1, p. 258] for the definitions of linking and non-linking and a proof of the basic result.

3. MINIMAL TOTAL ABSOLUTE CURVATURE IMMERSIONS.

In this section, we discuss the differential geometric formulation of tightness in terms of immersions with minimal possible total absolute curvature, due to Chern and Lashof [1].

Let $f:M^n \to E^m$, $m = n + k$, be a smooth immersion. Let S^{m-1} denote the sphere of unit vectors at the origin in E^m. For $p \in S^{m-1}$, the linear height function $\ell_p:E^m \to R$ is defined by

$$\ell_p(q) = \langle p,q\rangle,$$

where $\langle p,q\rangle$ is the Euclidean inner product. The restriction of ℓ_p to M, defined by $\ell_p(x) = \langle p, f(x)\rangle$ is a real-valued function on M.

We first recall some fundamental terminology for submanifolds of Euclidean space. Since f is an immersion, f is an embedding on a suitably small neighborhood of any point $x \in M$. For local calculations, we thus may identify T_xM and $f_*(T_xM)$. Suppose that $X \in T_xM$ and ξ is a field of unit normal vectors to f(M) in a neighborhood of x. Let D denote Euclidean covariant differentiation in E^m. One has the equation

$$D_X\xi = -A_\xi X + \nabla_X^\perp \xi,$$

where $-A_\xi X$ is the component of $D_X\xi$ tangent to M, and $\nabla_X^\perp\xi$ is the component normal to M. A_ξ is a tensor of type (1,1) on M, which depends only on ξ at the point x, whereas $\nabla^\perp$ is a differential operator depending on ξ in a neighborhood of x. A_ξ is called the <u>shape operator determined by ξ</u>, and $\nabla^\perp$ is called the <u>normal connection</u>.

Let BM denote the bundle of unit normal vectors to M. BM is an (m-1)-dimensional manifold. Let π:BM → M be the natural projection. Recall the Gauss map

(3.1) $$\nu : BM \to S^{m-1},$$

which takes a point $\xi \in T_x^\perp M$ for some $x \in M$, to the vector parallel to ξ at the origin. In the case where M is an oriented hypersurface, BM is just a covering of M with two disjoint sheets. The following result is well-known and is proven by a direct calculation, which we omit.

<u>Theorem 3.1</u>: <u>The nullity of the Gauss map ν at a point $\xi \in BM$ is equal to the nullity of A_ξ. In particular, ξ is a critical point of ν if and only if A_ξ is singular</u>.

We next examine the critical point theory of linear height functions.

<u>Theorem 3.2</u>: <u>Let $f:M \to E^m$ be an immersion and let $p \in S^{m-1}$</u>.

(a) <u>ℓ_p has a critical point at $x \in M$ if and only if p is orthogonal to T_xM</u>.

(b) <u>Suppose ℓ_p has a critical point at x. Then for X, Y $\in T_xM$, the Hessian H_x of ℓ_p at x satisfies</u>

$$H_x(X,Y) = \langle A_p X, Y\rangle.$$

Proof: (a) For the purpose of this local calculation, consider an embedded neighborhood U of x in M. Take $X \in T_xM$, and let $\gamma(t)$ be a curve in U with $\gamma(0) = x$, and initial tangent vector X. By definition

$$X\ell_p(x) = \frac{d}{dt}\,\ell_p(\gamma(t))\big|_{t=0} = \langle p, \frac{d\gamma}{dt}\rangle\big|_{t=0} = \langle p, X\rangle.$$

Thus, $X\ell_p(x) = 0$ if and only if $\langle p, X\rangle = 0$, and x is a critical point of ℓ_p precisely when p is orthogonal to T_xM.

(b) Extend Y to a vector field tangent to U. It is easy to verify that $H_x(X,Y) = XY\ell_p(x)$. Using (a), we get

$$H_x(X,Y) = X(Y\ell_p) = X\langle Y,p\rangle = \langle D_XY,p\rangle. \tag{3.2}$$

Let ξ be a field of unit normals to U with $\xi(x) = p$. Then $\langle Y,\xi\rangle = 0$ on U, so

$$\begin{aligned} 0 &= D_X\langle Y,\xi\rangle = \langle D_XY,\xi\rangle + \langle Y, D_X\xi\rangle \\ &= \langle D_XY,\xi\rangle + \langle Y, -A_\xi X\rangle \\ &= \langle D_XY,p\rangle + \langle Y, -A_pX\rangle. \end{aligned} \tag{3.3}$$

Thus from (3.2) and (3.3) we have $H_x(X,Y) = \langle A_pX, Y\rangle$.

Q.E.D.

The following results are immediate consequences of Theorem 3.2.

Corollary 3.3: Suppose $p \in T_x^{\perp}M$. Then

(a) ℓ_p has a degenerate critical point at x if and only if A_p is singular.

(b) If ℓ_p has a non-degenerate critical point at x, then the index of ℓ_p at x is equal to the number of negative eigenvalues of A_p.

From Theorem 3.1 and Corollary 3.3 one sees immediately:

Theorem 3.4: For $p \in S^{m-1}$, ℓ_p is a Morse function if and only if p is a regular value of the Gauss map ν.

Since BM and S^{m-1} are manifolds of the same dimension, Sard's theorem (see, for example, Milnor [4, p. 10]) implies

Corollary 3.5:

(a) For almost all $p \in S^{m-1}$, ℓ_p is a Morse function.

(b) Suppose ℓ_p has a non-degenerate critical point of index j at $x \in M$. Then there is a Morse function ℓ_q having a critical point $y \in M$ of index j (q and y may be chosen as close to p and x, respectively, as desired).

Proof: (a) This follows from Theorem 3.4 and Sard's Theorem.

(b) By Theorem 3.2 (a), we know that $p = \nu(\xi)$ with $\pi(\xi) = x$. Since ℓ_p has a non-degenerate critical point of index j at x, ν_* is non-singular and A_ξ has exactly j negative eigenvalues and n-j positive eigenvalues. Thus ν is a diffeomorphism of a neighborhood V of (x,ξ) in BM onto a neighborhood U of p in S^{m-1}. Let $q \in U$ be a regular value of ν. Then ℓ_q is a Morse function having a critical point y in $\pi(V)$. Moreover, since ν_* is non-singular on V, the number of negative eigenvalues of A_η must remain constant as η ranges over V. Thus, the index of ℓ_q at y is also equal to j.

Q.E.D.

We now turn to the question of total absolute curvature. For a regular value p of the Gauss map ν, let $\mu(p)$ be the number of critical points of the Morse function ℓ_p. The point p has a neighborhood U such that $\nu^{-1}U$ is the disjoint union of connected open sets $V_1,\ldots,V_k$, $k = \mu(p)$, on which $\nu: V_i \to U$ is a diffeomorphism. The Lipschitz-Killing curvature G on BM is defined by the equation

$$\nu^*(da) = G\, dA,$$

where da is the volume element on S^{m-1} normalized so that

$$\int_{S^{m-1}} da = 1,$$

and dA is the volume element on BM. Thus on V_i we have

$$\int_{V_i} G\, dA = \int_U da, \text{ if } G > 0 \text{ on } V_i$$

and

$$\int_{V_i} G\, dA = -\int_U da, \text{ if } G < 0 \text{ on } V_i.$$

That is,

$$\int_{V_i} |G|\, dA = \int_U da. \tag{3.4}$$

Let C denote the set of measure zero of critical values of ν. The function $\mu(p)$ is locally constant and so continuous on $S^{m-1} - C$. The total absolute curvature or total area on S^{m-1} covered by the image of the Gauss map is certainly

$$\tau(M,f) = \int_{S^{m-1}-C} \mu(p)\, da.$$

On the other hand, from equation (3.4) and the discussion preceding it, and from the theory of integration, we see

$$\int_{S^{m-1}-C} \mu(p)\, da = \int_{BM} |G|\, dA.$$

Thus, the study of total absolute curvature can now be formulated in terms of the function μ.

Since each Morse function ℓ_p certainly satisfies $\mu(p) \geqslant \gamma$, where γ is the Morse number of M, we have

$$\tau(M,f) \geqslant \gamma.$$

Kuiper [1] discovered the following fact, although the lemma which is necessary to complete the proof in full detail is due to Wilson [1].

Theorem 3.6: Let M be a manifold. The greatest lower bound of $\tau(M,f)$ for all smooth immersions f of M into all Euclidean spaces is equal to γ.

Proof: Let $f:M \to E^m$ be an immersion and ϕ a non-degenerate function on M with γ critical points. Define for $\lambda \in (0,\infty)$, the map $h_\lambda: M \to E^m \times E^1 = E^{m+1}$ by

$$h_\lambda(x) = (f(x), \lambda\phi(x)).$$

Clearly, each h_λ is an immersion since rank $h_\lambda \geq$ rank f = dim M at each point. The proof is accomplished by showing that $\tau(M,h_\lambda) \to \gamma$ as $\lambda \to \infty$.

Let $p = (0,0,\ldots,1)$. Then $\ell_p(h_\lambda(x)) = \lambda\phi(x)$, and so ℓ_p is non-degenerate and has γ critical points. The intuition behind the proof is that as λ gets large, the immersion gets very elongated in the direction of p, and more and more height functions have the same critical point behavior as ϕ, i.e., they have γ critical points. Thus the integral of $\mu(q)$ over the unit sphere S^m in E^{m+1} will get arbitrarily close to γ. We now give the details of the proof.

We first fix our attention on the immersion $h_1(x) = (f(x), \phi(x))$. As in the proof of Corollary 3.5, since p is a regular value of the Gauss map of h_1, there is a neighborhood U of p in S^m such that if $q \in U$, then $\ell_q \circ h_1$ is non-degenerate and has γ critical points on M. For the first part of this calculation, it is convenient to parametrize the linear height functions near ℓ_p by vectors with last coordinate equal to 1, rather than by unit vectors. So let $q = w + p$, where $w \in E^m$. There is an $\varepsilon > 0$ such that if $|w|^2 < \varepsilon$, then $\ell_q \circ h_1$ is non-degenerate and has γ critical points on M.

Note that

$$\ell_q(h_1(x)) = \langle w, f(x)\rangle + \phi(x).$$

Next we consider the immersion $h_\lambda(x) = (f(x), \lambda\phi(x))$. Again for $q = w + p$,

$$\ell_q(h_\lambda(x)) = \langle w, f(x)\rangle + \lambda\phi(x).$$

If we write $w = \lambda u$ for $u \in E^m$, then

$$\ell_q(h_\lambda(x)) = \langle \lambda u, f(x)\rangle + \lambda\phi(x) = \lambda(\langle u, f(x)\rangle + \phi(x)).$$

Hence, $\ell_q \circ h_\lambda$ is non-degenerate with γ critical points on M if $|u|^2 < \varepsilon$. We now return to parametrizing the height functions by unit vectors. Suppose $q = w + \beta p$ is a unit vector. Then ℓ_q has the same critical point behavior as $\ell_{q'}$, where $q' = (w/\beta) + p$. Hence ℓ_q is non-degenerate on $h_\lambda(M)$ with γ critical points if $(|w|/\beta)^2 < \lambda^2\varepsilon$.

Noting that $|w|^2 = 1-\beta^2$, the inequality can be written as

$$\mu_\lambda(q) = \gamma, \text{ if } |\beta| > (1+\lambda^2\varepsilon)^{-1/2},$$

where $q = w + \beta p$ is a unit vector and $\mu_\lambda(q)$ denotes the number of critical points of ℓ_q on $h_\lambda(M)$. Let $\alpha = (1 + \lambda^2\varepsilon)^{-1/2}$ and let

$$W_\lambda = (E^m \times [-\alpha, \alpha]) \cap S^m.$$

So if $q \in S^m - W_\lambda$, then $\mu_\lambda(q) = \gamma$. Now

$$\tau(M, h_\lambda) = \int_{S^m - W_\lambda} \mu_\lambda(q)\, da + \int_{W_\lambda} \mu_\lambda(q)\, da.$$

As $\lambda \to \infty$, the interval $[-\alpha, \alpha]$ tends to the set $\{0\}$, and hence W_λ tends to the set $(E^m \times \{0\}) \cap S^m = S^{m-1}$, which has measure zero in S^m. Hence

$$\lim_{\lambda\to\infty} \int_{S^m - W_\lambda} \mu_\lambda(q)\, da = \lim_{\lambda\to\infty} \int_{S^m - W_\lambda} \gamma da = \int_{S^m - S^{m-1}} \gamma da = \gamma.$$

The proof of the theorem is completed by the following lemma due to Wilson [1, p. 365].

<u>Lemma 3.7</u>: <u>Let</u> $I_\lambda = \int_{W_\lambda} \mu_\lambda(q)\, da$. <u>Then</u> $\lim_{\lambda\to\infty} I_\lambda = 0$.

Proof: The proof is accomplished by expressing each I_λ as an integral over the cylinder $S^{m-1} \times [-\varepsilon^{-1/2}, \varepsilon^{-1/2}]$, and then comparing I_λ with I_1. For $\lambda = 1$, since $\mu_1(q) > 0$ for all q,

$$0 \leq I_1 = \int_{W_1} \mu_1(q)\, da < \int_{S^m} \mu_1(q)\, da < \infty.$$

We now wish to parametrize W_λ by the cylinder $V = S^{m-1} \times [-\varepsilon^{-1/2}, \varepsilon^{-1/2}]$. Let $q = w + \beta p$ be a vector in S^m. Then q is in W_λ if and only if $|\beta| \leq \alpha = (1+\lambda^2\varepsilon)^{-1/2}$. Noting that $|w|^2 = 1-\beta^2$, a computation shows that q is in W_λ if and only if $\lambda\beta/|w|^2$ is in $[-\varepsilon^{-1/2}, \varepsilon^{-1/2}]$. Let $q' = w' + \eta p$, where $w' = w/|w|$ and $\eta = \lambda\beta/|w|$. Then q is in W_λ if and only if $q' \in V$. Let da^m denote the volume element of S^m (previously denoted by da), and let da^{m-1} denote the induced volume element on the equatorial sphere S^{m-1}. At the point $q = w + \beta p$ on S^m,

$$da^m = |w|^{m-2} da^{m-1} d\beta = \frac{|w|^{m-1}}{\lambda}\, da^{m-1} d\eta.$$

Since

$$\ell_\lambda(h_\lambda(x)) = \langle w, f(x)\rangle + \lambda\beta\phi(x) = |w|(\langle w', f(x)\rangle + \eta\phi(x))$$

$$= |w|\; \ell_{q'}(h_1(x)),$$

we have that $\mu_\lambda(q) = \mu_1(q')$, where, of course, the correspondence between q and q' depends on λ. Thus, in the following calculation, it is worth recalling that the vector w as a function of q' on V is also dependent on λ, and will be denoted by $w_\lambda(q')$. Thus,

$$I_\lambda = \int_{W_\lambda} \mu_\lambda(q)\, da^m = \int_V \mu_1(q') \frac{|w_\lambda(q')|^{m-1}}{\lambda}\, da^{m-1} d\eta$$

$$\leq \frac{1}{\lambda} \int_V \mu_1(q') da^{m-1} d\eta, \text{ since } |w_\lambda(q')| \leq 1,$$

$$< \frac{1}{\lambda} \left(\frac{1+\varepsilon}{\varepsilon}\right)^{1/2(m-1)} I_1,$$

since on W_1,

$$|w_1(q')|^2 \geq \frac{\varepsilon}{1+\varepsilon} \text{ and } I_1 = \int_V \mu_1(q')|w_1(q')|^{m-1} da^{m-1} d\eta.$$

Since $I_1 < \infty$, we conclude that $\lim_{\lambda \to \infty} I_\lambda = 0$.

Q.E.D.

Having established γ as the greatest lower bound for $\tau(M,f)$, we make the following definition.

Definition 3.8: An immersion $f:M \to E^m$ is said to have minimal total absolute curvature if $\tau(M,f) = \gamma$.

We close this section with the following obvious result.

Theorem 3.9: An immersion $f:M \to E^m$ has minimal total absolute curvature if and only if every non-degenerate linear height function ℓ_p has γ critical points.

Proof: If every non-degenerate ℓ_p has γ critical points, then clearly $\tau(M,f) = \gamma$. Conversely, suppose $\mu(p) > \gamma$ for a certain non-degenerate ℓ_p. Since μ is constant on a neighborhood U of p in S^{m-1}, we have $\mu > \gamma$ on a set of positive measure. Since $\mu \geq \gamma$ on $S^{m-1}-C$, we conclude that $\tau(M,f) > \gamma$.

4. CONVEX SETS.

Convex hypersurfaces are the first examples of immersions with minimal total absolute curvature. Moreover, the more topological (and less differential geometric) approach to tight immersions and mappings developed primarily by Kuiper and Banchoff has much of the flavor of the classical study of convex sets and uses many results from that theory as important tools. Thus, we now list some results which will be needed. These are well-known and can be found, for example, in the first seven pages of Busemann's Convex Surfaces [1].

Definition 4.1: A non-empty set K in E^m is called convex if for every two points, x, y in K, the closed line segment [x,y] is contained in K.

The dimension of a convex set K in E^m is the smallest integer r such that K is contained in an r-dimensional affine subspace of E^m. If K has dimension m, then one can show that K has interior points in E^m. A very important result for our study is the following.

Theorem 4.2: Let K be a compact m-dimensional convex subset of E^m. Then the boundary of K is homeomorphic to an (m-1)-sphere.

Proof: We merely outline the proof which is found in Busemann [1, p. 3].

Let p be a point in the interior of K. One shows that each ray γ emanating from p intersects ∂K in exactly one point $y(\gamma)$. Let $\delta(\gamma) = |p - y(\gamma)|$, and prove that $\delta(\gamma)$ is a continuous function on the set of rays emanating from p which is, of course, homeomorphic to S^{m-1}. Thus the continuous function δ has a positive minimum $\varepsilon > 0$. Define a map σ from ∂K to the (m-1)-sphere S^{m-1} centered at p of radius $\varepsilon/2$ by

$$\sigma(y) = p + (\varepsilon/2)\,\frac{y - p}{|y - p|}.$$

One easily shows that σ is a homeomorphism of ∂K onto S^{m-1}.

Q.E.D.

The boundary of an m-dimensional convex set in E^m is called a convex hypersurface in E^m. In differential geometry, one often defines a convex hypersurface in E^m in terms of support planes, as is done in Kobayashi-Nomizu [1, Vol. 2. p. 40]. Both formulations are useful in this study, so we will discuss the relationship between them.

Definition 4.3: Let V be a subset of E^m. A hyperplane π in E^m through a point $x \in V$ is called a support plane of V at x if V lies entirely in one of the closed half-spaces determined by π. The half-space is called a <u>support half-space</u>.

It is well-known from calculus that if $f:M^n \to E^{n+1}$ is an immersion, then the only possible support plane (even locally) to $f(M)$ at a point $f(x)$ is the tangent plane $f_*(T_xM)$. The differential geometric definition of a convex hypersurface is the following.

<u>Definition 4.4</u>: A smooth embedding $f:M^n \to E^{n+1}$ is called a <u>convex hypersurface</u> if for each $x \in M$, the tangent hyperplane to $f(M)$ at $f(x)$ is a support plane of $f(M)$.

To relate this definition to the one which arises from the theory of convex sets, we need the following basic result Busemann [1, pp. 4-5].

<u>Proposition 4.5</u>: <u>Let $K \neq E^m$ be a closed convex set in E^m.</u>

(a) <u>K is the intersection of all of its support half-spaces.</u>

(b) <u>Every point on ∂K has at least one support plane.</u>

(c) <u>A point in the interior of K has no support plane.</u>

<u>Definition 4.6</u>: Let V be a subset of E^m. The <u>convex hull</u> of V, denoted HV, is the smallest closed convex set which contains V.

If V is compact, then HV is compact. Also, a support plane of V is a support plane of HV.

<u>Theorem 4.7</u>: <u>Let $f:M \to E^{n+1}$ $(n \geq 2)$ be an immersion of a compact n-dimensional manifold. Then f embeds M onto $\partial Hf(M)$ if and only if every tangent plane to $f(M)$ is a support plane of $f(M)$.</u>

<u>Proof</u>: Suppose there is a point $x \in M$ such that $f_*(T_xM)$ is not a support plane of $f(M)$. Then $f(M)$ has no support plane at $f(x)$, and by Proposition 4.5 (b), $f(x)$ is not in $\partial Hf(M)$. Thus $f(M)$ is not embedded onto $\partial Hf(M)$.

Conversely, suppose each $f_*(T_xM)$ is a support plane of $f(M)$ and thus a support plane of $\partial Hf(M)$. By Proposition 4.5, each $f(x)$ lies in $\partial Hf(M)$.

Since $f(M)$ is compact, it is a closed subset of $\partial Hf(M)$. We now show $f(M)$ is also open in $\partial Hf(M)$.

Suppose $p = f(x)$ is in $f(M)$. Since f is an immersion, there is a neighborhood U of x in M homeomorphic to an open n-disk which is embedded by f into E^{n+1}. Since $f(M)$ lies in $\partial Hf(M)$, $f(U)$ is a subset of the n-sphere $\partial Hf(M)$ which is homeomorphic to an open n-disk. By the Brouwer invariance of domain theorem, Eilenberg-Steenrod [1, p. 303], $f(U)$ is an open set in $\partial Hf(M)$. Consequently, $f(M)$ is open in $\partial Hf(M)$, and by connectedness, $f(M) = \partial Hf(M)$. This argument actually implies that M is a covering space of $\partial Hf(M)$ with covering map f. Since $\partial Hf(M)$ is simply connected, f must be injective, and thus f is an embedding of M onto $\partial Hf(M)$.

Q.E.D.

It follows from Theorem 4.7 that if an embedding $f:M \to E^{n+1}$ is a compact, convex hypersurface then M must be diffeomorphic to S^n, the diffeomorphism being provided by $\sigma \circ f$, where σ is the map defined in Theorem 4.2.

<u>Theorem 4.8: If $f:S^n \to E^{n+1}$ is a compact convex hypersurface, then $\tau(S^n,f) = 2$. In other words, f embeds S^n with minimal total absolute curvature.</u>

<u>Proof</u>: Suppose a non-degenerate function ℓ_p has three critical points on S^n, call them x_1, x_2, x_3. The tangent hyperplanes at these three points are all parallel and they must all be support planes of $f(S^n)$. Clearly, if ℓ_p assumes three distinct values at x_1, x_2, x_3, then one of the planes does not support, contradicting convexity. On the other hand, if $\ell_p(x_i) = \ell_p(x_j)$ for $i \neq j$, then the tangent planes coincide. By convexity of $Hf(S^n)$, the closed segment $[x_i, x_j]$ lies on $Hf(S^n)$. But $Hf(S^n) = f(S^n)$ since f is a convex hypersurface. We conclude that, x_i and x_j are not isolated critical points of ℓ_p, contradicting the assumption that ℓ_p is non-degenerate.

Q.E.D.

The proof above is simply a proof of the well-known fact that a convex hypersurface has exactly two support planes in every direction. Its converse is far deeper and was proven by Chern and Lashof [1] in 1957 for smooth immersions. In 1970, Kuiper [8] proved this result (after necessary modifications of the definitions) for continuous immersions, and recently he has proven a version for tight maps of the sphere [13]. These latter proofs rely on analysis of the top-sets which we will discuss in Section 7. We will postpone our proof of the Chern-Lashof theorem until Section 8 of this chapter, so that we can make use of these techniques.

Chern and Lashof also obtained the following.

Theorem 4.9 (Chern-Lashof [1]): Let $f:M \to E^m$ be an immersion of a compact n-manifold. Then

(a) $\tau(M,f) \geq 2$,

(b) If $\tau(M,f) < 3$, then M is homeomorphic to an n-sphere.

Proof: (a) Every ℓ_p has a maximum and a minimum on M, so $\mu(p) \geq 2$ for all p.

(b) If $\tau(M,f) < 3$, there must be a set U of positive measure on which $\mu(p) = 2$. So there is a non-degenerate ℓ_p with two critical points, and M is homeomorphic to S^n by Reeb's theorem (see, for example, Milnor [3, p. 24]).

Q.E.D.

Remark 4.10: If $\tau(M,f) = 3$, then M need not be homeomorphic to S^n. The well-known Veronese embedding f of the real projective plane P^2 into E^5 satisfies $\tau(P^2,f) = 3$, and thus it has minimal total absolute curvature. See Example 7.3 and Section 9.

Remark 4.11: An example of a manifold which cannot be immersed with minimal total absolute curvature in any E^m (Kuiper [1, p. 86]):

If M is homeomorphic to S^7, a tight immersion of $f:M \to E^m$ would embed M as a convex hypersurface in an $E^8 \subset E^m$, and M would be diffeomorphic to the standard 7-sphere. Hence, the exotic 7-spheres of Milnor [5] cannot be immersed with minimal total absolute curvature in any E^m. Further, Ferus [1] proved that every embedding of an exotic n-sphere ($n \geq 5$) in E^{n+2} has total absolute curvature $\tau \geq 4$.

5. TIGHT MAPS AND IMMERSIONS.

In the first years of following Chern and Lashof's paper in 1957, the theory of minimal total absolute curvature immersions underwent substantial development and reformulation. Since this notion is a generalization of convexity, Kuiper [5]-[6] at first called these "convex immersions". Banchoff [1] first used the designation tight in conjunction with his introduction of the two-piece property. In 1962, Kuiper [6] formulated tightness in terms of intersections with half-spaces and injectivity of induced maps on homology. This has obvious advantages in that it lends itself to the top-set approach, and that it makes tightness a property which is invariant under projective transformations. Kuiper [6] noted that his formulation is equivalent to minimal total absolute curvature for manifolds which satisfy his

Condition 3A: The Morse number γ of M is equal to the sum of the Betti numbers $\beta(M;F)$ for some field F.

He observed that all two-dimensional surfaces satisfy this condition with $F = Z_2$. Suppose that $f:M \to E^m$ is a minimal total absolute curvature immersion of a manifold satisfying condition 3A with $F = Z_2$. If the hyperplane π is a level set of a non-degenerate height function and h is a closed half-space determined by π, then by Theorem 2.1, the map

$$H_*(f^{-1}h) \to H_*(M) \quad (Z_2\text{-homology})$$

is injective. This motivates the following definition of tightness, Kuiper [13]. (Unless stated otherwise, all topological spaces are assumed to be Hausdorff).

Definition 5.1: A map f of a compact topological space X into E^m is called tight if for every closed half-space h in E^m, the induced homomorphism

$$H_*(f^{-1}h) \to H_*(X)$$

in Cech homology with Z_2 coefficients is injective. A subset of E^m is called a tight set if the inclusion map is a tight map.

Remark 5.2 (a): The use of Cech homology is motivated by its continuity property which is necessary to eliminate the requirement that the half-space h be determined by a non-degenerate height function. Of course for triangulable spaces (and thus smooth manifolds), Cech homology agrees with singular homology. A second obvious remark is that this definition extends the notion of tightness beyond the category of smooth immersions of manifolds.

Remark 5.2 (b): Tightness is a projective property. From the definition, it is clear that only the underlying affine space A^m is needed to verify if a map $f:X \to E^m$ is tight, and not the metric space E^m. The affine space A^m can be closed by adding the classes of parallel lines as new points, thus obtaining real projective space P^m, and A^m is naturally embedded as the complement of a hyperplane P^{m-1} in P^m. If $f:X \to A^m$ is tight and $\sigma:P^m \to P^m$ is a projective transformation such that the image $\sigma(fX)$ lies in A^m, then $\sigma f:X \to A^m$ is also tight. This follows immediately from the definition, since for every half-space h in A^m, $(\sigma f)^{-1}(h) = f^{-1}(\sigma^{-1}h) = f^{-1}h'$, for the appropriate half-space h'.

Remark 5.2 (c): Orthogonal projections of tight maps are tight. Suppose $f:X \to E^m$ is tight and $\phi:E^m \to E^k$ is orthogonal projection onto a Euclidean subspace of E^m. Then $\phi \circ f:X \to E^k$ is tight. To see this, let h be the closed half-space in E^k given by $\ell_p \leq r$ for $p \in E^k$. Then the same inequality in E^m gives a half-space h' in E^m such that $\phi^{-1}h = h'$. Thus, $(\phi \circ f)^{-1}h = f^{-1}h'$, and the tightness of $\phi \circ f$ follows from the tightness of f. Similarly, if $f:X \to E^m$ is tight and $i:E^m \to E^{m+j}$ is inclusion of E^m into a higher-dimensional Euclidean space, then $i \circ f:X \to E^{m+j}$ is tight. Here if h is a half-space in E^{m+j} and $h' = h \cap E^m$, then $(i \circ f)^{-1}h = f^{-1}h'$.

Our next objective is to prove that tightness is equivalent to minimal total absolute curvature for immersions of manifolds which satisfy condition 3A with $F = Z_2$. The important point here is to show that for an immersion with minimal total absolute curvature, the injectivity condition on homology holds for half-spaces determined by degenerate height functions, as well as by non-degenerate ones. This fact greatly simplifies many arguments. The main ingredients are the continuity property of Cech homology and the following lemma.

Lemma 5.3: Let $f:M \to E^m$ be an immersion of a compact manifold. Suppose U is an open subset of M containing $M_r(p)$ for some $p \in S^{m-1}$, $r \in R$. Then there exists a non-degenerate height function ℓ_p and a real number s such that

$$M_r(p) \subset M_s^-(q) \subset M_s(q) \subset U.$$

Proof: Since $M_r(p)$ is compact and U is open, one easily shows that there exists $\varepsilon > 0$ such that $M_{r+\varepsilon}(p) \subset U$. Let K be the maximum value which any linear height function assumes on M. Suppose a unit vector p' is orthogonal to p, and let

$$q = \cos\alpha\, p + \sin\alpha\, p'.$$

Then

$$p = \sec\alpha\, q - \tan\alpha\, p'.$$

For any $x \in M$,

$$|\ell_p(x)-\ell_q(x)| = |\langle p-q,f(x)\rangle| = |\langle(\sec\alpha - 1)q - \tan\alpha\, p',f(x)\rangle|$$

$$\leq |\sec\alpha - 1|\,|\ell_q(x)| + |\tan\alpha|\,|\ell_{p'}(x)| \leq (|\sec\alpha - 1| + |\tan\alpha|)K.$$

Choose α near 0 so that

$$|\sec\alpha - 1| < \varepsilon/2K \quad \text{and} \quad |\tan\alpha| < \varepsilon/2K.$$

Then

$$|\ell_p(x)-\ell_q(x)| < \varepsilon$$

for any $x \in M$, and therefore

$$M_r(p) \subset M^-_{r+\varepsilon/2}(q) \subset M_{r+\varepsilon/2}(q) \subset M_{r+\varepsilon}(p) \subset U$$

for all q in the neighborhood of p in S^{m-1} determined by α, and thus for some q with ℓ_q non-degenerate.

Q.E.D.

We now prove the main result of this section, due to Kuiper [13].

Theorem 5.4: An immersion $f:M \to E^m$ of a compact manifold is tight if and only if the Morse number $\gamma(M) = \beta(M;Z_2)$ and f has minimal total absolute curvature.

Proof: If f is tight and ℓ_p is any non-degenerate height function, then it follows from Theorem 2.1 that ℓ_p has exactly $\beta = \beta(M;Z_2)$ critical points. Hence $\beta \geq \gamma$ and $\tau(M,f) = \beta$. However, the Morse inequalities imply that $\gamma \geq \beta$, and f has minimal total absolute curvature γ.

Conversely, if f has minimal total absolute curvature and $\gamma = \beta(M;Z_2)$, then, by Theorem 2.1 the homomorphism

$$H_*(f^{-1}h) \to H_*(M) \quad (Z_2\text{-homology})$$

is injective for each half-space h determined by a non-degenerate height function. Suppose now that $f^{-1}h = M_r(p)$ for a degenerate height function ℓ_p and some real number r. We must show that the induced homomorphism on homology is still injective in this case. Here we need to use the continuity property of Cech homology. We wish to produce a nested sequence of half-spaces h_i, i = 1,2,3,... satisfying

$$(5.1) \qquad f^{-1}(h_i) \supset f^{-1}(h_{i+1}) \supset \cdots \supset \bigcap_{j=1}^{\infty} f^{-1}(h_j) = M_r(p), \; i = 1,2,\ldots$$

such that the homomorphism in Z_2-homology

$$(5.2) \qquad H_*(f^{-1}(h_i)) \to H_*(M) \text{ is injective, } i = 1,2,\ldots$$

If (5.1) and (5.2) are satisfied, then

(5.3) $H_*(f^{-1}(h_i)) \to H_*(f^{-1}(h_j))$ is injective for all $i > j$.

The continuity property of Cech homology (see Eilenberg-Steenrod [1, p. 261]) says that

$$H_*(M_r(p)) = \overleftarrow{\lim_{i\to\infty}} H_*(f^{-1}(h_i)) .$$

Equations (5.3) and Theorem 3.4 of Eilenberg-Steenrod [1, p. 216] on inverse limits imply that the map

$$H_*(M_r(p)) \to H_*(f^{-1}(h_i))$$

is injective for each i. Thus, so is the map

$$H_*(M_r(p)) \to H_*(M).$$

We produce the sequence $\{h_i\}$ inductively using Lemma 5.3 to find a non-degenerate height function ℓ_q and a real number s satisfying

$$M_r(p) \subset M_s^-(q) \subset M_s(q) \subset M_{r+(1/i)}^-(p).$$

The set U in Lemma 5.3 should be taken to be the previous $M_s^-(q)$ (except for $i = 1$ when $U = M_{r+1}^-(p)$). Then h_i is the half-space $\ell_q \leq s$ constructed at the i-th step. Note that (5.2) is satisfied since each ℓ_q is non-degenerate and that the h_i are nested. Finally, since

$$f^{-1}(h_{i+1}) \subset M_{r+(1/i)}(p),$$

it is clear that

$$\bigcap_{j=1}^{\infty} f^{-1}(h_j) = M_r(p),$$

and the theorem is proven.

Q.E.D.

Remark 5.5: Until recently, every known example $f:M \to E^m$ of an immersion with minimal total absolute curvature satisfied $\gamma(M) = \beta(M;Z_2)$. (See Kuiper [13], p. 107). However, Kuiper and Meeks [1] have produced for each $n \geq 4$, a hypersurface M in E^{n+1} with minimal total absolute curvature $\tau = 12$ and $\beta(M;Z_2) = 8$.

For a compact connected manifold M, Morse [1, p. 383] showed that if a non-degenerate function ϕ has more than one local maximum or minimum, one can produce a non-degenerate function ϕ which is polar (i.e., has exactly one local maximum and one local minimum) and which has fewer critical points than ϕ. It follows that if $\mu(\phi) = \gamma(M)$, then ϕ is polar.

Proposition 5.6: Let M be a connected, compact 2-dimensional surface. Then

(a) $\gamma(M) = \beta(M;Z_2) = 4-\chi(M)$, where $\chi(M)$ is the Euler characteristic of M.

(b) For a non-degenerate function ϕ on M, $\mu(\phi) = \gamma$ if and only if ϕ is polar.

Proof: We prove both statements simultaneously. As noted above, if $\mu(\phi) = \gamma$ for some non-degenerate ϕ, then ϕ is polar. Conversely, suppose ϕ is polar. Then $\mu_0(\phi) = \beta_0(M;Z_2) = 1$ and $\mu_2(\phi) = \beta_2(M;Z_2) = 1$. This and the well-known Morse relation (see, for example Milnor [3, p. 29])

$$\sum_{k=0}^{2} (-1)^k \mu_k(\phi) = \sum_{k=0}^{2} (-1)^k \beta_k(M;Z_2) = \chi(M),$$

implies that $\mu_1(\phi) = \beta_1(M;Z_2)$ also. Hence if ϕ is any polar non-degenerate function on M, then $\mu(\phi) = \beta(M;Z_2)$. Thus, any non-degenerate function on M has at least $\beta(M;Z_2)$ critical points. We conclude that $\gamma = \beta(M;Z_2) = 4 - \chi$, and that if ϕ is polar, then $\mu(\phi) = \gamma$.

Q.E.D.

This and Theorem 5.4 yield the following corollary.

Corollary 5.7: Let $f:M \to E^m$ be an immersion of a connected, compact two-dimensional surface. Then the following are equivalent:

(a) f is tight.

(b) f has minimal total absolute curvature.

(c) Every non-degenerate linear height function is polar.

Banchoff [1] introduced the notion of the two-piece property largely to take advantage of the above corollary.

Definition 5.8: A map $f:X \to E^m$ of a compact, connected topological space X is said to have the two-piece property (TPP) if $f^{-1}h$ is connected for every closed half-space h in E^m.

For an embedded space X in E^m, the TPP means that every hyperplane π cuts X into at most two pieces. Since $\beta_0(f^{-1}h;Z_2)$ is merely the number of connected components of $f^{-1}h$, we see that if f is tight, then $\beta_0(f^{-1}h;Z_2) \leq 1$ for all half-spaces h, and so f has the TPP. Thus, we have

Theorem 5.9: Let f be a continuous map of a compact connected space X into E^m. If f is tight, then f has the TPP.

In Example 5.23, we will exhibit a smooth TPP embedding due to Kuiper of the 3-sphere into E^4 which is not tight. As Corollary 5.12 will show, the TPP and tightness are equivalent for smooth immersions of compact 2-manifolds (without boundary). In fact, Kuiper [13, p. 10] pointed out that these two properties are equivalent for continuous maps of compact 2-manifolds (see Lastufka [1, p. 396] for a proof). However, the TPP is weaker than tightness for embeddings of 2-manifolds with boundary (see Example 5.23).

While the TPP is weaker than tightness, it has the virtue of simplicity. Moreover, it is sufficient to prove Little and Pohl's result on tight immersions of maximal codimension, which we will discuss in Section 10 of this chapter.

It is useful to observe that the TPP can be formulated equally well in terms of open rather than closed half-spaces. The proof here is due to Lastufka [1].

Proposition 5.10: Let $f:X \to E^m$ be a continuous map of a compact space. Then f has the TPP if and only if $f^{-1}(\text{int } h)$ is connected for every open half-space int h in E^m.

Proof: Suppose first that f has the TPP. Let int h be the open half-space given by the equality $\ell_p < r$, and assume that $f^{-1}(\text{int } h)$ is non-empty. Choose $x \in f^{-1}(\text{int } h)$. Then $\ell_p(x) = r - \varepsilon$ for some $\varepsilon > 0$. Consider the sequence of closed half-spaces h_k, $k = 1,2,\ldots$ given by the inequalities $\ell_p \leq r - (\varepsilon/k)$. By the TPP, each $f^{-1}(h_k)$ is connected. Further, $\bigcap f^{-1}(h_k)$ contains the point x and is, therefore, non-empty. Hence the union of connected sets,

$$\bigcup f^{-1}(h_k) = f^{-1}(\text{int } h)$$

is connected.

Conversely, assume that $f^{-1}(\text{int } h)$ is connected for every open half-space. Suppose $f^{-1}h$ is not connected for the closed half-space h given by the inequality $\ell_p \leq r$. Then, there exist two disjoint, non-empty, relatively closed subsets A and B of $f^{-1}h$ such that $f^{-1}h = A \cup B$. Since $f^{-1}h$ is closed in X, so are A and B. Since X is a normal topological space, there exist disjoint open subsets V and W in X such that $A \subset V$ and $B \subset W$. Let $U = V \cup W$. Then $f^{-1}h$ lies in U and from the compactness of $f^{-1}h$, we can find an $\varepsilon > 0$ such that U contains $f^{-1}(\text{int } h')$ where int h' is given by the inequality $\ell_p < r + \varepsilon$. This implies that $f^{-1}(\text{int } h')$ is not connected, a contradiction.

Q.E.D.

Theorem 5.11: Let $f:M \to E^m$ be a immersion of a compact, connected manifold. Then f has the TPP if and only if every non-degenerate linear height function is polar (i.e., has one maximum, one minimum).

This and Corollary 5.7 yield the following:

Corollary 5.12: A TPP immersion $f:M \to E^m$ of a compact, connected two-dimensional surface is tight.

We prove Theorem 5.11 via the following sequence of elementary propositions.

Proposition 5.13: Let $f:M \to E^m$ be an immersion of a connected, compact manifold. A non-degenerate linear height function ℓ_p is polar if and only if $f^{-1}h$ is connected for every half-space h determined by ℓ_p.

Proof: Suppose

$$h = \{y \in E^m \mid \ell_p(y) \geq r\}$$

is such that $f^{-1}h$ has at least two components V_1 and V_2. Then ℓ_p has at least two local maxima, one on V_1, and one on V_2.

Conversely, suppose a non-degenerate ℓ_p has two local maxima x_1 and x_2 (if ℓ_p has two local minima, then ℓ_{-p} has two local maxima) at heights r_1 and r_2 with $r_1 \leq r_2$. By the lemma of Morse (Lemma 1.2), there is a neighborhood U of x_1 not containing x_2 such that for sufficiently small $\varepsilon > 0$, the closed set

$$V = \{x \in \overline{U} \mid \ell_p(x) \geq r_1 - \varepsilon\}$$

is contained in the interior of U. Let h be the half-space determined by the equation $\ell_p \geq r_1 - \varepsilon$. Then V is a component of $f^{-1}h$, but x_2 is in $f^{-1}h - V$. So, $f^{-1}h$ is not connected.

Q.E.D.

In order to complete the proof of Theorem 5.11, we must also show that if every non-degenerate height function is polar then $f^{-1}h$ is connected for half-spaces determined by degenerate height functions. This could be handled using Cech homology as in the proof of Theorem 5.4, but the following lemma enables one to give a much more elementary proof.

Lemma 5.14 (Kuiper [4, p. 14]: Let $f:M \to E^m$ be an immersion. Suppose $W \subset U \subset \overline{U} \subset M$, with U an open set such that for some $p \in S^{m-1}$ we have $\ell_p(x) = \gamma$ for $x \in W$ and $\ell_p(x) < \gamma$ for $x \in \overline{U} - W$. Then there is a neighborhood N of p in S^{m-1} such that if $q \in N$ then ℓ_q assumes its maximum on $\overline{U}$ in the interior of U.

Proof: Let $\gamma - 3\varepsilon$, $\varepsilon > 0$, be the maximum value of ℓ_p on ∂U. There is a neighborhood N of p in S^{m-1} such that for all $q \in N$,

$$|\ell_q(x) - \ell_p(x)| < \varepsilon \quad \text{for all} \quad x \in \overline{U}.$$

Thus, for $x \in W$, $\ell_q(x) \geqslant \gamma - \varepsilon$, so ℓ_q assumes its maximum on $\overline{U}$ in the interior of U. Note that if ℓ_q is non-degenerate, this means that ℓ_q has a local maximum at a point in U.

Q.E.D.

Proof of Theorem 5.11:

If f has the TPP, then by Proposition 5.13, every non-degenerate ℓ_p is polar. Conversely, suppose f does not have the TPP. Then there is a half-space h such that $f^{-1}h$ is not connected. Suppose h is determined by the inequality $\ell_p \geqslant r$. Let V_1 and V_2 be two components of $f^{-1}h$. Since V_1 and V_2 are compact, there exist disjoint open sets U_1 and U_2 with $V_i \subset U_i$. The function ℓ_p assumes its maximum value on U_i at points in the set V_i. Applying Lemma 5.14 to U_1 and U_2, there is a non-degenerate function ℓ_q with q near p, which has local maxima on both U_1 and U_2. Hence, ℓ_q is not polar.

Q.E.D.

It is sometimes useful to phrase the two-piece property in terms of support planes. Part (a) of the following definition agrees with Definition 4.3 in the case where f is an embedding.

Definition 5.15: Let $f:M \to E^m$ be a smooth immersion of a compact manifold.

(a) a hyperplane π in E^m through a point f(x) is called a support plane of f(M) at x if f(M) lies entirely in one of the closed half-spaces determined by π.

(b) a hyperplane π through $f(x)$ is called a local support plane of $f(M)$ at x if there exists a neighborhood U of x such that $f(U)$ lies in a closed half-space determined by π and $f(\partial U)$ lies in the interior of this half-space.

The last condition in the definition of a local support plane, i.e. $f(\partial U)$ lies in the interior of the half-space, is necessitated by examples like the tight torus of Kuiper shown in Figure 7.3. If this condition were omitted, then the plane of the face F_3 would be a local support plane at any of the points in the interior of F_3.

Theorem 5.16: If $f:M \to E^m$ is a TPP immersion of a compact manifold, then every local support plane of $f(M)$ is a support plane of $f(M)$.

Proof: Suppose that a local support plane π is given by the equation $\ell_p = \gamma$, so that $\ell_p \leq \gamma$ on U and $\ell_p < \gamma$ on ∂U. If π is not a support plane, then there exists a point $y \in M$ with $\ell_p(y) > \gamma$. If h is the half-space given by $\ell_p \geq \gamma$, then $f^{-1}h$ has at least two connected components, one contained in U and one in $M - \overline{U}$, contradicting the TPP.

Q.E.D.

Note that the converse of Theorem 5.16 is not true. If $f:S^1 \to E^2$ is an immersion which wraps twice around a Euclidean circle at constant speed, then every local support plane is a global support plane, but f does not have the TPP. A similar example for surfaces is obtained by taking $f:T^2 \to E^3$ to be a double covering of a torus of revolution. However, the converse of Theorem 5.16 does hold if f is assumed to be an embedding.

Theorem 5.17: Let $f:M \to E^m$ be a smooth embedding of a compact manifold. Then the following are equivalent.

(a) f has the TPP.

(b) every local support plane of $f(M)$ is a support plane of $f(M)$.

(c) every local extremum of a non-degenerate height function is an absolute extremum.

Proof: (a) $\to$ (b) by Theorem 5.16.

(b) $\to$ (c): If a non-degenerate ℓ_p has a local extremum which is not an absolute extremum at x, then the tangent plane at f(x) is a local support plane which is not a global support plane, by the lemma of Morse.

(c) $\to$ (a): Suppose f does not have the TPP. By Theorem 5.11, there exists a non-degenerate ℓ_p with at least two maxima. If the maxima are at different heights, then the proof is finished. So suppose $\ell_p(x_1) = \ell_p(x_2)$ is the absolute maximum of the non-degenerate height function ℓ_p on M. By taking a Euclidean translation, we can assume that $f(x_1)$ is the origin so that $\ell_p(x_1) = \ell_p(x_2) = 0$ is the maximum of ℓ_p on M. Let $\varepsilon = |f(x_2)|$, and let q be the unit vector $f(x_2)/\varepsilon$. Let $U_1 = f^{-1}(B_{\varepsilon/3}f(x_1))$ and $U_2 = f^{-1}(B_{\varepsilon/3}f(x_2))$. Then the maximum of ℓ_q on U_1 is $\varepsilon/3$ and the minimum of ℓ_q on U_2 is $2\varepsilon/3$. Let N_1 and N_2 be the respective neighborhoods of p in S^{m-1} given by applying Lemma 5.14 to U_1 and U_2. Let

$$p' = \cos\alpha\, p + \sin\alpha\, q$$

be a point in $N_1 \cap N_2$ for some α, $0 < \alpha < \pi/2$. Then, by Lemma 5.14, $\ell_{p'}$ has a local maximum value in U_1 and one in U_2. But if $y \in U_1$, then

$$\ell_{p'}(y) = \cos\alpha\, \ell_p(y) + \sin\alpha\, \ell_q(y) \leq \cos\alpha\,(0) + \sin\alpha\,(\tfrac{\varepsilon}{3}) = \tfrac{\varepsilon}{3}\sin\alpha.$$

On the other hand, $\ell_{p'}(x_2) = \varepsilon\sin\alpha$. Thus, the maxima of $\ell_{p'}$ in U_1 and U_2 are at different heights. If $\ell_{p'}$ is degenerate, then there exists a non-degenerate height function nearby which has local maxima in U_1 and U_2 at different heights.

Q.E.D.

Recall that a map f of a topological space X into E^m is called substantial if f(X) is not contained in any hyperplane in E^m. Kuiper [8] showed that the two-piece property imposes an upper bound on the codimension of a substantial smooth immersion of a manifold.

Theorem 5.18: Let $f:M \to E^m$ be a substantial smooth immersion of a compact n-dimensional manifold. If f has the TPP, then $m \leq \frac{1}{2} n(n+3)$.

Proof: Let ℓ_p be a non-degenerate linear height function on M with absolute maximum at $x \in M$. After a translation of coordinates, we may assume that $f(x)$ is the origin of E^m, and hence $\ell_p(x) = 0$. By Theorem 3.2, we know that p is normal to $f(M)$ at $f(x)$ and that the Hessian $H(X,Y) = \langle A_p X, Y\rangle$ of ℓ_p at x is negative definite. Let $T_x^{\perp}M$ denote the normal space to $f(M)$ at $f(x)$, and let V be the vector space of symmetric bilinear forms on T_xM. Define a linear map $\Phi: T_x^{\perp}M \to V$ by $\Phi(q) = A_q$, i.e.

$$\Phi(q)(X,Y) = \langle A_q X, Y\rangle, \quad X,Y \in T_xM.$$

The dimension of $T_x^{\perp}M$ is $m-n$, while V has dimension $\frac{1}{2} n(n+1)$. Thus, if

$$m-n > \frac{1}{2} n(n+1), \text{ i.e. } m > \frac{1}{2} n(n+3),$$

then the kernel of Φ is non-trivial. We now complete the proof by showing that the TPP implies that Φ is injective. Suppose there exists $q \neq 0$ in $T_x^{\perp}M$ with $A_q = 0$. Let $z(t) = p + tq$. Then $z(t) \in T_x^{\perp}M$ for all t, and

$$A_{z(t)} = A_p + t\, A_q = A_p \quad \text{for all } t.$$

Thus, $\ell_{z(t)}$ has a non-degenerate maximum at x for all t. On the other hand, since f is substantial, there exists a point $y \in M$ such that $\ell_q(y) \neq 0$. Then,

$$\ell_{z(t)}(y) = \ell_p(y) + t\, \ell_q(y),$$

and for a suitable choice of t, we have $\ell_{z(t)}(y) > 0$. Therefore, this $\ell_{z(t)}$ does not assume its absolute maximum value at x, and f does not have the TPP by Theorem 5.16. Contradiction.

Q.E.D.

In Section 9, we will show that the Veronese embeddings $f: P^n \to E^m$,

$m = \frac{1}{2}n(n+3)$, are tight and substantial. Thus, the bound in Theorem 5.18 cannot be decreased. A remarkable result, due to Kuiper [6] for n=2, and Little and Pohl [1] for higher dimensions, is that up to projective transformation, these Veronese embeddings are the only TPP embeddings of any manifold with the maximal codimension. The proof will be given in Section 10. Finally, for every integer k satisfying $1 < k < n(n+1)/2$, there exists a tight substantial embedding of an n-dimensional manifold into E^{n+k} (see Theorem 9.6).

The method of proof in Theorem 5.18 yields even stronger results if f is assumed to be tight. For example (Chern-Lashof [1]):

<u>Theorem 5.19</u>: <u>If $f:S^n \to E^m$ is a tight smooth substantial immersion, then $m=n+1$.</u>

<u>Proof</u>: Let Φ be the map from $T_x^{\perp}S^n$ to V, as in Theorem 5.18. As the proof of that theorem shows, the TPP (and hence tightness) implies that Φ is injective. Let p and q be linearly independent in $T_x^{\perp}S^n$. Then A_p and A_q are not scalar multiples of one another. This implies, after some linear algebra, that there exists a real number t such that for $z(t) = p + tq$, the shape operator $A_{z(t)}$ is non-degenerate with index k different from 0 and n. Hence, $\ell_{z(t)}$ has a non-degenerate critical point of index k at x, and by Corollary 3.5, there exists a Morse function $\ell_{p'}$ on S^n which has a critical point of index k, contradicting tightness. We conclude that $T_x^{\perp}S^n$ must have dimension 1, i.e. $m=n+1$.

Q.E.D.

Of course, a much stronger conclusion actually holds, namely that f embeds S^n onto a convex hypersurface in E^{n+1}. Further, one only needs to assume that f is a tight topological immersion to obtain this conclusion (see Section 8).

Next, we give a more general version of these results which has applications, in particular, to the case where M is a projective space (see Theorem 9.4). This result is also due to Kuiper [1].

Let $(\beta_0,\ldots,\beta_n)$ be an (n+1)-tuple of non-negative integers. Let $c(\beta_0,\ldots,\beta_n)$ be the maximal dimension of a linear family of symmetric bilinear forms in n variables which contains a positive definite form and such that no form of the family has index k if $\beta_k = 0$. Of course, in our context, β_k will be the k-th Z_2-Betti number of a manifold M. In the case where M is a projective plane over R, C or the quaternions Q, the number $c(\beta_0,\ldots,\beta_n)$ can be computed (see Kuiper [8, p. 232]).

Theorem 5.20: Let $f:M \to E^m$ be a tight smooth substantial immersion of a compact connected n-dimensional manifold and let β_j be the j-th Z_2-Betti number of M. Then

$$m-n \leq c(\beta_0,\ldots,\beta_n) \leq \frac{1}{2}\, n(n+1)$$

Proof: With the notation of Theorem 5.18, we know that $\Phi:T_x{}^{\perp}M \to V$ is injective, and so the codimension m-n is equal to the dimension of the image of Φ. This image is a vector space which contains a positive definite bilinear form. Further, if $\beta_k = 0$, then no bilinear form in the image of Φ can have index k, for that would imply the existence of a non-degenerate linear height function having a critical point of index k, contradicting tightness. Thus,

$$m-n = \dim(\text{Image } \Phi) \leq c(\beta_0,\ldots,\beta_n).$$

Q.E.D

In [5], Banchoff showed that for all $n \geq 3$ and for any $m > n$, there exist substantial polyhedral embeddings of S^n into E^m which have the TPP. This clearly demonstrates that smoothness is necessary in Theorem 5.18. Even if one assumes that f is tight in Theorem 5.18, the assumption of smoothness is still necessary. Banchoff [1] showed that if M is a compact surface, then the upper bound on m for a tight substantial polyhedral embedding of M into E^m is

$$m = \frac{1}{2}\left(5 + \sqrt{49 - 24\chi(M)}\right)$$

Moreover, this upper bound can be attained for every surface except a Klein bottle (see Kühnel [3]).

We now describe Banchoff's tight polyhedral embedding of P^2 into E^5.

Example 5.21: A tight substantial polyhedral embedding of P^2 into E^5. Let $v_1, \ldots, v_6$ be the standard unit basis vectors in E^6. Then the ten triangles indicated in Figure 5.1 fit together to form a projective plane P^2 lying in the 5-plane

$$E^5 = \{x \in E^6 \mid \sum_{i=1}^{6} x_i = 1\}.$$

This is a realization of the well-known 6-vertex triangulation of P^2. Since this surface contains all 15 segments $[v_i, v_j]$, its intersection with any half-space is connected and thus it has the TPP. See Banchoff [1] for a similar substantial TPP embedding of T^2 into E^6 using the triangulation with 7 vertices shown in Figure 5.2. Kühnel [3] has shown further that if M is a compact surface, then there exists a tight substantial polyhedral embedding of M into E^m if and only if

(i) $3 \leq m \leq \frac{1}{2}(5 + \sqrt{49 - 24\chi(M)})$ if M is orientable;

(ii) $4 \leq m \leq \frac{1}{2}(5 + \sqrt{49 - 24\chi(M)})$ if M is non-orientable and $\chi(M) \neq 0$;

(iii) $4 \leq m \leq 5$ if M is a Klein bottle.

Kühnel also has similar results for compact surfaces with boundary. Finally, using their 9-vertex triangulation of the complex projective plane CP^2, Kühnel and Banchoff [1] have produced a tight polyhedral embedding of CP^2 into E^8.

Banchoff's [5] TPP polyhedral embeddings of S^n into E^m for $m > n + 1$ are not tight. In fact, as noted earlier, Kuiper's version of the Chern-Lashof theorem shows that a tight continuous immersion of S^n into E^m must be a convex hypersurface. In [8], Kuiper modified Banchoff's examples to give a smooth TPP embedding of S^3 into E^4 which is not tight. We will conclude this section with that example and an even simpler example of an embedded

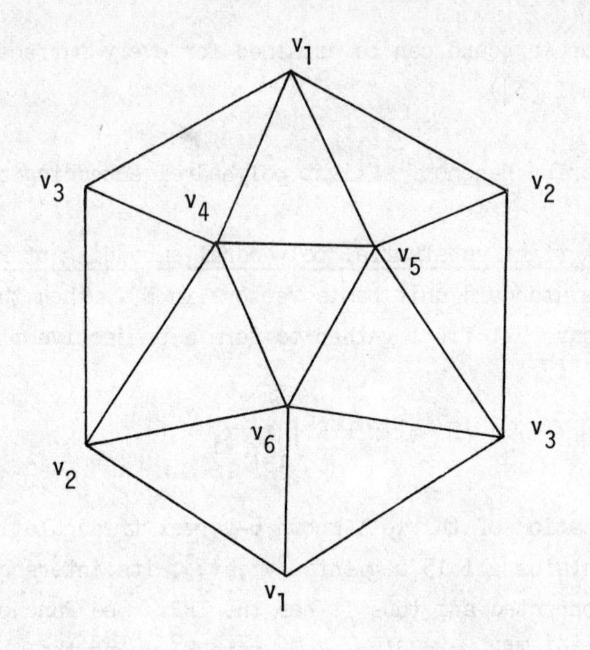

Figure 5.1 The 6-vertex triangulation of P^2

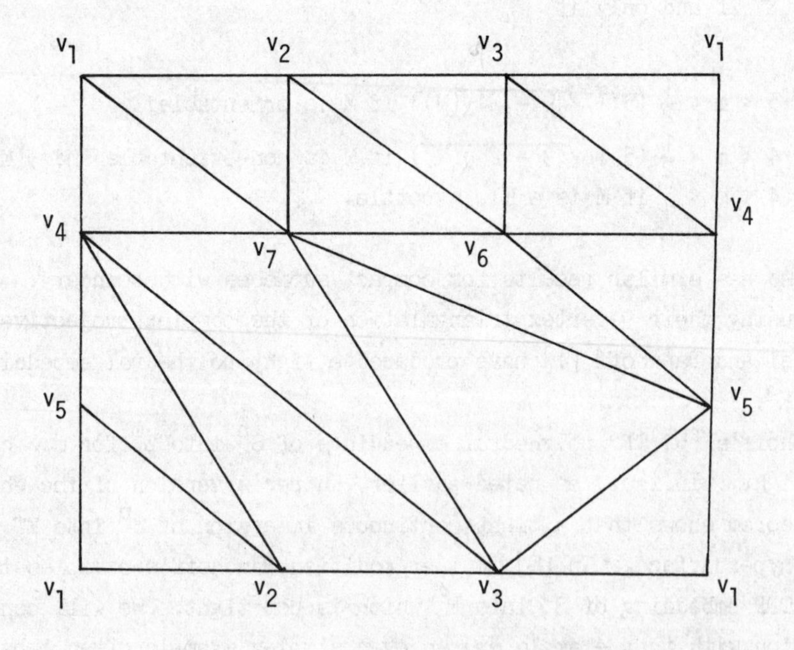

Figure 5.2 The 7-vertex triangulation of T^2

surface with boundary which has the TPP but is not tight. First, we introduce one more useful definition.

Definition 5.22: A map f of a compact connected topological space X into E^m is called k-tight if for every closed half-space h in E^m and for every integer $i \leq k$, the induced homomorphism

$$H_i(f^{-1}h) \to H_i(X), \; Z_2\text{-Cech homology},$$

is injective.

Of course, 0-tightness is just the two-piece property. If $f:M \to E^m$ is a smooth immersion of a compact manifold, then f is k-tight if and only if every non-degenerate linear height function ℓ_p has $\beta_i(M;Z_2)$ critical points of index i for $0 \leq i \leq k$.

5.23 Examples with the two-piece property which are not tight.
The first example, due to Kuiper [13], is a TPP embedding of an annulus in E^3 which is not tight (see Figure 5.3). Let X be the annulus given by the relations:

$$x^2 + y^2 + z^2 = 1, \quad -\frac{1}{\sqrt{2}} \leq z \leq \frac{1}{\sqrt{2}}.$$

The set X clearly has the TPP. On the other hand, let h be the half-space given by the inequality $x \geq -\frac{3}{4}$. Then

$$H_1(X \cap h) = Z_2 \times Z_2 \text{ and } H_1(X) = Z_2,$$

so that injectivity on the 1-dimensional homology cannot hold, i.e. X is not 1-tight. For more discussion of tight and TPP manifolds with boundary see Banchoff [4], Kühnel [2]-[5], Lastufka [1], Rodriguez [1]-[2].

We now present Kuiper's [8] smooth embedding of S^3 into E^4 which has the TPP but is not tight. An example is given by the equation

$$\frac{x^2 + w^2}{16} + y^2 + (z - (x^2 + w^2))^2 = 1.$$

Algebraic examples of this type are motivated by the following geometrical construction. Begin with a standard torus of revolution in E^3 with y-axis as axis of rotation (see Figure 5.4). Take the lower half of the torus and smoothly cap the open ends with convex surfaces. Of course, there must be a smooth cylinder of transition from the torus to the convex cap. Both the smoothing and the capping can be done while preserving symmetry with respect to the xz and yz planes. This gives a smooth surface S of the type

$$\phi(x^2, y^2, z) = 0. \tag{5.4}$$

An algebraic example is

$$\frac{x^2}{16} + y^2 + (z - x^2)^2 = 1. \tag{5.5}$$

The desired embedding M of the 3-sphere in E^4 is obtained by rotating the surface S given by (5.4) or (5.5) about the yz-plane in 4-space, i.e. replace x^2 by $x^2 + w^2$. There are clearly linear height functions which have non-degenerate critical points of index 1 or 2 on M. Thus, M is not a tightly embedded 3-sphere. We will show that M has the TPP by demonstrating that all local extrema of height functions lie on the boundary of the convex hull of M and are thus global extrema (see Theorem 5.17). In view of the rotational symmetry of M, it is sufficient to study the behavior of non-degenerate linear height functions ℓ_q for $q \in E^3$. Furthermore, if q is normal to M at a point Q, then Q must lie in E^3. If Q lies on ∂HS, then Q also lies on ∂HM. Thus, the problem is reduced to showing that if Q is not in ∂HS, then ℓ_q does not have a local extremum at Q. The convex hull of S is the union of the convex hulls of the horizontal sections in the planes z = constant. The various possibilities for the sections are shown in Figure 5.5, and the boundaries of the convex hulls are indicated by the dotted lines. Suppose Q is a point at level 2 or level 3 which does not lie in the boundary of the convex hull of the cross-section. Then, the surface S does not have a local support plane at Q, i.e. the restriction of ℓ_q to S has a critical point of index 1 at q. Thus, the height function ℓ_q on M

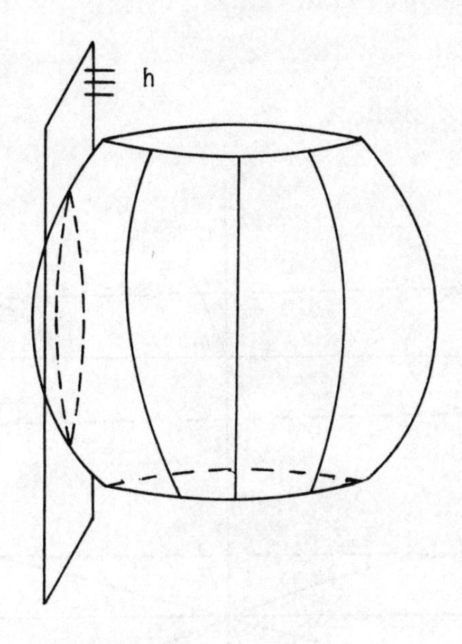

Figure 5.3 TPP annulus which is not tight

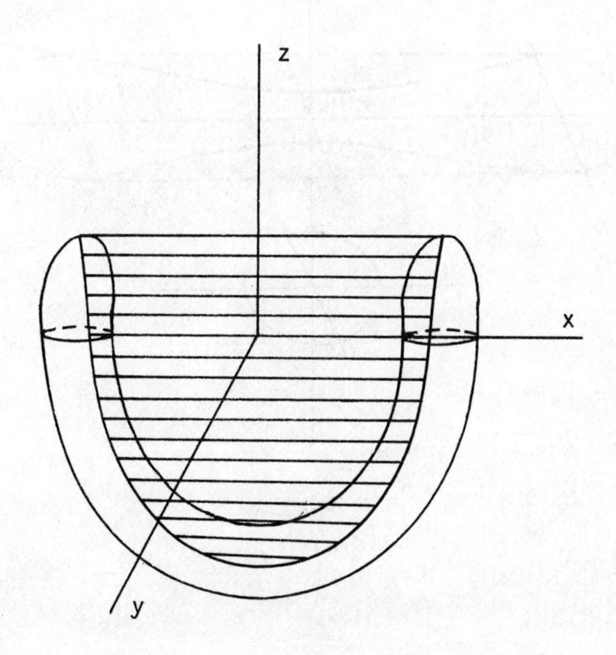

Figure 5.4 The profile surface S

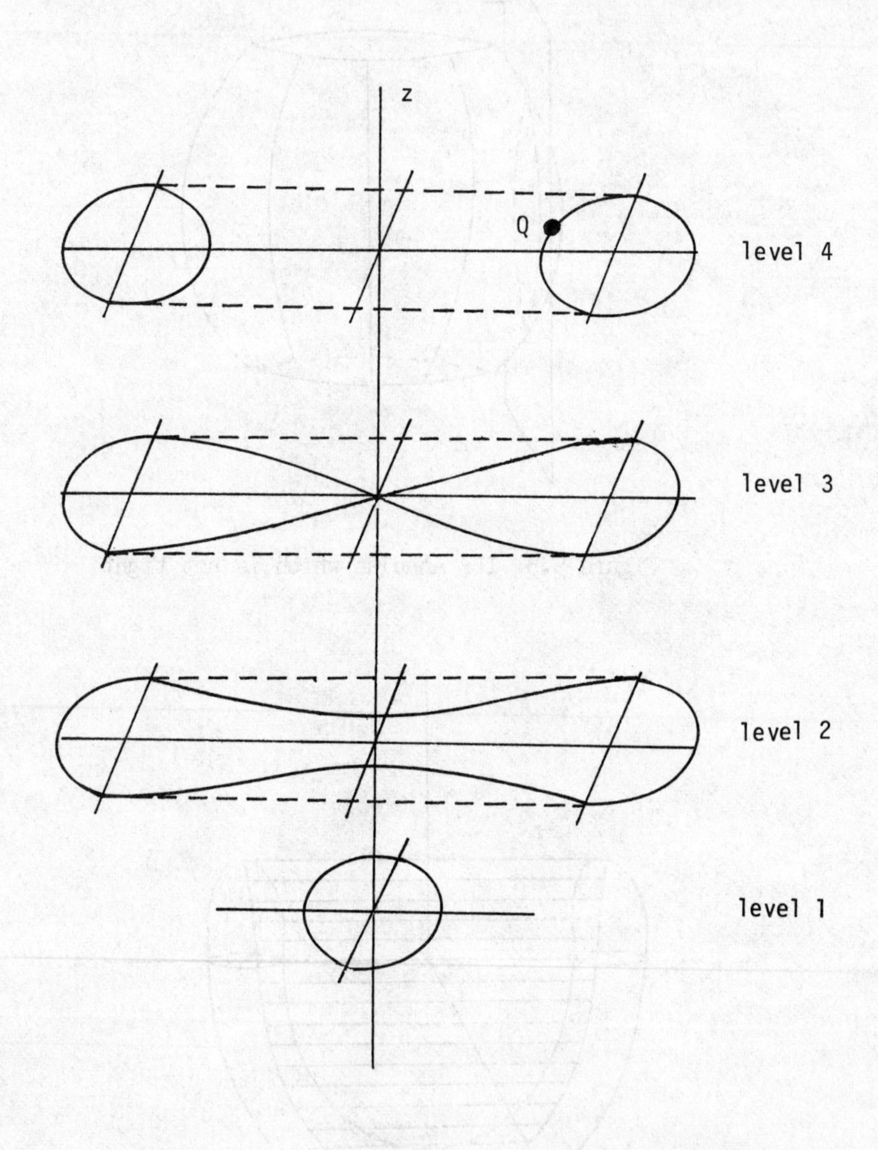

Figure 5.5 Horizontal sections of the surface S

does not have a local extremum at Q. Next, suppose that Q is a point at level 4 which is not in ∂HS. The surface S has positive Gaussian curvature at Q, so that the restriction of ℓ_q to S does have a local extremum at Q. However, under the rotation about the yz-plane in E^4, the two convex curves in level 4 sweep out a torus T in xyw-space which does not have a local support plane at Q. Thus again, the function ℓ_q on M cannot have a local extremum at Q. Therefore, every local extremum of a non-degenerate height function on M lies on ∂HM, and M has the TPP by Theorem 5.17. We know that M is not 1-tight since there exist non-degenerate height functions which have critical points of index 1 and $\beta_1(S^3) = 0$. In terms of intersections with half-spaces, it is also easy to see that M is not 1-tight. If h is the half-space $z \geq 0$ for the surface (5.4) constructed by rotating Figure 5.3, then $h \cap M$ is diffeomorphic to $S^1 \times D^2$, and the homomorphism

$$H_1(S^1 \times D^2) = Z_2 \rightarrow H_1(S^3) = 0$$

is not injective.

More generally, Kuiper [8, p. 221] showed, in the same manner, that the hypersurface with equation

$$\frac{1}{16}(x_1^2 + \ldots + x_{p+1}^2) + (y_1^2 + \ldots + y_q^2) + (z - (x_1^2 + \ldots + x_{p+1}^2))^2 = 1$$

in E^{p+q+2}, $q = p$ or $q = p + 1$, is an embedded S^{p+q+1} which is $(p-1)$-tight but not tight.

6. PRODUCTS OF TIGHT IMMERSIONS; TAUT IMMERSIONS.

We begin this section by showing that the product of two tight immersions is tight. This provides new examples, namely the product of two convex hypersurfaces. In particular, we study products of two Euclidean spheres and their conformal images known as the cyclides of Dupin. These turn out to satisfy the stronger condition of tautness, which we introduce here. Taut immersions will be studied much more extensively in Chapter 2.

Suppose $f: M \rightarrow E^m$, $g: M' \rightarrow E^{m'}$ are immersions of compact manifolds. One defines the product immersion

$$f \times g : M \times M' \to E^m \times E^{m'} = E^{m+m'}$$

by

$$(f \times g)\,(x,y) = (f(x), g(y)).$$

Kuiper [1, p. 76] showed that $\tau(f \times g) = \tau(f)\,\tau(g)$ and one can approach the subject from that viewpoint. However, we will use critical point theory instead. Let p be a unit vector in $E^{m+m'}$. We can write p uniquely as

$$p = \cos\theta\, q + \sin\theta\, q'$$

for $q \in E^m$, $q' \in E^{m'}$, $0 \leq \theta \leq \pi/2$.

<u>Lemma 6.1</u>: <u>Let $p = \cos\theta\, q + \sin\theta\, q'$, $0 \leq \theta \leq \pi/2$.</u>

(a) ℓ_p <u>is non-degenerate on $M \times M'$ if and only if $0 < \theta < \pi/2$ and</u> ℓ_q, $\ell_{q'}$, <u>are non-degenerate on M, M', respectively.</u>

(b) <u>If</u> ℓ_p <u>is non-degenerate, then</u>

$$\underline{\mu}(\ell_p) = \underline{\mu}(\ell_q)\,\underline{\mu}(\ell_{q'}).$$

<u>Proof</u>: Let $(x,y) \in M \times M'$ and $X \in T_x M$, $Y \in T_y M'$. Then an easy computation yields

$$(\ell_p)_*(X,Y) = \cos\theta\,(\ell_q)_* X + \sin\theta\,(\ell_{q'})_* Y. \tag{6.1}$$

If $\theta = 0$ or $\pi/2$, respectively, the critical points of ℓ_p are all points of the form $\{x\} \times M'$, where x is a critical point of ℓ_q or $M \times \{y\}$, where y is a critical point of $\ell_{q'}$, respectively. Since the critical points are not isolated in these cases, ℓ_p is clearly a degenerate function.

If $0 < \theta < \pi/2$, then it is evident from equation (6.1) that ℓ_p has a critical point at (x,y) if and only if ℓ_q has a critical point at x and $\ell_{q'}$ has a critical point at y.

Let $(u^1,\ldots,u^n)$ and $(v^1,\ldots,v^{n'})$ be local coordinates in neighborhoods of x in M and y in M', respectively. Clearly, with respect to the local coordinates $(u^1,\ldots,u^n, v^1,\ldots,v^{n'})$ in a neighborhood of (x,y) in $M \times M'$, the Hessian of ℓ_p has the form

$$(6.2) \qquad H_{(x,y)}(\ell_p) = \left[\begin{array}{c|c} \cos\theta\, H_x(\ell_q) & 0 \\ \hline 0 & \sin\theta\, H_y(\ell_{q'}) \end{array}\right].$$

Consequently, $H_{(x,y)}(\ell_p)$ is non-singular if and only if $H_x(\ell_q)$ and $H_y(\ell_{q'})$ are non-singular. Thus ℓ_p is non-degenerate and only if ℓ_q and $\ell_{q'}$ are non-degenerate.

Moreover, it is clear from (6.2) that the index of ℓ_p at a non-degenerate critical point (x,y) is equal to the sum of the indices of ℓ_q at x and $\ell_{q'}$ at y. Thus, for $0 \leq k \leq n+n'$,

$$\mu_k(\ell_p) = \sum_{i+j=k} \mu_i(\ell_q)\, \mu_j(\ell_{q'}).$$

From this, one computes

$$\mu(\ell_p) = \sum_{k=0}^{n+m} \mu_k(\ell_p)$$

$$= \sum_{k=0}^{n+m} \sum_{i+j=k} \mu_i(\ell_q)\, \mu_j(\ell_{q'})$$

$$= \mu(\ell_q)\, \mu(\ell_{q'}).$$

Q.E.D.

<u>Theorem 6.2</u>: <u>Suppose $f:M \to E^m$ and $g:M' \to E^{m'}$ are tight immersions of compact manifolds. Then $f \times g$ is a tight immersion of $M \times M'$ into $E^{m+m'}$.</u>

<u>Proof</u>: In the notation of Lemma 6.1, for any non-degenerate function ℓ_p on $M \times M'$,

$$\mu(\ell_p) = \mu(\ell_q)\, \mu(\ell_{q'})$$

for appropriate q, q'. Since f and g are tight,

$$\mu(\ell_q) = \beta(M;Z_2) \text{ and } \mu(\ell_{q'}) = \beta(M';Z_2)$$

for every such pair of non-degenerate functions ℓ_q, $\ell_{q'}$. Thus

$$\mu(\ell_p) = \beta(M;Z_2)\ \beta(M';Z_2) = \beta(M \times M';\ Z_2),$$

with the last equality due to the Künneth formula (see, for example, Greenberg [1, p. 198]). Hence $f \times g$ is tight.

Q.E.D.

T. Ozawa [4] has recently shown that this result still holds if X and Y are only assumed to be compact topological spaces and f and g are tight continuous maps.

From the preceding theorem we immediately obtain an important class of tight immersions. Suppose

$$f_i : S^{n_i} \to E^{n_i+1}, \qquad 1 \leq i \leq k,$$

is an embedding of S^{n_i} as a convex hypersurface in E^{n_i+1}. By Theorem 4.8, f_i is tight, and thus by Theorem 6.2,

$$f_1 \times \ldots \times f_k : S^{n_1} \times \ldots \times S^{n_k} \to E^{n_1 + \ldots + n_k + k}$$

is a tight immersion. An interesting special case of this example is the following.

Suppose $S^k(a)$ in E^{k+1} is the Euclidean hypersphere of radius a centered at the origin. By Theorem 6.2,

$$M = S^k(a) \times S^{n-k}(b) \subset E^{k+1} \times E^{n-k+1} = E^{n+2}$$

is tightly embedded. Moreover, if $r = (a^2 + b^2)^{1/2}$, then M lies in $S^{n+1}(r)$. An immersion whose image lies in a Euclidean sphere is called <u>spherical</u>, and one formulation of the definition of a taut immersion f is simply to require that f be tight and spherical. We will soon formulate the definition in an equivalent way which arises after stereographic projection.

For $q \in S^{n+1}(r)$, let P_q be stereographic projection from $S^{n+1}(r) - \{q\}$

onto the set

$$E_q^{n+1} = \{x \in E^{n+2} \mid \langle x,q \rangle = 0\}.$$

The images of M under stereographic projection belong to a collection of hypersurfaces called the cyclides of Dupin. For the moment, we will concentrate on the case where the point q is not on M, and thus $P_q(M)$ is compact and diffeomorphic to M. If the point q lies in one of the factors E^k or E^{n-k}, the image is called a round cyclide (or torus of revolution in the case $M^2 = S^1 \times S^1$). Otherwise, the cyclide appears as in Figure 5.2 of Chapter 2 and is called a ring cyclide. A ring cyclide can also be obtained from a round cyclide by an inversion of E^{n+1} in a sphere. (Recall that the inversion ϕ of E^{n+1} in the sphere centered at p of radius ρ is the map of $E^{n+1}-\{p\}$ onto itself which sends a point q to the point $\phi(q)$ on the ray from p through q satisfying $|q-p|\ |\phi(q)-p| = \rho^2$.)

The 2-dimensional cyclides were originally defined by Dupin in 1822 as the envelope of the family of spheres tangent to three fixed spheres in E^3. These surfaces were studied extensively by many prominent mathematicians of the nineteenth century including Cayley, Darboux, Klein, Liouville, and Maxwell (see Lilienthal [1, pp. 290-293] for a list of their respective contributions). The full collection of cyclides includes some surfaces with singularities and some which are unbounded (see Section 5 of Chapter 2).

From Dupin's definition, one derives that the cyclides are those surfaces in E^3 which are simultaneously the envelopes of two one-parameter families of spheres whose centers lie along a pair of so-called focal conics (see Section 5 of Chapter 2). These conics are, in fact, the focal set of the surface, and classically the cyclides were characterized as the only surfaces with the property that both sheets of the focal set are curves. This characterization in terms of the focal set was generalized to dimensions greater than two (see Cecil-Ryan [2], [5] and Pinkall [3]), and will be studied in Section 6 of Chapter 2.

We wish to show that the ring cyclides are tight in E^{n+1}, and moreover, that they are also taut in the sense we are about to define. Suppose now that $f:M \to S^m \subset E^{m+1}$ is a tight spherical immersion of a compact manifold M. A hyperplane π in E^{m+1} intersects S^m in either a single point or in an (m-1)- sphere S^{m-1}. In either case, each half-space h determined by π intersects S^m in a closed ball B in the metric of S^m. (Note that B could have radius 0 and consist of a single point.) Hence, the tightness of f implies that the map

$$H_*(f^{-1}B) \to H_*(M) \ (Z_2\text{-homology})$$

is injective for every closed ball B in S^m.

Stereographic projection P_q takes the sphere S^{m-1} to an Euclidean hypersphere or hyperplane in E^m_q, the latter case occurring when q lies on S^{m-1}. Thus, a closed ball determined by S^{m-1} in S^m is taken by P_q to either a closed ball, the complement of an open ball or a closed half-space in E^m. Therefore, we state the definition of a taut map into Euclidean space as follows.

Definition 6.3: A map f of a compact topological space X into E^m is called taut if for every closed ball, complement of an open ball, or closed half-space Ω in E^m, the induced homomorphism

$$H_*(f^{-1}\Omega) \to H_*(X)$$

in Z_2-Cech homology is injective.

It follows immediately from Definitions 5.1 and 6.3 that,

Proposition 6.4: A taut map of a compact topological space X into E^m is tight.

Moreover, the definition of tautness was designed to make the following true.

Theorem 6.5: Let $f:X \to S^m \subset E^{m+1}$ be a tight spherical map of a compact topological space X and let P_q be stereographic projection from $S^m - \{q\}$ to E^m with pole q not in f(X). Then $P_q \circ f$ is a taut map of X into E^m.

Therefore, the cyclides are taut, and hence tight, in E^m. Finally, we show that a tight spherical map $f:X \to S^m$ is also taut when considered as a map into E^{m+1}.

Theorem 6.6: Let $f:X \to S^m \subset E^{m+1}$ be a tight, spherical map of a compact topological space X. Then f is a taut map of X into E^{m+1}.

Proof: Let Ω be a closed ball or the complement of an open ball in E^{m+1}. Then $\Omega \cap S^m = h \cap S^m$ for some closed half space h in E^{m+1}. Since $f(X) \subset S^m$, we have

$$f^{-1}\Omega = f^{-1}(\Omega \cap S^m) = f^{-1}(h \cap S^m) = f^{-1}h.$$

By tightness, the map

$$H_*(f^{-1}h) \to H_*(X)$$

is injective, and so f is taut.

Q.E.D.

7. TIGHT SURFACES.

In this section, we cover the known results on tight immersions of surfaces into E^3, which are primarily contained in Kuiper [4]-[5]. Kuiper produced tight smooth embeddings of all orientable surfaces into E^3, and tight smooth immersions of non-orientable surfaces with Euler characteristic $\chi < -1$. At the same time, he showed that tight smooth immersions of the projective plane P^2 or the Klein bottle K^2 into E^3 were impossible. The question of existence of a tight immersion in the case $M = P^2 \# T^2$ $(\chi = -1)$ is still open. In addition, Kuiper [16] has shown that no continuous tight immersion of P^2 or K^2 into E^3 exists.

The principal tool in these results is the analysis of the top-sets. We will prove several general results on top-sets which have gradually distilled into Kuiper's Fundamental Theorem [13]: a top-set of a tight set is tight. This is not only essential for the results of this section, but also for the proof of the Chern-Lashof theorem and for many other results in the subject.

Examples of Smooth Tight Surfaces

We already have examples of tight embeddings of the 2-sphere, namely convex surfaces, and of the torus, the cyclides of Dupin. In [5], Kuiper produced the following examples of tight smooth embeddings of orientable surfaces, and tight immersions of non-orientable surfaces with even Euler characteristic $\chi < -1$. These examples are smooth, but not analytic because of the flat portions at top and bottom. In fact, such analytic examples were only recently constructed by Banchoff and Kuiper [1] for orientable surfaces and by Kuiper [15] for non-orientable surfaces with even $\chi < -1$ (see Remark 7.29).

We first introduce two basic constructions: a convex surface with planar pieces and a handle of negative Gaussian curvature which can be smoothly attached to the planar pieces. Both can be obtained as surfaces of revolution of the curve in Figure 7.1. The convex surface can also be obtained by smoothing a rectangular parallelpiped. (This is the best model for illustrations.) In Figure 7.1, the function $x = g(z)$ is smooth for $0 < z < 1$ with horizontal tangents at the points $(0,0)$ and $(0,1)$. Further, $g(z)$ is strictly increasing on $[0,1/2]$ and strictly decreasing on $[1/2,1]$, reaching a maximum value $x = a$ at $z = 1/2$. Its local inverses, of the form $z = h(x)$, are assumed to be C^∞-flat at the points $(0,0)$ and $(0,1)$, i.e., all the derivatives of h are zero there. If one rotates the curve around the axis $x = -b$, $b > 0$, and fills in the holes with planar disks, one obtains a smooth closed convex surface with disks

$$(x + b)^2 + y^2 \leq b^2, \quad z = 0 \quad \text{or} \quad z = 1,$$

in the top and bottom planes. On the other hand, rotating the curve about the axis $x = b$, $b > a$, yields a cylindrical surface with negative Gaussian curvature on its interior and a boundary consisting of the circles

$$(x - b)^2 + y^2 = b^2, \quad z = 0, \ z = 1.$$

If a surface has two planar pieces covering the respective circles, we can remove the interiors of the disks bounded by the circles and insert the cylindrical surface without destroying smoothness. Moreover, it follows from Theorem 3.2 that no height function has a local extremum on a surface

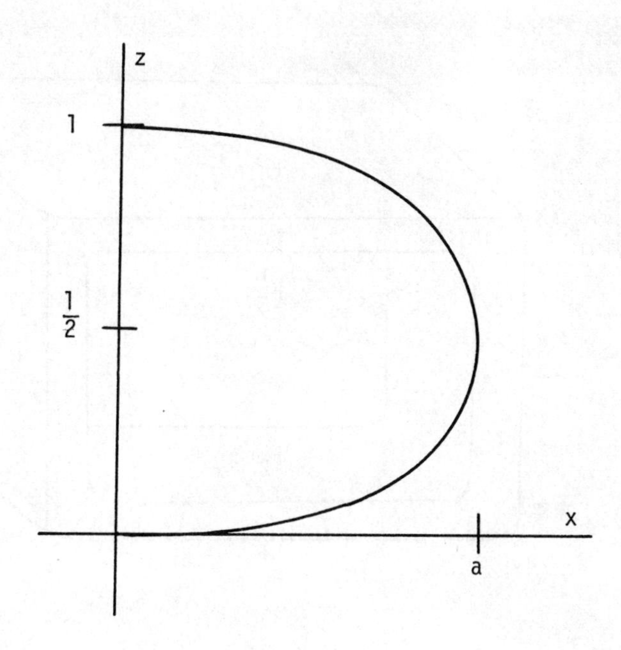

Figure 7.1 Profile Curve

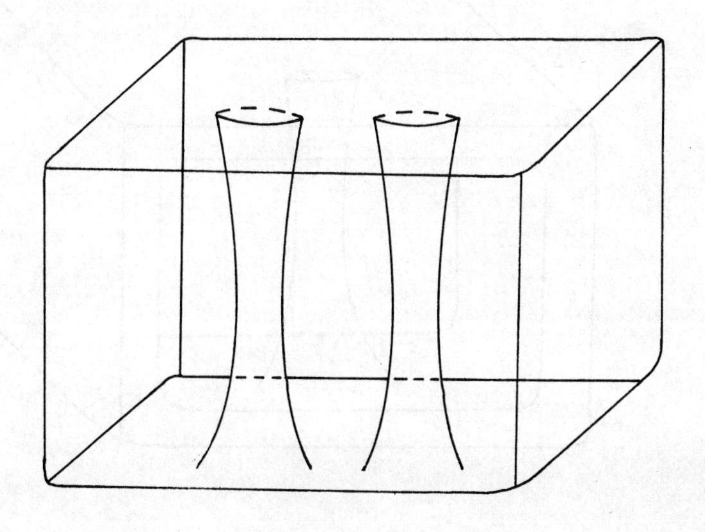

Figure 7.2 Tight embedding of a two-holed torus

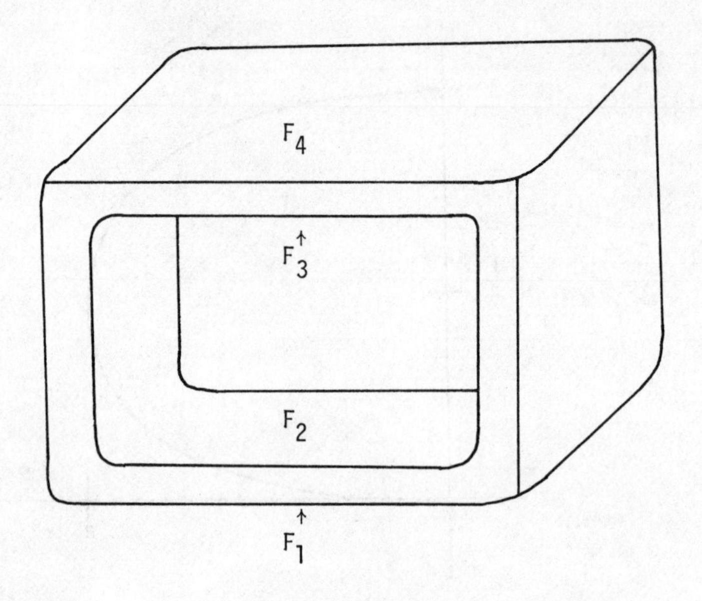

Figure 7.3 A tight torus

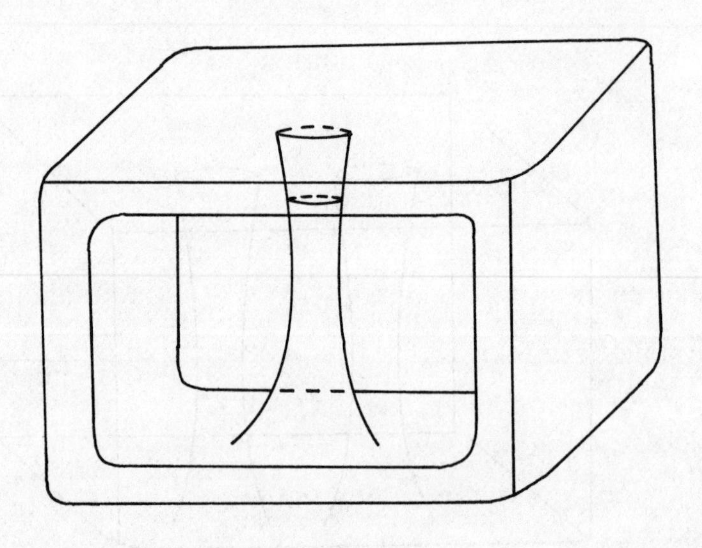

Figure 7.4 Tight immersion of $T^2 \# K^2$

of negative curvature. Thus, by Corollary 5.7, if the original surface is tight, then so is the resulting surface, i.e., each non-degenerate height function is still polar.

Example 7.1: Tight smooth embeddings of orientable surfaces

One begins with a convex surface M which has two parallel planar pieces. To construct a tight orientable surface M of genus g, remove g circular disks from the top and bottom planar pieces in identical fashion, and join each pair by a smooth handle of negative Gaussian curvature (see Figure 7.2 for $g = 2$).

An alternate construction begins with the tight smooth torus shown in Figure 7.3 with four horizontal planar pieces denoted by F_1, F_2, F_3, F_4 in ascending order of height. To get a tight embedding of an orientable surface with genus $g \geq 2$, add $g-1$ handles of non-positive curvature from F_2 to F_3. To get a tight smooth immersion which is not an embedding, add the $g-1$ handles from F_1 to F_4, cutting through the faces F_2 and F_3. For the torus, however, it is not at all obvious how to construct a tight immersion into E^3 which is not an embedding. An ingenious construction of such an immersion due to Kuiper is given in detail in the paper of Langevin and Rosenberg [1, p. 409].

Example 7.2: Tight smooth immersions of non-orientable surfaces with even χ

This example is a slight variation of the preceding one. One begins with the tight torus, as shown in Figure 7.3. To construct a tight immersion of a Klein bottle with a handle, $T^2 \# K^2$, remove circular disks from F_2 and F_4, and join the boundary circles by a handle of negative curvature which cuts through F_3 (see Figure 7.4). To construct a tight immersion of a non-orientable surface with $\chi = -2k$, $k > 1$, adjoin $k-1$ more smooth handles from F_2 to F_3 to the surface in Figure 7.4. Algebraic examples have been constructed by Kuiper [15]. Tight immersions of non-orientable surfaces with odd $\chi \leq -3$ have been constructed by Kuiper [5] also, but these are much more complicated (see Remark 7.23).

Example 7.3: Veronese surface

Let S^2 be the unit sphere in E^3 given by the equation

$$x^2 + y^2 + z^2 = 1.$$

Consider the map from S^2 into E^6 given by

$$(x,y,z) \to (x^2, y^2, z^2, \sqrt{2}\, yz, \sqrt{2}\, zx, \sqrt{2}\, xy).$$

This map takes the same value on antipodal points, so it induces a map $\phi: P^2 \to E^6$. One can show by an elementary direct computation that ϕ is injective on P^2 and that ϕ is an immersion (see Sections 9 and 10 for more detail and generalizations to higher dimensions). Hence, ϕ is a smooth embedding. If $(u_1, \ldots, u_6)$ are the standard coordinates in E^6, then it follows from $x^2 + y^2 + z^2 = 1$, that

$$u_1 + u_2 + u_3 = 1,$$

and thus $\phi(P^2)$ lies in the subspace E^5 given by this equation. It is easy to show directly that ϕ is a substantial embedding into E^5. Further, $\phi(P^2)$ lies on the unit sphere S^5 in E^6 given by the equation

$$u_1^2 + \ldots + u_6^2 = 1.$$

Hence, ϕ is a spherical embedding into the 4-sphere $S^4 = S^5 \cap E^5$. To see that ϕ has the TPP, note that a hyperplane in E^5, given by an equation

$$a_1 u_1 + \ldots + a_6 u_6 = c,$$

cuts P^2 in a conic. Such a conic does not separate P^2 into more than two pieces, and so ϕ has the TPP. Since P^2 is 2-dimensional, we conclude from Corollary 5.12 that ϕ is tight. Further, since ϕ is tight and spherical, it follows from Theorem 6.6 that ϕ is taut.

There are two standard ways to obtain a tight substantial embedding of

P^2 into E^4 from the Veronese embedding ϕ. First let P be stereographic projection from $S^4 - \{q\}$ to E^4 with pole q not on $\phi(P^2)$. Then, by Theorem 6.5, $P \circ \phi$ is a taut (and hence tight) embedding of P^2 into E^4. Secondly, we can take orthogonal projection of E^6 onto the 4-space E^4 spanned by the vectors $(e_1-e_2)/\sqrt{2}$, e_4, e_5, e_6, where $e_1,\ldots,e_6$ is the standard orthonormal basis for E^6. This gives a parametrization

$$f:(x,y,z) \to \left(\frac{x^2 - y^2}{\sqrt{2}}, \sqrt{2}\, yz, \sqrt{2}\, zx, \sqrt{2}\, xy\right)$$

which is also an embedding on P^2. Since tightness is perserved under orthogonal projection (see Remark 5.2(c)), f is a tight embedding of P^2 into E^4.

We now define the notion of a top-set.

<u>Definition 7.4</u>: Let $f:X \to E^m$ be a continuous map of a compact space. For a unit vector $p \in E^m$, the <u>top-set $\Omega(p)$</u> is the set

$$\Omega(p) = \{x \in X \mid \ell_p(x) = \max_{y \in X} \ell_p(y)\}.$$

If f is a tight smooth immersion of a smooth manifold and ℓ_p is non-degenerate, then $\Omega(p)$ is the single point where ℓ_p attains its maximum. It is precisely the top-sets of degenerate height functions which are most important, however.

<u>Definition 7.5</u>: Let $f:X \to E^m$ be a continuous map of a compact space. For p_1, p_2 an orthonormal 2-frame in E^m, the <u>top^2-set</u> $\Omega(p_1,p_2)$ is defined by

$$\Omega(p_1,p_2) = \{x \in \Omega(p_1) \mid \ell_{p_2}(x) = \max_{y \in \Omega(p_1)} \ell_{p_2}(y)\}.$$

By induction, for an orthonormal k-frame $p_1,\ldots,p_k$, we define the <u>top^k-set</u>

$$\Omega(p_1,\ldots,p_k) = \{x \in \Omega(p_1,\ldots,p_{k-1}) \mid \ell_{p_k}(x) = \max_{y \in \Omega(p_1,\ldots,p_{k-1})} \ell_{p_k}(y)\}.$$

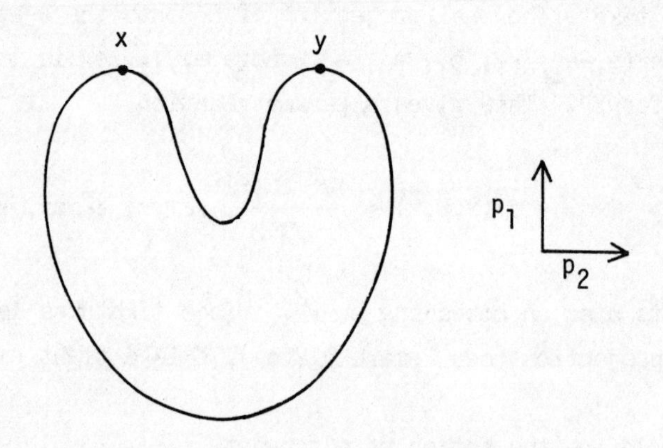

Figure 7.5 Top-set of an embedded circle

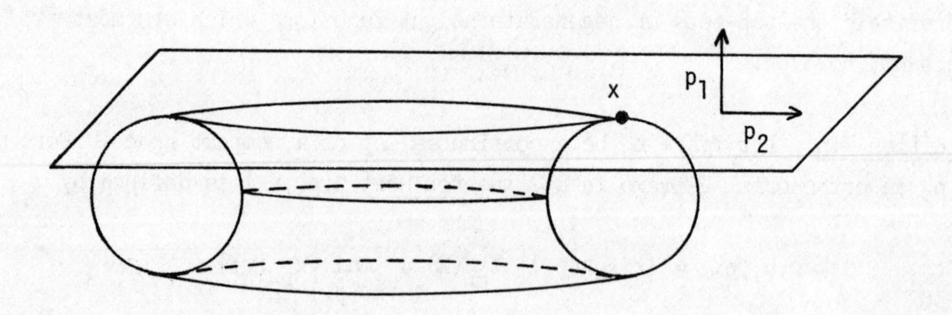

Figure 7.6 Top-set of a torus of revolution

A top*-set is a top^k-set for some $k \geqslant 1$.

We now consider some examples of top-sets.

Example 7.6: f is the embedding of S^1 as the closed plane curve in the Figure 7.5. Then, $\Omega(p_1) = \{x,y\}$ and $\Omega(p_1,p_2) = \{y\}$.

Example 7.7: $f:T^2 \to E^3$ is an embedding as a torus of revolution. $\Omega(p_1)$ is the circle in T^2 which is embedded onto the highest horizontal circle. $\Omega(p_1,p_2)$ is the point x (see Figure 7.6).

Example 7.8: f is the embedding of the torus T^2 onto the surface in Figure 7.3. For p normal to the front plane, $f\Omega(p)$ is an annulus. For q normal to the top plane, $f\Omega(q)$ is a convex disk.

For tight immersions $f:M \to E^3$ of surfaces, the important case to study occurs when the convex hull $Hf\Omega(p)$ is 2-dimensional. Kuiper [4] showed that the boundary curve $\partial Hf\Omega(p)$ is always contained in the image of f, and that f is an embedding on $\gamma = f^{-1}(\partial Hf\Omega(p))$. The crucial distinction is between those γ which separate M and those which do not. If γ separates M, then we will see that f must embed the top-set onto the convex disk bounded by $f(\gamma)$, as with $\Omega(q)$ in Example 7.8. Thus, the interesting case occurs when γ does not separate M, e.g., the top and bottom circles of a torus of revolution, the boundary curve of the top-set $\Omega(p)$ in Example 7.8 and the boundary curve of the annulus in the top plane of the immersion shown in Figure 7.4.

Kuiper's Fundamental Theorem has evolved since 1960, when he showed [4] that a top^2-set of a tight surface in E^3 must be connected, i.e., a top-set of a tight surface is also tight. In the 1970 Inventiones Math. paper [8], the result took the form of the "Key Lemma." In 1972 [10], he showed that every top*-set of a TPP embedding has the TPP. Finally in the 1979 Chern symposium paper [13], the result was further generalized, with the use of Cech homology, to the

Fundamental Theorem: A top*-set of a tight set $X \subset E^m$ is tight.

We first prove a version of the Key Lemma as it appeared in Kuiper-Pohl

[1, p. 180]. Let M be a compact connected topological space and $f:M \to E^m$ a continuous map. We will use the notation,

$$M_r^+(p) = \{x \in M | \ell_p(x) > r\},$$

and

$$\bar{M}_r^+(p) = \{x \in M | \ell_p(x) \geq r\},$$

will denote the closure of $M_r^+(p)$.

<u>Lemma 7.9</u>: <u>(Key Lemma): Let $f:M \to E^m$ be a continuous map of a compact topological space. Suppose Ω is a top^*-set of f and h is a closed half-space in E^m such that $\Omega \cap f^{-1}h$ is contained in an open set U in M. Then there exists a unit vector q and a real number r such that</u>

$$\Omega \cap f^{-1}h \subset M_r^+(q) \subset \bar{M}_r^+(q) \subset U.$$

<u>Moreover, the same result holds for all q' sufficiently near q</u>.

<u>Proof</u>: Suppose Ω is the top^{k-1}-set $\Omega(p_1,\ldots,p_{k-1})$ and h is given by the inequality $\langle x, p_k\rangle \geq 0$. The choice of origin in E^m can be made so that the height function determined by p_j takes the maximum value 0 on the top^j-set $\Omega(p_1,\ldots,p_j)$ for $1 \leq j \leq k-1$. Thus $f(\Omega \cap h)$ lies in the closed (m-k+1)-half-plane K_1 determined by the equations

$$\ell_{p_1} = \ldots = \ell_{p_{k-1}} = 0, \ \ell_{p_k} \geq 0.$$

On the other hand, $f(M-(\Omega \cap f^{-1}h))$ lies in the convex, not closed, set K_2 which is the union of the sets $S_1,\ldots,S_k$ with equations:

$$S_1: \ \ell_{p_1} < 0$$
$$S_2: \ \ell_{p_1} = 0, \ \ell_{p_2} < 0 \ldots$$
$$S_k: \ \ell_{p_1} = \ldots = \ell_{p_{k-1}} = 0, \ \ell_{p_k} < 0.$$

The convex sets K_1 and K_2 are disjoint, and K_1 contains the compact convex set $H(f\Omega \cap h)$, while K_2 contains the compact set $Hf(M-U)$. There exists a hyperplane $\ell_q = r$ which separates the disjoint compact convex sets $H(f\Omega \cap h)$ and $Hf(M-U)$ and is disjoint from both. It follows that

$$\Omega \cap f^{-1}h \subset M_r^{\ +}(q) \subset \overline{M}_r^{\ +}(q) \subset U,$$

and this also holds for the same r, for all q' sufficiently near q.

Q.E.D.

We first show that if f has the TPP, then so also do all of its top^*-sets. We then prove the Fundamental Theorem.

<u>Corollary 7.10</u>: <u>Let $f:M \to E^m$ be a continuous map of a compact space. If f has the TPP, then every top^*-set of f has the TPP.</u>

<u>Proof</u>: We show that every top^k-set is connected. This implies that every top^{k-1}-set has the TPP. Suppose that a top^k-set Ω is not connected. Let U_1 and U_2 be disjoint open sets in M, each having non-empty intersection with Ω. Apply the Key Lemma to $U = U_1 \cup U_2$ with the half-space h chosen so that $\Omega \cap f^{-1}h = \Omega$. One gets a unit vector q and real number r so that

$$\Omega \subset \overline{M}_q^{\ +}(r) \subset U.$$

Thus $\overline{M}_q^{\ +}(r) \cap U_i$ is non-empty for i = 1,2, and we conclude that $\overline{M}_q^{\ +}(r)$ is not connected, contradicting the assumption that f has the TPP.

Q.E.D

Theorem 7.11: (Fundamental Theorem). Let $f:M \to E^m$ be a continuous map of a compact space. If f is tight and Ω is a top^*-set of f, then the map $f|_\Omega$ is also tight.

Proof: The proof is accomplished through the continuity property of Cech homology, where the approximating sequence is constructed using the Key Lemma. We do the proof for top^1-sets and the result for top^k-sets follows by induction. Finally, we will do the proof only for the case when f is an embedding. Therefore, we can identify M with f(M) and greatly simplify the notation. The generalization of the proof to the case where f is a map is straightforward, although cumbersome.

Let Ω be a top^1-set of f, say $\Omega = \Omega(p)$. Note that the tightness of f implies that the map

$$H_*(\Omega) \to H_*(M), \ (Z_2 - \text{homology})$$

is injective, since $\Omega = M \cap h'$, where h' is determined by the inequality $\ell_p \geq r_0 = \max \ell_p$.

Let h be an arbitrary half-space in E^m. We must show that the map

$$\alpha: H_*(\Omega \cap h) \to H_*(\Omega)$$

is injective. We now use the Key Lemma to produce a sequence of half-spaces h_i, i = 1,2,... such that

$$(7.1) \qquad h_i \cap M \supset h_{i+1} \cap M \supset \ldots \supset \bigcap_{j=1}^{\infty} h_j \cap M = h \cap \Omega.$$

The construction resembles that used in Theorem 5.4. At the ith step of the induction, we use the Key Lemma to choose a unit vector q and number r such that

$$h \cap \Omega \subset M_r^+(q) \subset \bar{M}_r^+(q) \cap U,$$

where U is the set of points of the previous $M_r^+(q)$ whose distance to $h \cap \Omega$

is less than $1/i$. (For $i = 1$, use M for the previous $M_r^+(q)$.) Then h_i is the half-space $\ell_q > r$, constructed at the ith step and the sequence $\{h_i\}$ satisfies (7.1).

The continuity property of Cech homology says that

$$H_*(h \cap \Omega) = \varprojlim_{i\to\infty} H_*(h_i \cap M).$$

We now have obtained the following commutative diagram of morphisms in homology. The maps other than α, β, γ are injective because the embedding $f:M \to E^m$ is tight.

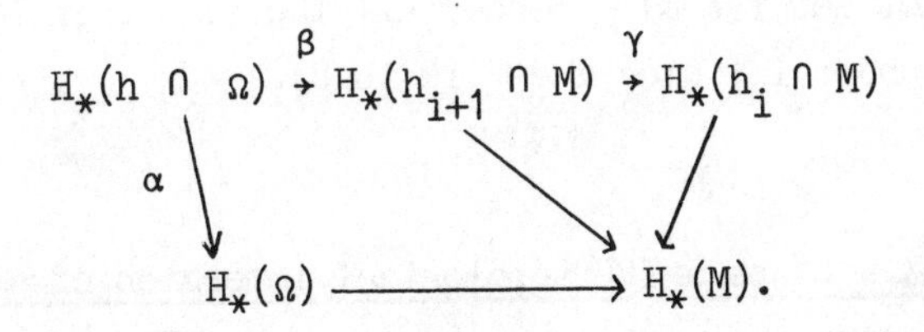

Then every map of the form γ in the diagram is injective and we conclude from Theorem 3.4 of Eilenberg-Steenrod [1, p. 216] on inverse limits that each map of the form β is injective. The commutativity of the above diagram then implies that α is injective, as desired.

Q.E.D.

We now turn to the subject of tight immersions of surfaces into E^3. Later, we will consider smooth immersions, but at present we will only assume continuity. We recall the definitions of embedding and immersion in the topological category. A <u>topological embedding</u> of a topological space X into E^m is a map $f:X \to E^m$ such that f is a homeomorphism of X onto $f(X)$, considered as a subspace of E^m. It is well known that if f is a continuous injective map and X is compact, then f is an embedding. A <u>topological immersion</u> $f:X \to E^m$ of a compact topological space is a map such that every point in X has a neighborhood on which f restricts to an embedding. The reader is referred to Lastufka [1] for a rather complete treatment of tight topological immersions. The following lemma is useful for handling immersions in this context.

Lemma 7.12: Let $f:X \to E^m$ be a topological immersion of a compact space. The set of points in E^m with more than one pre-image is closed.

Proof: We show that the complement of the set in question is open. Let Q be a point in E^m such that $f^{-1}(Q) = \{x\}$. Let U be a neighborhood of x in X on which f is an embedding. Then Q is not in f(X-U), so there exists $\varepsilon > 0$ such that the open ball $B_\varepsilon(Q)$ in E^m centered at Q with radius ε does not intersect the compact set f(X-U). Hence, all the points in $B_\varepsilon(Q)$ have at most one pre-image under f (which must lie in U).

Q.E.D.

Lemma 7.13: Let $f:X \to E^m$ be a TPP topological immersion of a compact space. Let Ω be a top*-set such that $f(\Omega)$ is a convex set. Then f is injective on Ω.

Proof: The proof is by induction on the dimension j of the convex set $f(\Omega)$. First suppose that j = 0, i.e. $f(\Omega)$ is a point P. Since f is an immersion and X is compact, Ω consists of a finite number of points. But f has the TPP, so Ω is connected by Corollary 7.10, i.e. Ω consists of just one point.

Suppose now that the lemma is true for all top*-sets whose image is a convex set of dimension less than j. Suppose that $f(\Omega)$ is a closed j-dimensional convex disk D in a j-plane E^j. Note that $\Sigma = \partial D$ is the union of the sets $\pi \cap D$, where π is a support plane for D in E^j (see, for example, Lastufka [1, p. 378] for a proof). Each $\pi \cap D$ is equal to $f(\Omega')$, where Ω' is a top-set of Ω. Since $f(\Omega')$ is a convex set of dimension less than j, f is injective on Ω' by the induction hypothesis. Hence f is an embedding on $\gamma = f^{-1}\Sigma$. Let

$$K = \{x \in \Omega \mid f(x) = f(y) \text{ for some } y \neq x \text{ in } \Omega\}.$$

Suppose that K is non-empty. By Lemma 7.12, K is closed in X and hence compact. Since f is an embedding on γ, $f(K)$ is contained in the interior of D, and so also is $Hf(K)$. The set $Hf(K)$ is a closed convex disk of dimension less than or equal to j. One can find a point Q in $\partial Hf(K)$ such that $Q \in f(K)$ and $Hf(K)$ has a strict support $(j-1)$-plane π in E^j. As before, $f^{-1}(Q)$ consists of a finite number as before of points $x_1,\ldots,x_\ell$ with $\ell \geq 2$. Let h be a closed half-space in E^m whose boundary hyperplane intersects E^j in the $(j-1)$-plane π with $h \cap f(K) = \{Q\}$. All points in $h \cap D$ have exactly one pre-image except Q. Thus, f is a homeomorphism from the set $V = \Omega \cap f^{-1}(h-\{Q\})$ onto $(h \cap D)-\{Q\}$.

We claim that only one point in $f^{-1}(Q)$ is a limit point of V. To see this, let $x \in f^{-1}(Q)$, and let W be a neighborhood of x in X on which f is an embedding. Let $U = (f|_W)^{-1}(B_\varepsilon(Q))$ and $S = V \cap f^{-1}(B_\varepsilon(Q))$. It is important to note that S is connected, as f embeds S onto the set $(h \cap D \cap B_\varepsilon(Q))-Q$ (the shaded region in Figure 7.7). (Note that this is what fails if one tries to duplicate this proof for the example in Remark 7.14, where one only knows that $f(\Omega)$ is a closed convex disk with a number of disjoint open convex disks removed.)

We next show that $U \cap S$ is both open and closed in S. It is obviously open, since U is open in X and S has the relative topology. Suppose that a sequence $\{y_n\}$ in $U \cap S$ converges to a point $y \in S$. Then $d(Q,f(y)) < \varepsilon$, and for all n sufficiently large, $f(y_n) \in \overline{B}_\delta(Q)$ for some $\delta < \varepsilon$. Thus, these y_n all belong to the compact subset $L = (f|_W)^{-1}(\overline{B}_\delta(Q))$ of U. Hence, the limit point y is in L and so $y \in U$, as desired. Thus, if $U \cap S$ is non-empty, it equals S.

Now suppose $U_1,\ldots,U_\ell$ are disjoint neighborhoods of the points $x_1,\ldots,x_\ell$ in $f^{-1}(Q)$, all constructed as the set U above with the same ε. At least one of the points in $f^{-1}(Q)$, say x_1, is a limit point of V. Since $U_1 \cap S$ is non-empty, $U_1 \cap S = S$. Consequently, $U_i \cap S = \emptyset$ for $2 \leq i \leq \ell$, and x_1 is the only limit point of V in $f^{-1}(Q)$. We now see that the set

$$\Omega \cap f^{-1}h = \overline{V} \cup \{x_2,\ldots,x_\ell\}$$

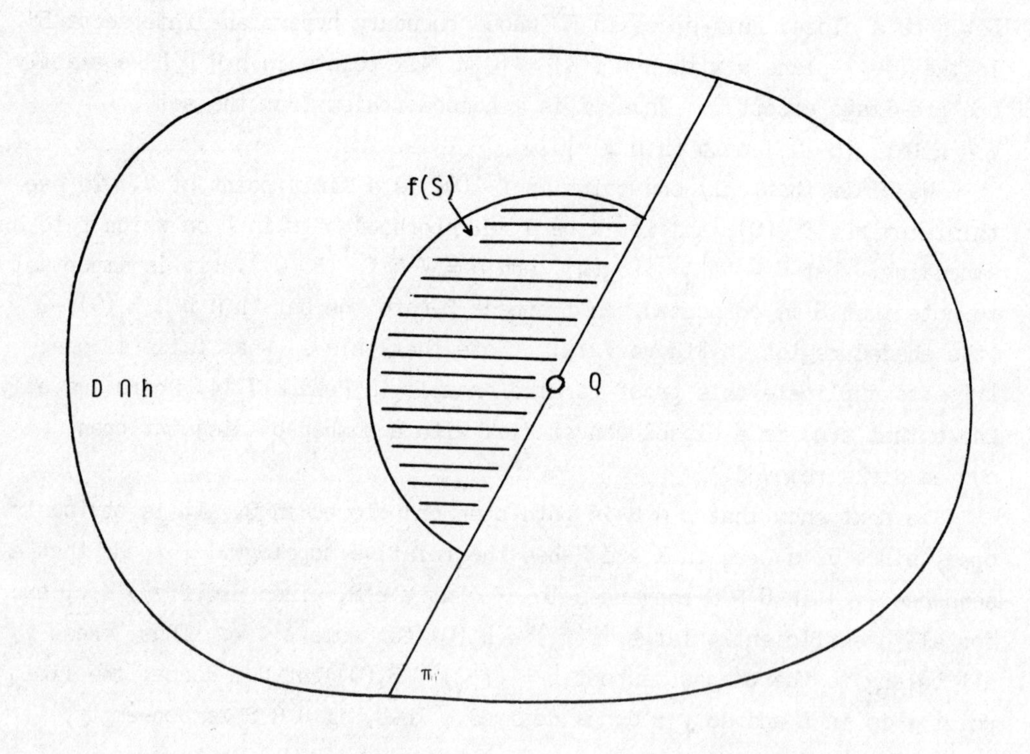

Figure 7.7 The Image f(S) in the disk D

is disconnected, contradicting the fact that Ω has the TPP by Corollary 7.10.

Q.E.D.

Given the preceding Lemma 7.13, it is worthwhile to note that a top-set of an arbitrary tight smooth immersion is not necessarily embedded. The following example is due to T. Banchoff and was published in Cecil-Ryan [8]. In his thesis, L. Rodriguez [1] gave a similar TPP immersion of a surface with boundary into E^3 which has the same non-embedded top-set. On the other hand, Thorbergsson [3] has recently shown that if a top*-set is a Z_2-Cech cohomology point, then the corresponding top*-map is injective and its image is convex.

<u>Remark 7.14</u>: <u>A top-set of a tight immersion is not necessarily embedded</u>. For the sake of clarity, we first give a polyhedral version of the example. One can think of the tight rectangular torus T in Figure 7.8 as having been constructed by taking two identical annuli (which lie in the front and back planes of Figure 7.8) and joining corresponding points of their boundaries by line segments. If we apply the same construction to the plane figure obtained by adding a rectangular strip with vertices P, Q, R, S (see Figure 7.9) across the annulus, the result is a tight topological embedding of the 2-holed torus. This may also be thought of as adding a handle (based on PQRS) to the torus T. We can also add such a handle based on P'Q'R'S'. Of course, this will intersect the other handle. However, the result is a tight topological immersion of f of a 3-holed torus M whose top-set in the front plane is not embedded (see Figure 7.9). Its image is the outer annulus with two strips added to form a cross. The points in the center square are covered twice. Topologically, the front top-set Ω is a torus with two disks removed. The boundary of one disk is embedded by f onto the outer boundary curve of $f(\Omega)$. The boundary of the other disk is immersed onto the curve indicated by the arrows in Figure 7.9 which has rotation index 3. The image $f(\Omega)$ is a TPP plane set, a closed convex disk with 4 disjoint open convex disks removed. However, if C is the boundary of one of these disks, then $f^{-1}C$ is not a closed curve in M. The construction can be modified to provide a smooth tight immersion of the 3-holed torus with a non-embedded

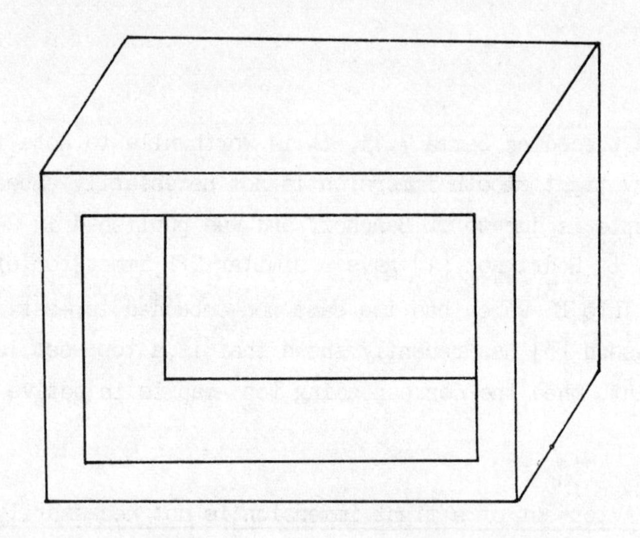

Figure 7.8 Tight rectangular torus

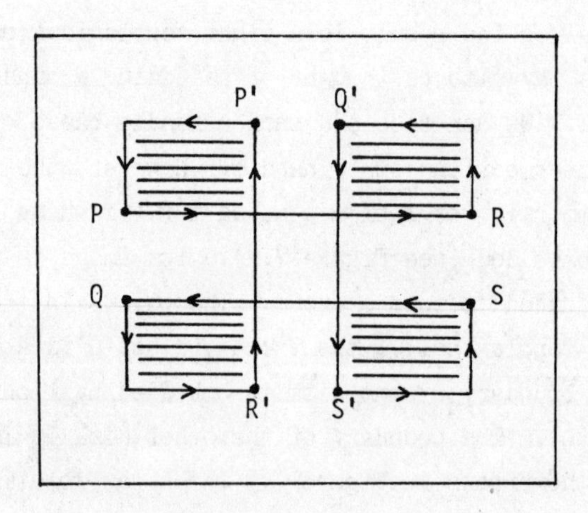

Figure 7.9 Non-embedded top-set

top-set. Begin with the smooth torus in Figure 7.3. The handles described above can be smoothed in such a way that they have non-positive Gaussian curvature, and the front (and back) top-set still has the same topological properties. The curve with rotation index 3 is now a smooth locally convex immersed curve. As noted earlier, the method of proof in Lemma 7.13 breaks down in this example if the point Q (in that proof) is taken to be one of the four corner points of the doubly covered square.

A second example of a non-embedded top-set in higher dimensions is the following due to G. Thorbergsson. Let $f:M \to E^3$ be a tight immersion of a compact surface which is not an embedding, e.g. the surface in Figure 7.4. Let $g:S^1 \to E^2$ embed S^1 as a circle. Then by Theorem 6.2, the immersion

$$f \times g : M \times S^1 \to E^3 \times E^2 = E^5$$

is tight. If one considers a unit vector of the form $p = (0,q)$ in E^5, then $\Omega(p) = M \times \{q\}$, and the restriction of $f \times g$ to $\Omega(p)$ is not an embedding.

Recall that tightness and the TPP are equivalent for continuous maps of surfaces.

Lemma 7.15: *Let $f:M \to E^3$ be a TPP topological immersion of a compact surface M. For any top* - set Ω,*

(a) *if $Hf(\Omega)$ has dimension 0 or 1, then f is an embedding of Ω onto $Hf(\Omega)$.*

(b) *if $Hf(\Omega)$ is 2-dimensional, then $\partial Hf(\Omega) \subset f(\Omega)$, and f is an embedding on $\gamma = f^{-1}(\partial Hf(\Omega))$. Further, if γ separates M, then Ω is the closure of one of the components of $M - \gamma$, and f embeds Ω onto the disk $Hf(\Omega)$.*

Proof: (a) If $Hf(\Omega)$ has dimension 0, then f is injective on Ω by Lemma 7.13. Next, suppose $Hf(\Omega)$ is the segment $[Q_1,Q_2]$ in E^3. Since every point in $Hf(\Omega)$ is a convex combination of points in $f(\Omega)$, the points Q_1 and Q_2 are in $f(\Omega)$. Since Ω is connected, the connected set $f(\Omega)$ must be all of $[Q_1,Q_2]$, and again f is injective on Ω by Lemma 7.13.

(b) Suppose $Hf(\Omega)$ lies in the 2-plane E^2. Every support line ℓ of

$Hf(\Omega)$ in E^2 is also a support line of $f(\Omega)$. For such a support line ℓ, the set $\Omega' = f^{-1}\ell \cap \Omega$ is a top-set of Ω, and $Hf(\Omega')$ has dimension 0 or 1. By part (a), f is an embedding of Ω' onto $Hf(\Omega') = \ell \cap Hf(\Omega)$. Now $\partial Hf(\Omega)$ is the union of these intersections $\ell \cap Hf(\Omega)$, where ℓ is a support line of $f(\Omega)$ in E^2. Hence, $\partial Hf(\Omega) \subset f(\Omega)$ and f is an embedding on $\gamma = f^{-1}(\partial Hf(\Omega))$.

Finally, suppose γ separates M and that the equation of the plane of $f(\gamma)$ is $\ell_p = r$. Then, one of the components of $M - \gamma$, call it V, is mapped into the convex disk $Hf(\Omega)$. For otherwise, $M_r^-(p)$ is disconnected in violation of the TPP. Since f is an immersion, the Invariance of Domain theorem (see Eilenberg-Steenrod [1, p. 303]) implies that f(V) is an open subset of the open convex disk, interior $Hf(\Omega)$. We next show that f(V) is also closed in the interior of $Hf(\Omega)$. Suppose Q is a limit point of f(V) in the interior of $Hf(\Omega)$, and let $Q_1, \ldots, Q_n, \ldots$ be a sequence in f(V) converging to Q. Choose $x_i \in V$ such that $f(x_i) = Q_i$. The sequence $\{x_i\}$ has a limit point x in the compact set $\overline{V} = V \cup \gamma$. By continuity, $f(x) = Q$, and x cannot be in γ, since $f(\gamma)$ is disjoint from the interior of $Hf(\Omega)$. Thus $x \in V$, and so $Q \in f(V)$ and f(V) is closed in interior $Hf(\Omega)$. Therefore f(V) = interior $Hf(\Omega)$, and $f(\Omega)$ is a convex set. By Lemma 7.13, f embeds Ω onto the convex set $Hf(\Omega)$, and $\Omega = \overline{V}$.

Q.E.D.

As a consequence, we get Kuiper's [4] proof of the Chern-Lashof theorem in E^3.

<u>Theorem 7.16</u>: <u>Let $f: S^2 \to E^3$ be a tight topological immersion. Then f embeds S^2 onto the convex surface $\partial Hf(S^2)$.</u>

<u>Proof</u>: As before, we note that $\partial Hf(S^2)$ is the union of the $Hf(\Omega)$, where Ω is a top-set of f. Since every closed curve γ separates S^2, Lemma 7.15 implies that f embeds each top-set Ω onto $Hf(\Omega)$. Thus, $f^{-1}(\partial Hf(S^2))$ is a 2-sphere embedded in S^2, and it must therefore be equal to S^2.

Q.E.D.

We see from Lemma 7.15 that if $Hf(\Omega)$ is 2-dimensional and $Hf(\Omega) \neq f(\Omega)$, then $\gamma = f^{-1}(\partial Hf(\Omega))$ does not separate M. Such a curve γ is called a <u>top-cycle</u> of the immersion.

Examples of top-cycles

(a) the top and bottom circles of a torus of revolution;

(b) the tight immersion of $T^2 \# K^2$ in Figure 7.4 has three top-cycles, the outer boundary curves of the annuli in the front, back, and top planes.

Lemma 7.17: Let $f:M \to E^3$ be a tight topological immersion of a compact, connected surface. Then the number of top-cycles is finite.

Proof: We show that the set of points p in S^2 such that $\Omega(p)$ contains a top-cycle is isolated and hence finite. Suppose that $\Omega(p)$ contains a top-cycle. By Lemma 7.12, there exists an $\varepsilon > 0$ such that f restricts to an embedding on the inverse image of the Euclidean tube of radius ε around $f(\gamma)$. Let U be an annular neighborhood of γ which is embedded inside this tube. Then one can show, as in the proof of Lemma 5.14, that there exists a neighborhood N of p in S^2 such that if $q \in N$, then $\Omega(q) \subset U$. Now suppose that $q \in N$ and write $q = \cos\alpha\, p + \sin\alpha\, p'$ with p' orthogonal to p and $\sin\alpha > 0$. Let r be the value of $\ell_{p'}$ on the top^2-set $\Omega(p,p')$. The top-set $\Omega(q)$ is contained in the intersection of U with the half-space $\ell_{p'} \geq r$. Hence, $\Omega(q)$ lies in a contractible neighborhood of $\Omega(p,p')$ (which is a point or segment), and $\Omega(q)$ cannot contain a top-cycle.

Q.E.D.

Lemmas 7.15 and 7.17 can be summarized in the following structure theorem of Kuiper [5]. Note that if f is smooth, then any two top-cycles must be disjoint since the normal to f(M) at a point in $\Omega(p)$ must be p. However, if f is only assumed to be continuous, then two top-cycles can intersect in a point or a closed line segment.

Theorem 7.18: Let $f:M \to E^3$ be a tight topological immersion of a compact connected surface M not homeomorphic to S^2. Then M is the union of two non-empty disjoint open sets U and V and the finite number of top-cycles $\gamma_1,\ldots,\gamma_k$, such that:

(a) The set U is embedded onto the complement in $\partial Hf(M)$ of a finite

<u>number of plane closed convex disks</u> $D_1, \ldots, D_k$, <u>where</u> $\gamma_i = f^{-1}(\partial D_i)$ <u>for</u> $1 \leq i \leq k$.

(b) <u>If f is smooth, then the Gaussian curvature satisfies</u> $K \geq 0$ <u>on U and</u> $K \leq 0$ <u>on V.</u>

<u>Proof</u>: (a) There must exist some top-cycles, for otherwise, f embeds M onto $\partial Hf(M)$, and M is a sphere. Let $\gamma_1, \ldots, \gamma_k$ be the finite number of top-cycles, and let $D_1, \ldots, D_k$ be the plane closed convex disks such that $f(\gamma_i) = \partial D_i$, $1 \leq i \leq k$. We know that $\partial Hf(M)$ is the union of the $Hf(\Omega)$ as Ω ranges over the top-sets of f. If a point p is in the set

$$W = \partial Hf(M) - (D_1 \cup \ldots \cup D_k),$$

then by Lemma 7.15, $p \in f(\Omega) = Hf(\Omega)$, where Ω is a top-set such that $Hf(\Omega)$ has dimension 0 or 1, or such that $Hf(\Omega)$ is 2-dimensional and f embeds Ω onto $Hf(\Omega)$. Thus, f is an embedding on $U = f^{-1}W$. Furthermore, it also follows from Lemma 7.15(b) that f is an embedding on the closed set A which is defined as the union of U and the top-cycles. Let V = M-A and the proof is finished.

(b) The Lemma 1.2 of Morse and Theorem 3.2 show that a smooth immersion has a strict local support plane at points where $K > 0$ and has no local support plane at points where $K < 0$. If $x \in U$, then f has a global support plane at f(x), and thus $K(x) \geq 0$. On the other hand, suppose $K(x) > 0$ at a point $x \in V$. Then, the unit normal p to the tangent plane of f(x) can be chosen so that ℓ_p has a non-degenerate maximum at x. By Corollary 3.5, there exists a non-degenerate ℓ_q which has a local maximum near x and hence in V. However, ℓ_q assumes its global maximum at a point in U. Thus, ℓ_q is not polar, contradicting tightness (Corollary 5.7).

Q.E.D.

In the case where the restriction of f to every top-set is injective we can obtain another natural decomposition of M which is slightly different than Theorem 7.18. Even though top-sets are not always embedded, as Remark 7.14 shows, they are embedded in most known examples. First, as a consequence of the Fundamental Theorem and Banchoff's [4, p. 261] determination of TPP subsets of the plane, we get

Lemma 7.19: Let $f:M \to E^3$ be a tight topological immersion of a compact, connected surface and suppose that f is injective on a top-set Ω such that $Hf(\Omega)$ is 2-dimensional. Then f embeds Ω onto a closed convex plane disk with k disjoint open convex disks removed, where $0 \leq k \leq \beta_1(M;Z_2)$.

Proof: Since f is injective on Ω and Ω is compact, f is an embedding on Ω, and thus $f(\Omega)$ has the TPP by the Fundamental Theorem. Hence, $f(\Omega)$ is a closed convex disk with a possibly infinite number of disjoint open convex disks removed. The boundary of each disk removed carries a generator for $H_1(\Omega)$. Note that $\Omega = f^{-1}h$, where h is one of the half-spaces determined by the plane of $f(\Omega)$. By tightness, the homomorphism $H_1(\Omega) \to H_1(M)$ is injective, and so the number of disks removed satisfies $0 \leq k \leq \beta_1(M;Z_2)$.

Q.E.D.

Suppose f is a tight topological immersion into E^3 of a compact surface M not homeomorphic to S^2 which is injective on every top-set. By Lemma 7.15 and 7.19, the intersection of f(M) with $\partial Hf(M)$ is the complement in $\partial Hf(M)$ of a finite number of open disks (some of which may be coplanar). In this case, one could reformulate Theorem 7.18 using these disks rather than the possibly larger disks bounded by the images of the top-cycles. As with the top-cycles, if $\gamma = f^{-1}(\partial D)$ for one of these disks D, then γ cannot separate M. For example, let S be a convex surface with parallel planar pieces on top and bottom. These are necessarily closed convex disks. Form a new tight surface M with genus g by removing g open convex disks from each of the larger convex disks and adjoining g smooth cylinders with non-positive Gaussian curvature from the top to the bottom. There are only two top-cycles, the boundary curves of the original two closed convex disks. The set U of Theorem 7.18 is embedded by f onto the complement of these two convex disks in $\partial Hf(M)$. The top and bottom sets, except for the top-cycles themselves, lie in V. The intersection of f(M) with $\partial Hf(M)$ is the complement of the 2g smaller convex disks in $\partial Hf(M)$. If top-sets were always embedded, it would probably be preferable to call the boundaries of these smaller disks top-cycles, rather than the outer boundary curves. However, when a 2-dimensional top-set is not embedded, only the first definition continues to make sense as noted in Remark 7.14. We note that the <u>bounds given in Theorems 7.20 and 7.25 for the number of top-cycles also hold for the number of disks in this decomposition</u>.

We now obtain a result concerning the number of top-cycles. We again exclude the case $M = S^2$, since there are no top-cycles. In the case where M is non-orientable, we will be able to sharpen this result further, after we have shown that there does not exist a tight immersion of the Klein bottle into E^3. (See Theorems 7.25 and 7.26). These results can also be found in Cecil-Ryan [8].

<u>Theorem 7.20</u>: <u>Let $f: M \to E^3$ be a tight topological immersion of a compact connected surface M not homeomorphic to S^2, and let $\alpha(f)$ be the number of top-cycles of f. Then</u>

$$2 \leq \alpha(f) \leq 2 - \chi(M).$$

<u>Moreover, if $\alpha(f) = 2 - \chi(M)$, then the top-cycles come in pairs, each joined by a topological cylinder</u>.

<u>Proof</u>: We recall that there must exist some top-cycles, for otherwise f embeds M onto the convex surface $\partial Hf(M)$. There cannot be exactly one top-cycle, for it would then separate M by Theorem 7.18, contradicting the definition of top-cycle. Thus $2 \leq \alpha(f)$. Now let U and V be as in Theorem 7.18. Suppose first that all the top-cycles are disjoint. Let $V_1, \ldots, V_k$ be the connected components of V. Each $\overline{V}_i$ is a manifold with boundary consisting of at least two top-cycles, since no single top-cycle bounds.
So $\chi(\overline{V}_i) \leq 0$ and $\chi(\overline{V}_i) = 0$ if and only if the boundary of $\overline{V}_i$ is precisely two top-cycles, and $\overline{V}_i$ has no handles or cross-caps. Let A be the union of U and the top-cycles. The Euler characteristic of A is $2 - \alpha(f)$. Since $M = A \cup \overline{V}_1 \cup \ldots \cup \overline{V}_k$, and A and $\overline{V}_i$ intersect in a disjoint union of curves, we get

$$\chi(M) = \chi(A) + \chi(\overline{V}_1) + \ldots + \chi(\overline{V}_k).$$

Thus
$$\chi(M) \leq 2 - \alpha(f),$$

and
$$\alpha(f) \leq 2 - \chi(M).$$

Equality is possible only if each $\chi(\overline{V}_i) = 0$, i.e., V_i is a cylinder.

If the top-cycles are not disjoint, this argument must be modified slightly. Since f is an embedding on a neighborhood of A, we can replace A by a larger closed set Z whose boundary consists of $\alpha(f)$ disjoint curves near the top-cycles. Let the components of M-Z play the role of the V_i in the argument above. This shows that even in this case, if $\alpha(f) = 2-\chi(M)$, then each component V_i of V is an open cylinder. (An example of a tight surface whose top-cycles intersect is shown in Figure 1b of Banchoff-Kuiper [1].)

Q.E.D.

We get immediately the following corollary.

<u>Corollary 7.21</u>: <u>There does not exist a tight topological immersion of the projective plane into E^3.</u>

We next prove the more difficult result that there does not exist a tight immersion of the Klein bottle K^2 into E^3. We will assume that f is smooth as Kuiper did in his original paper. However, the result holds even if f is only assumed to be a topological immersion, but the proof will not be given here (see Kuiper [16]).

<u>Theorem 7.22</u>: <u>There does not exist a tight smooth immersion of the Klein bottle into E^3.</u>

<u>Proof</u>: Suppose M is a compact surface with $\chi(M) = 0$, i.e., a torus or Klein bottle. We show that if there exists a tight smooth immersion $f: M \to E^3$, then M can be oriented. Hence, no tight immersion of the Klein bottle into E^3 exists. By Theorem 7.20, since $\chi(M) = 0$, there must be exactly two top-

cycles, γ_1 and γ_2, and they are connected by a cylinder V of non-positive curvature. The two curves γ_1 and γ_2 must be disjoint, since f is smooth. By taking a projective transformation, we can assume that $f(\gamma_1)$ and $f(\gamma_2)$ lie in parallel planes, say $\ell_p \equiv r_1$ and $\ell_p \equiv r_2$, respectively, with $r_1 > r_2$. The top-set $\Omega(p)$ is connected, so there exists a smooth choice of unit normal to M on $\Omega(p)$, namely, the constant vector field p itself. We can extend this choice of unit normal to produce an oriented neighborhood of $\Omega(p)$ in M which has a well-defined Gauss map into S^2. Let U_1 be the inverse image under this Gauss map of a neighborhood N_1 of p in S^2 which is small enough that no two vectors in N_1 are perpendicular to each other. (We need this property later.) Clearly U_1 is a neighborhood of $\Omega(p)$. Choose a neighborhood U_2 of the top-set $\Omega(-p)$ containing γ_2 in exactly the same way. Applying Lemma 7.9 to U_1 and U_2, there exists a non-degenerate ℓ_q with q near p such that

$$\Omega(p) \subset \overline{M}_b^{+}(q) \subset U_1, \quad \Omega(-p) \subset \overline{M}_a^{-}(q) \subset U_2,$$

for appropriate non-critical values $a < b$. The compact manifolds with boundary $\overline{M}_b^{+}(q)$ and $\overline{M}_a^{-}(q)$ contain the top-cycles γ_1 and γ_2, respectively. Hence, neither is homotopic to a point. Since ℓ_q has four critical points on M, it must have a maximum and a saddle on $\overline{M}_b^{+}(q)$ and a minimum and a saddle on $\overline{M}_a^{-}(q)$. Thus, each of these manifolds with boundary has Euler characteristic zero, and since they are orientable, they are cylinders.

Let U and V be as in Theorem 7.18. By the intermediate value theorem, the plane $\ell_q \equiv s$, for $a < s < b$, must intersect both f(U) and f(V). Since there are no critical values of ℓ_q in the interval $[a,b]$, the set $\overline{M}_s^{-}(q)$ can be retracted onto $\overline{M}_a^{-}(q)$, and so it is also a cylinder with boundary consisting of two smooth curves, α_s in U and β_s in V.

The set $U \cup \overline{M}_b^{+}(q) \cup \overline{M}_a^{-}(q)$ is a topological cylinder. Put an orientation on a neighborhood W of this cylinder and on the parallel planes $\ell_q \equiv s$, $a < s < b$. This induces an orientation on each of the smooth plane curves α_s. By identifying the parallel planes, we have a smooth family of embedded smooth curves α_s, all of which must have the same rotation index, say +1.

This orientation of W also induces an orientation on the curves β_b and β_a. The oriented curve β_b is immersed into the same plane as α_b. We now show that these two curves have the same rotation index. The curves α_b and β_b form the boundary of the oriented cylinder $\overline{M}_b^{\,+}(q)$ and so there exists a smooth homotopy of regular curves ϕ_t, $0 \leq t \leq 1$, in $\overline{M}_b^{\,+}(q)$ with $\phi_0 = \alpha_b$ and $\phi_1 = \beta_b$. Referring to the construction of the neighborhood U_1, we know that $q \in N_1$, since ℓ_q has a critical point in U_1. By the definition of N_1, q is not tangent to M anywhere on U_1. Therefore, if π is orthogonal projection of E^3 onto the plane $\ell_q \equiv b$, then $\pi \circ f$ is an immersion on U_1. Hence, the family of curves $\pi \circ f\ (\phi_t)$ is a regular homotopy of regular curves in the plane $\ell_q \equiv b$, and the curves $\pi \circ f\ (\phi_0) = f(\alpha_b)$ and $\pi \circ f\ (\phi_1) = f(\beta_b)$ have the same rotation index, +1. The same argument shows that $f(\alpha_a)$ and $f(\beta_a)$ have the same rotation index +1 in the plane $\ell_q = a$.

Let C be the cylinder which is the union of the curves β_s, $a \leq s \leq b$. A choice of orientation of C makes $\{\beta_s\}$, into a smooth family of oriented regular plane curves (identifying the oriented parallel planes again), all of which must, therefore, have the same rotation index. From above, we know that the rotation index of β_b would have to be 1 or −1. Now choose the orientation for C which gives the rotation index +1 to all of the β_s. Then β_b and β_a have the same orientation as they were given by the orientation on W. Hence, we have given a consistent orientation to all of M, which is thus a torus and not a Klein bottle.

Q.E.D.

Remark 7.23: Tight immersions of surfaces with odd χ into E^3. Kuiper [5] produced tight immersions of compact surfaces with odd $\chi \leq -3$ into E^3. The existence of a tight immersion into E^3 in the case $\chi = -1$ ($M = P^2 \# T^2$) is still an open question. Kuiper's construction for $\chi = -3$ (also see Kuiper [11]) is rather involved, and we will not give the details here, but we will mention the main steps. He begins by trying to construct a tight immersion of $P^2 \# T^2$ into E^3. Start with a convex surface S with two identical convex disks in parallel planes. Let γ_1 and γ_2 be the boundaries of the convex disks. They will be top-cycles of the final immersion. Kuiper then defines an immersion of $P^2 \# T^2$ into E^3 by removing the two open convex disks and hanging a cylinder with a cross-cap W between γ_1 and γ_2 in the interior of the compact convex region in E^3 bounded by S. He is not able to do this in such a way that the Gaussian curvature is non-positive on W, but the region of positive curvature in W is contained in a small disk D. He then produces a tight immersion of $P^2 \# T^2 \# T^2$ by joining a cylinder of non-positive curvature from the boundary of a slightly larger disk containing D to a flat piece which can be suitably situated on the convex surface S. Now we have $K \leq 0$ except at the points of S, and thus the new immersion is tight. This immersion has 3 top-cycles, γ_1, γ_2 and a third in the flat piece of S where the new handle is adjoined. One can produce a similar tight immersion of $P^2 \# T^2 \# T^2$ with only two top-cyles, γ_1 and γ_2, by adjoining the end of the new handle to a flat piece in the region W near γ_1 or γ_2, rather than to a flat piece in S.

Remark 7.24: Tight C^∞-stable maps of surfaces into E^3. Although there do not exist tight smooth immersions of P^2 and K^2 into E^3 and the case $M = P^2 \# T^2$ is not settled, Kuiper [11] has shown that there exist tight C^∞-stable maps of all compact surfaces into E^3. A map $f:M \to E^m$ of a smooth manifold is said to be C^∞-stable if for any g sufficiently near to f in the C^∞-topology there exist diffeomorphisms $\phi:M \to M$ and $\psi:E^m \to E^m$ such that $g = \psi \circ f \circ \phi^{-1}$ (see Thom [1]). Immersions are C^∞-stable. Two well-known examples of tight C^∞-stable maps of P^2 into E^3 are obtained as projections of the Veronese surface (see Example 7.3) into 3-dimensional subspaces. Recall the embedding f of P^2 into E^4 obtained as a projection of the Veronese surface in Example 7.3,

$$f: (x,y,z) \to \left(\frac{(x^2 - y^2)}{\sqrt{2}}, \sqrt{2}\, yz, \sqrt{2}\, zx, \sqrt{2}\, xy)\right).$$

Projection of f onto the first 3 coordinates gives a cross-cap which is an immersion except at two Whitney pinch points (see Hilbert-Cohn-Vossen [1, pp. 313-317] and Banchoff [7] for more detail). A small disk neighborhood of each pinch point is mapped onto a cone over a figure 8. Thom [1] proved that this is the only kind of singularity possible for C^∞-stable maps from E^2 to E^3. On the other hand, projection of the map f onto the last three coordinates gives Steiner's Roman surface. This well-known mapping of P^2

into E^3 has six pinch points which are the endpoints of three segments of double points which intersect in a triple point (see Hilbert-Cohn-Vossen [1, pp. 302-304] and Banchoff [7, p. 1011]). Since both the cross-cap and the Roman surface are projections of the Veronese surfaces, they are tight maps by Remark 5.2(c). Banchoff [7] discusses the family of maps obtained as one rotates from the cross-cap to the Roman surface. All of the C^∞-stable maps in the family have either 2 or 6 pinch points which led him to conjecture that any C^∞-stable tight mapping of P^2 into E^3 must have either 2 or 6 pinch points. Kuiper also produced a tight C^∞-stable map of K^2 into E^3. Finally, one can produce a C^∞-stable tight map of $P^2 \# T^2$ into E^3 by modifying the cross-cap mapping $g:P^2 \to E^3$ according to the following often-used procedure which we will call <u>adding a handle</u>. First flatten the surface $g(P^2)$ in disk neighborhoods of two points of $g(P^2)$ on the boundary of its convex hull $\partial Hg(P^2)$. This can be done without destroying the convexity of these neighborhoods. Then apply a projective transformation of E^3 which takes the two planar disks into parallel planes such that some line orthogonal to the planes intersects the interiors of both disks. Now remove an open circular disk from each of the two planar faces and join the two boundary circles by a smooth cylinder of non-positive Gaussian curvature (obtained as a surface of revolution around the line orthogonal to the two planes). The result is a tight C^∞-stable map of $P^2 \# T^2$ into E^3 which has only two singularities, the two pinch points of g. See Coghlan [1] for further work on tight stable maps.

Using Kuiper's topological version of Theorem 7.22, we can obtain a sharper bound on the number of top-cycles of a tight immersion of a non-orientable surface in E^3. As in the proof of Theorem 7.20, we will assume that the top-cycles are disjoint in the proofs of the following two theorems. If they are not disjoint, the proofs can be modified similarly.

<u>Theorem 7.25</u>: <u>Let $f:M \to E^3$ be a tight topological immersion of a compact connected non-orientable surface, and let $\alpha(f)$ be the number of top-cycles of f. Then $2 \leq \alpha(f) \leq 1 - \chi(M)$.</u>

<u>Proof</u>: In Theorem 7.20, we showed that $\alpha(f) \leq 2 - \chi(M)$. If equality holds, then $\alpha(f)$ is twice the number of components of V (using the decomposition of M in Theorem 7.18), and hence $\chi(M)$ is even. So if $\chi(M)$ is odd, we have

immediately that $\alpha(f) \leq 1 - \chi(M)$. For the case when $\chi(M)$ is even, a different proof is required. Suppose $\alpha(f) = 2 - \chi(M)$. Then there exist $k = \alpha(f)/2$ cylinders V_i joined to corresponding pairs of top-cycles by Theorem 7.20. Let A be the union of U and the top-cycles. One can put an orientation on f(A), which induces an orientation on each of the plane convex curves which are images of the top-cycles. One can also orient each of the cylinders V_i, thereby inducing an orientation on the top-cycles. For at least one of the cylinders, say V_1, neither possible orientation of V_1 is consistent with the orientation of A on both boundary curves. Otherwise, M is orientable. Now, replace the cylinders $V_2, \ldots, V_k$ by the convex disks bounded by all of their bounding top-cycles. The new surface still has the two-piece property. Thus, it is a tight continuous immersion of the Klein bottle contradicting Kuiper's topological version of Theorem 7.22. Therefore, $\alpha(f) = 2 - \chi(M)$ is not possible, and we conclude that $\alpha(f) \leq 1 - \chi(M)$.

Q.E.D.

Finally we find the following equivalent condition to the existence of a tight immersion in the open case of $M = P^2 \# T^2$.

Theorem 7.26: Let M be a compact, connected surface with odd Euler characteristic $\chi(M) < -1$. Then there exists a tight topological, resp. smooth, immersion $f: M \to E^3$ with $\alpha(f) = 1 - \chi(M)$ if and only if there exists a tight topological, resp. smooth, immersion of $P^2 \# T^2$ into E^3.

Proof: Suppose $\chi(M) < -1$ and there exists a tight immersion $f: M \to E^3$ with $\alpha(f) = 1 - \chi(M)$, which is an even number. The argument with the Euler characteristic in part (a) of Theorem 7.20 shows that all $k = (1 - \chi(M))/2$ closures of components of V satisfy $\chi(\overline{V}_i) = 0$, except one component, say V_1, with $\chi(\overline{V}_1) = -1$. Thus each $\overline{V}_i$, $2 \leq i \leq k$, is a cylinder while $\overline{V}_1$ must have two boundary curves and one cross-cap. As in Theorem 7.25, if we replace the cylinders $V_2, \ldots, V_k$ by the convex disks bounded by their top-cycles, we obtain a tight immersion of $P^2 \# T^2$ into E^3. If the original immersion f is assumed to be smooth, then the new immersion can be made smooth by modifying f in a small neighborhood of the top-cycles.

Conversely, suppose there exists a tight immersion $g: P^2 \# T^2 \to E^3$. By

Theorem 7.25, it must have exactly two top-cyles. Let U be as in Theorem 7.18. To produce a tight immersion f of the non-orientable surface M with $\chi = -3$ having $\alpha(f) = 4$, add a handle as in Remark 7.24. If the original immersion is smooth, this can be done preserving smoothness. Treat the case when $\chi(M) < -3$ simply by repeating this construction the necessary number of times.

Q.E.D.

We conclude this discussion of the number of top-cycles by constructing examples with $\alpha(f)$ equal to each number between the bounds in Theorems 7.20 and 7.25, except, of course, $\alpha(f) = 1 - \chi(M)$ in the case of odd χ.

Examples 7.27: Tight immersion with various values of $\alpha(f)$.

(a) orientable surfaces: One proceeds inductively. To get all possible values of $\alpha(f)$ for the two-holed torus, begin with the tight torus of Figure 7.3. To get $\alpha(f) = 2$, add a handle of negative curvature from the face F_2 to the face F_3. To get $\alpha(f) = 3$, add the handle from F_1 to F_2, and to get $\alpha(f) = 4$, add it from F_1 to F_4. For a 3-holed torus, begin with a suitable choice of one of the above three surfaces and adjoin a handle as in Remark 7.24. We remark that the algebraic tight embeddings of Banchoff and Kuiper [1] all have $\alpha(f) = 2 - \chi(M)$ (see Remark 7.29).

(b) non-orientable surfaces, even χ: Example 7.2 (Figure 7.4) for the case $T^2 \# K^2$ has three top-cycles. For $\chi(M) < -2$, follow the inductive procedure described in (a) to add handles to the surface in Figure 7.4 and get $\alpha(f)$ equal to any number between three and $1-\chi(M)$. To get $\alpha(f) = 2$ for $T^2 \# K^2$, first modify the torus in Figure 7.3, by slanting the front annulus slightly toward the back, so that the front top-set is just the boundary curve. Then add the handle from F_3 through F_2 to a circle cut from the slanted annulus. Finally, to get $\alpha(f) = 2$ in the case of $\chi < -2$, one simply adjoins the required number of handles from F_2 to F_3 to this model.

(c) non-orientable surfaces, odd χ: The basic building block is Kuiper's rather complicated example [5, pp. 89-91] with $\chi = -3$ and $\alpha(f) = 3$. As above, one can arrange for flat pieces on the boundary of the convex hull and on the inside in order to add handles increasing $\alpha(f)$ by 0, 1 or 2. The maximum obtained by this procedure is $\alpha(f) = -\chi$ (see Theorem 7.26). To get $\alpha(f) = 2$ in the case $\chi = -3$, adjoin the handle, required to eliminate the "nap of positive curvature" [5, p. 89], from the region near the nap to a circle lying within a slanted top annulus, rather than to a circle lying in the set U (in the terminology of Theorem 7.18).

Remark 7.28: Tight substantial smooth immersions of surfaces into E^4. By Theorem 5.18, we know that if $f:M \to E^m$ is a tight substantial smooth immersion of a 2-dimensional surface, then $m \leq 5$. Further, if m=5, then M is P^2, and f is projectively equivalent to a Veronese surface (see Section 10). We now make a few comments about tight smooth substantial immersions of surfaces into E^4. In Example 7.3, we exhibited a tight smooth embedding of P^2 into E^4 obtained as a projection (or stereographic projection) of a Veronese surface. In Section 6, we gave a tight smooth substantial embedding of T^2 into E^4 as a product embedding of two convex curves. The procedure described below (due to Kuiper [8, p. 213]) enables one to produce tight smooth substantial embeddings into E^4 of all orientable surfaces (except, of course, S^2) and immersions of all non-orientable surfaces with $\chi \leq -4$. The key open question remaining is whether there exists a tight smooth immersion of the Klein bottle into E^4. We first construct a tight substantial embedding into E^4 of an orientable surface with genus g. Let C be an embedded smooth convex plane curve containing two parallel line segments L_1 and L_2. Then $C = f(S^1)$ is a tight embedding of S^1 into E^2. Therefore

$$\phi = f \times f: (x,y) \to (f(x), f(y)) \in E^2 \times E^2 = E^4$$

is a tight substantial embedding of T^2 into E^4. The plane sets

$$\Gamma_1 = \{(x,y) | \ x \in L_1, \ y \in L_1\} \text{ and } \Gamma_2 = \{(x,y) | \ x \in L_2, \ y \in L_2\}$$

are parallel, and thus they are contained in a 3-dimensional subspace E^3 in E^4. We next add g-1 handles of non-positive curvature from Γ_1 to Γ_2 as in Example 7.1. If a height function ℓ_p, $p \in E^4$, has a non-degenerate critical point on one of the handles, then the critical point must be of index 1. Thus, every non-degenerate ℓ_p is still polar, and f is tight by Corollary 5.7. This same method yields tight substantial immersions into E^4 of all non-orientable surfaces with $\chi \leq -4$. The only difference is what is attached between the two parallel faces Γ_1 and Γ_2. In the case $\chi = -4$, one removes convex disks from Γ_1 and Γ_2 and attaches the set V (as in Theorem 7.18) bounded by the two top-cycles of the tight immersion of $T^2 \# K^2$ given in Example 7.27(b). For even $\chi < -4$, one follows the same procedure, again attaching the set V from the appropriate example with higher genus in 7.27(b) having 2 top-cycles. For $\chi = -5$, attach the set V from the immersion of $P^2 \# T^2 \# T^2$ with 2 top-cycles in Example 7.27(c). For odd $\chi < -5$, use the appropriate example with higher genus from Example 7.27(c).

The most exhaustive attempt to classify tight substantial surfaces in E^4 is that of C.S. Chen and W.F. Pohl [1] in which they find a structure theorem similar to but necessarily more complicated than Theorem 7.18. Their work is based to some extent on the previous work of Chen [1]-[3] on tight immersions of codimension 2 and on the thesis of B. Hempstead [1]. Thorbergsson [3] has also obtained some structure theorems for tight immersions of codimension 2.

Remark 7.29: Tight analytic immersions of surfaces into E^3.

At the beginning of this section, we gave examples of tight immersions of orientable surfaces and of non-orientable surfaces with even $\chi \leq -2$. These are smooth, but they are clearly not analytic because they contain open planar pieces. While it was clear that the sphere and the torus admit tight analytic embeddings (a round sphere and a torus of revolution), it was not at all obvious how to produce tight analytic immersions of the other surfaces. By a rather ingenious method, Banchoff and Kuiper [1] produced tight analytic embeddings of all orientable surfaces into E^3, and Kuiper [15] subsequently produced tight analytic immersions of non-orientable surfaces with even $\chi \leq -2$. This method involves polynomial approximation of smooth functions. It would still be interesting if one could produce a simple explicit model of a tight analytic embedding of a surface with genus $g > 1$.

Alexandrov [1] proved that tight analytic embeddings of surfaces into E^3 are rigid, i.e., two isometric analytically embedded tight surfaces are congruent in E^3. This generalizes the well-known fact that convex surfaces are rigid. Nirenberg [1] obtained rigidity of tight smooth immersions of compact orientable surfaces under the additional assumptions that grad $K \neq 0$ at every boundary point of the set M^- where $K < 0$, and that each component of M^- contains at most one closed asymptotic curve. He shows, incidentally, that the first assumption implies that the immersion has the maximum number 2g top-cycles, where g is the genus of M.

Banchoff [8] showed that, in general, tight polyhedral surfaces in E^3 are not rigid.

<u>Remark 7.30</u>: <u>Total absolute curvature for knotted surfaces</u>.

In 1929, Fenchel [1] showed that for a smooth space curve $\gamma: S^1 \to E^3$, the total absolute curvature,

$$\tau(\gamma) = \frac{1}{2\pi} \int_{S^1} |\kappa(s)|\, ds,$$

satisfies $\tau(\gamma) \geq 2$, and that equality holds if and only if γ is a plane convex curve. Borsuk [1], in 1947, generalized Fenchel's result to closed curves in E^m, $m > 3$. In 1949, Fary [1] and Milnor [1,2] proved independently that if the space curve γ is a knot, then $\tau(\gamma) \geq 4$. Milnor showed further that $\tau(\gamma) = 4$ implies that γ unknotted. We now paraphrase the summary of the key arguments of Fary and Milnor given in Kuiper-Meeks [1].

If $\tau(\gamma) < 4$, then there exists some linear height function ℓ_p which has only two critical points on γ. This implies that γ is the boundary of an embedded disk D in E^3, which is ruled by straight line segments in the planes orthogonal to p. Since γ is the boundary of the disk D, it is unknotted.

Fary and Milnor observed that for any closed embedded polygon which is not in a plane, the total curvature, suitably defined, can be decreased by some small perturbation of the vertices. Milnor showed further that the addition of one new vertex to a closed polygon cannot decrease the total absolute curvature. It follows, for any curve γ, that $\tau(\gamma)$ is an upper bound for the total absolute curvature of all inscribed polygons. Then, for any nonplanar closed embedded curve γ, one can find a nearby isotopic curve with smaller total absolute curvature. Hence, Milnor's conclusion that $\tau(\gamma) = 4$ implies that γ is unknotted follows from the earlier result. Milnor's result is the best possible, since the trefoil knot can be represented by curves γ with $\tau(\gamma) = 4 + \varepsilon$ for any $\varepsilon > 0$.

A compact surface M embedded in E^3 is said to be <u>knotted</u> if it is not isotopic to the standard embedding. A model for the standard embedding of the orientable surface M_g of genus g is given in Example 7.1, i.e. begin with a convex surface with two parallel planar pieces on top and bottom and adjoin g handles of negative Gaussian curvature obtained as surfaces of revolution from the curve in Figure 7.1.

Langevin and Rosenberg [1] showed that if $\tau(f) < 8$ for a smooth embedding $f:T^2 \to E^3$, then f is unknotted. This was generalized to higher genus by Meeks [1, p. 400] and Morton [1] independently, who showed that if $\tau(f) < 2g + 6$ for a smooth embedding $f:M_g \to E^3$, then f is unknotted. More generally, they show that if there exists a non-degenerate height function ℓ_p with only one local maximum on an embedded compact surface in E^3, then the surface is unknotted.

In a recent paper, Kuiper and Meeks [1] handle the case $\tau(f) = 2g + 6$ and substantially develop the theory of total curvature for knotted surfaces. They first exhibit an embedding $f:T^2 \to E^3$ with $\tau(f) = 8$ such that $\tau(h) > 8$ for any h near f in the space of embeddings of T^2 into E^3. This shows that Milnor's τ-decreasing method, which was used in the case $\tau(\gamma) = 4$ for curves, will not work for surfaces. Then, using a different approach, they show that if g = 1 or 2 and $\tau(f) = 2g + 6$ then $f:M_g \to E^3$ is unknotted. On the other hand, for $g > 2$, they produce examples with $\tau(f) = 2g + 6$ which are knotted. In the course of the paper, they introduce the notion of isotopy tightness, i.e. an embedding $f:M \to E^3$ is called <u>isotopy tight</u> if $\tau(h) \geq \tau(f)$ for all embeddings h in the ambient isotopy class of f, i.e. the path component of f in the space of all embeddings of M into E^3. Their examples with $\tau(f) = 2g + 6$ for $g > 2$ are, in fact, isotopy tight. They go on to prove a significant number of results on isotopy tightness. At the same time, they pose many natural questions which remain to be answered.

Finally, we mention an earlier related result of Ferus [1] who showed that if $\tau(f) < 4$ for a smooth embedding $f:M \to E^m$ of a compact, connected, smooth n-manifold with even $\chi(M)$, then M is homeomorphic to S^n, and f(M) is topologically unknotted. More precisely, for each $x \in f(M)$, there exists a homeomorphism h of E^m which maps f(M) onto the standard sphere S^n such that $E^m - \{x\}$ is mapped diffeomorphically onto $E^m - \{h(x)\}$. In the case $m = n + 2$, $n = 1$ or $n \geq 5$, under the same hypotheses, Ferus is able to show that M is diffeomorphic to the standard sphere S^n, and that f is differentiably unknotted, i.e. there exists a diffeomorphism of E^{n+2} which maps $f(M)$ onto S^n. This result also holds for n = 4, if one assumes that M is diffeomorphic to S^4. As a corollary, Ferus obtains the result that an embedding of an exotic n-sphere $(n \geq 5)$ in E^{n+2} has total absolute curvature $\tau \geq 4$.

8. THE CHERN-LASHOF THEOREM.

In 1957, Chern and Lashof [1] proved that a smooth immersion $f:S^n \to E^m$ with minimal total absolute curvature must embed S^n as a convex hypersurface in a Euclidean subspace E^{n+1} in E^m. In 1970, Kuiper [8] proved the theorem under the assumption that f is a topological immersion. In 1979, he further generalized the result to show that if $f:S^n \to E^m$ is a continuous tight substantial map, then $m = n+1$ and $f(S^n) = \partial Hf(S^n)$ is the boundary of a convex (n+1)-dimensional body [13]. In this section, we present Kuiper's proof of the theorem for topological immersions. We first introduce some terminology.

Definition 8.1: A top^*-set is called an E^j-top^*-set if its image spans an affine j-dimensional plane in E^m.

Recall that a path-connected topological space X is said to be k-connected, $k \geq 1$, if the homotopy groups $\pi_i(X) = 0$ for $1 \leq i \leq k$; 0-connected means path-connected. If X is k-connected and $g:S^j \to X$ is a continuous map with $1 \leq j \leq k$, then g is homotopic to a constant map on S^j. For our purposes, the main applications of the following lemma are to n-spheres, which are (n-1)-connected.

Lemma 8.2: Let $f:M \to E^m$ be a topological immersion of a compact topological manifold. Assume that for almost all unit vectors p, the set $M_r(p)$ is k-connected for all $r \in R$. Then for any E^j-top^*-set Ω, we have:

(a) if $j \leq k + 1$, then f is an embedding of $f(\Omega)$ onto $Hf(\Omega)$.

(b) if $j = k + 2$, then $\partial Hf(\Omega) \subset f(\Omega)$, and f is an embedding on $\gamma = f^{-1}(\partial Hf(\Omega))$.

Remark 8.3: Note that the hypotheses are equivalent to assuming that f has the TPP in the case $k = 0$ and are stronger otherwise. Thus, this lemma is a generalization of Lemma 7.15.

Proof of Lemma 8.2: Fix a value of $k \geq 0$. The proof is by induction on j. For $j = 0$ and $j = 1$, the proof is exactly the same as the proof of Lemma 7.15 (a), which only assumes that f has the TPP.

We now do the induction step. Take j with $1 \leq j \leq k + 2$, and suppose (a) is true for all E^i-top*-sets with $i < j$. We prove that (a) and (b) hold for E^j-top*-sets.

Let Ω be an E^j-top*-set, and let E^j be the j-plane containing the closed convex disk $D = Hf(\Omega)$. Every support plane π of D in E^j is also a support plane of $f(\Omega)$. For such a support plane, $f^{-1}\pi = \Omega'$ is a top-set of Ω, and Ω' is an E^i-top*-set of f with $i < j$. By the induction hypothesis, f is an embedding of Ω' onto $Hf(\Omega') = \pi \cap D$. Now $\Sigma = \partial D$ is the union of the $\pi \cap D$, where π is a support plane of D in E^j. Hence Σ lies in $f(\Omega)$, and f is an embedding on $\gamma = f^{-1}\Sigma$. If $j = k + 2$, we have proven (b) and are finished.

Now suppose $j \leq k + 1$. To complete the proof of (a), we only need to show that $f(\Omega) = D$, since then f must be injective on Ω by Lemma 7.13. Suppose there exists a point P in D which is not in $f(\Omega)$. Since Ω is compact, there is a closed ball $\bar{B}_{2\delta}(P)$ of radius $2\delta > 0$ in E^m centered at P which is disjoint from $f(\Omega)$. Let Σ be the (j-1)-sphere ∂D. We have shown above that f is an embedding on $\gamma = f^{-1}\Sigma$. We next show that Σ is not contractible in the open set V in E^m,

$$V = \{Q \in E^m \mid d(Q,f(\Omega)) < \delta\} - \bar{B}_{2\delta}(P).$$

To see this, let ϕ be a orthogonal projection from E^m to E^j. Note that $\phi(V) \subset E^j - \bar{B}_\delta(P)$, by the triangle inequality. If $F:S^{j-1} \times [0,1] \to V$ were a homotopy of Σ to a point, then $\phi \circ F$ would be a homotopy of Σ to a point in $E^j - \bar{B}_\delta(P)$, which is impossible.

Now, let $U = f^{-1}V$. Then U is open, and it follows from what we have just shown that $\gamma = f^{-1}\Sigma$ is not contractible in U. However, by the Key Lemma 7.9, there exists a unit vector p and real number r with

$$\Omega \subset M_r(p) \subset U,$$

and p can be chosen so that $M_r(p)$ is k-connected. Thus, $\gamma \subset M_r(p)$ and since $M_r(p)$ is k-connected, the (j-1)-sphere γ is contractible in $M_r(p)$, and thus contractible in U. Contradiction.

Q.E.D.

We now give a proof of the Chern-Lashof theorem for weakly non-degenerate topological immersions. First recall some definitions from the topological category (see Kuiper [9] for details). A continuous real-valued function ϕ on a compact topological manifold M is called <u>non-critical</u> at $x \in M$ if there exists a coordinate chart in a neighborhood of x for which ϕ is one of the coordinates. Otherwise, x is called a <u>critical point</u> of ϕ. A topological immersion $f:M \to E^m$ is called <u>weakly non-degenerate</u> if almost all linear height functions ℓ_p have isolated critical points on M. If f is a C^2-immersion or a piece-wise linear immersion, then f is weakly non-degenerate. Using topological critical point theory, one can show that if $f:S^n \to E^m$ is tight and weakly non-degenerate, then almost all linear height functions have exactly 2 critical points.

<u>Theorem 8.4 (Chern-Lashof, Kuiper)</u>: <u>Let $f:S^n \to E^m$ be a substantial topological immersion such that almost all linear height functions have exactly two critical points. Then $m = n+1$ and f embeds S^n as a convex hypersurface</u>.

<u>Proof</u>: First note that if ℓ_p has exactly two critical points, then each $M_r(p)$ is (n-1)-connected. For suppose ℓ_p has a minimum value a, at a critical point x_1 and a maximum value b at x_2. If $a < r < b$, then there exists a deformation retraction of $M_r(p)$ onto $M_a(p) = \{x_1\}$ (see Kuiper [9, p. 80] for this result in the topological context). Thus $M_r(p)$ is clearly (n-1)-connected, while if $r > b$, $M_r(p) = S^n$.

Now suppose there exists an E^{n+1}-top*-set Ω. By Lemma 8.2(b), $\gamma = f^{-1}(\partial Hf(\Omega))$ is an n-sphere embedded in S^n. Hence $\gamma = S^n$ and f embeds S^n as the convex hypersurface $\partial Hf(\Omega)$ in E^{n+1}. Furthermore, since f is substantial, $E^{n+1} = E^m$.

Next suppose that all top^1-sets of f are E^j-top-sets for $j < n$. Then part (a) of Lemma 8.2 is applicable to all top^1-sets of f. We conclude that

$f(S^n)$ contains the (m-1)-sphere $\Sigma = \partial Hf(S^n)$, since Σ is the union of the $Hf(\Omega) = f(\Omega)$, as Ω ranges over the top^1-sets of f. Moreover, $\gamma = f^{-1}\Sigma$ is an embedded (m-1)-sphere in S^n. Thus, $m \leq n+1$. However, $m \geq n+1$, since f is an immersion, and we see that $m = n+1$ and f embeds S^n onto Σ.

We complete the proof by showing that there do not exist any E^j-top^*-sets for $j > n+1$. If such a top^*-set exists, then by repeating the process of taking top-sets, one eventually produces either an E^{n+1}-top^*-set or an E^q-top^*-set Ω with $q > n+1$ such that all of the top^1-sets of Ω are E^i-top^*-sets for $i \leq n$. In the latter case, one applies the above argument to get that S^n contains the embedded (q-1)-sphere $f^{-1}(\partial Hf(\Omega))$. Thus $q \leq n+1$, a contradiction. As to the former case, we have already noted that the existence of an E^{n+1}-top^*-set implies that $m = n+1$, thereby precluding the existence of an E^j-top^*-set with $j > n+1$.

Q.E.D.

9. VERONESE MANIFOLDS.

In this section, we discuss the well-known Veronese embeddings of projective spaces into Euclidean spaces. In [1], Kuiper showed that if $f:M^n \to E^m$ is a substantial TPP immersion, then $m \leq N = \frac{1}{2}n(n+3)$, (see Theorem 5.18). He also showed that the Veronese embedding of RP^n into E^N is tight, as are the embeddings due to H. Hopf [1] of RP^n substantially into E^m, $2n + 1 \leq m \leq N$, obtained from the Veronese embedding by projection onto E^m. Kuiper [2] noted that similar embeddings of CP^n, QP^n (Q = quaternions) and OP^2 (O = Cayley numbers) are also tight, and Tai [1] gave a detailed proof of this fact (also see Kuiper [8], [13]). These embeddings are examples of an extensive class of submanifolds, the standard embeddings of symmetric R-spaces, which were shown to be tight by Takeuchi and Kobayashi [1] (also see Kobayashi [1], [2] and Wilson [2]). The standard symmetric R-spaces are also minimal submanifolds of a hypersphere in Euclidean space (see Takeuchi-Kobayashi [1], Ferus [3]) and hence they are taut by Theorem 6.6. Kelly [2] showed that these are essentially all of the tight equivariant embeddings of irreducible locally symmetric homogeneous spaces. Approaching the subject from a different point of view, Ferus [3] characterized the standard symmetric R-spaces as the only compact (irreducible) extrinsically symmetric submanifolds of Euclidean space. Here a submanifold M in E^m is called extrinsically symmetric if for each $x \in M$, the manifold M is invariant under the reflection σ_x of E^m in the normal space to M at x. Ferus [5] used this characterization to give a different proof of the tightness of symmetric R-spaces. This is a rich subject, but we will not attempt to discuss it further here. (Remark: Blomstrom [1] has recently considered extrinsically symmetric submanifolds of pseudo-Riemannian space forms.)

On the other hand, Kuiper [6] proved that a substantial TPP immersion $f:M^2 \to E^5$ must be a Veronese surface (up to projective transformation). This was generalized by Little and Pohl [1], who showed that a substantial TPP immersion of an n-dimensional manifold M into E^N must be a Veronese embedding of RP^n. We will give their proof in Section 10. Later, Kuiper and Pohl [1] showed that if $f:RP^2 \to E^m$, $m \geq 5$ is a substantial TPP topological embedding, then $m = 5$ and f is either a smooth Veronese embedding or the polyhedral embedding of Banchoff given in Example 5.21. Aside from the papers already mentioned, the reader is referred to Little [1] - [2], Hempstead [1], and Chen-Pohl [1] for further discussion of Veronese manifolds and tightness.

We will follow the approach and use the notation of Tai [1] in the subsequent presentation of the Veronese embeddings. We let F denote one of the algebras R, C or Q. For a quaternion

$$q = r_0 + r_1 i + r_2 j + r_3 k$$

the conjugate $\bar{q}$, is defined by

$$\bar{q} = r_0 - r_1 i - r_2 j - r_3 k.$$

and its norm is given by $|q| = (q\bar{q})^{1/2}$. If $q \in C$, then $\bar{q}$ is the ordinary complex conjugate, and $\bar{q} = q$ for $q \in R$. Define $d = 1,2,4$ depending on whether $F = R$, C or Q, respectively. For matrices over F, let $A^* = \bar{A}^t$. Then it is easy to check that

(a) $(AB)^* = B^* A^*$

and

(b) $\operatorname{Re} \operatorname{trace} (AB) = \operatorname{Re} \operatorname{trace} (BA)$

whenever the indicated operations make sense. Now let $M(n+1,F)$ be the space of all $(n+1) \times (n+1)$ matrices over F. Let

$$H(n+1,F) = \{A \in M(n+1,F) | A^* = A\}.$$

Elements of $H(n+1,F)$ are called Hermitian matrices. If A is Hermitian, the off-diagonal entries of A are in F, while the diagonal entries must be real. Hence $H(n+1,F)$ is a real vector space of dimension $\frac{n}{2}(n+1)d + n + 1$. Let $U(n+1,F) = \{A \in M(n+1,F) | AA^* = I\}$. For $F = R, C, Q$ respectively, $U(n+1,F)$ is equal to $O(n+1)$, $U(n+1)$, $Sp(n+1)$, respectively.

F^{n+1} can be considered as a Euclidean space of dimension $(n+1)d$. The usual Euclidean inner product on $F^{n+1} = E^{(n+1)d}$ is given by

$$\langle x,y\rangle = \mathrm{Re}(x^* y),$$

where $x = (x_0,\ldots,x_n)$ and $y = (y_0,\ldots,y_n)$ in F^{n+1} are represented as column matrices. Then $\langle x, Ay\rangle = \langle A^* x, y\rangle$ for all $A \in M(n+1,F)$.

The space $M(n+1,F)$ can be considered as Euclidean space of dimension $(n+1)^2 d$, and the usual inner product takes the form

$$\langle A,B\rangle = \mathrm{Re\ trace}\ (AB^*),\ A,B \in M(n+1,F).$$

Of course, on the subspace $H(n+1,F)$, this simplifies to

$$\langle A,B\rangle = \mathrm{Re\ trace}\ (AB).$$

Let $S^{(n+1)d-1}$ be the unit sphere in F^{n+1} and FP^n the quotient space of $S^{(n+1)d-1}$ under the relation

$$(x_0,\ldots,x_n) \sim (x_0\lambda,\ldots,x_n\lambda),\ \lambda \in F,\ |\lambda| = 1.$$

Consider the map from $S^{(n+1)d-1}$ to $H(n+1,F)$ given by

$$(9.1) \qquad x \to xx^* = \begin{bmatrix} |x_0|^2 & x_0\bar{x}_1 & \cdots & x_0\bar{x}_n \\ x_1\bar{x}_0 & |x_1|^2 & \cdots & x_1\bar{x}_n \\ \cdots & \cdots & \cdots & \cdots \\ x_n\bar{x}_0 & x_n\bar{x}_1 & \cdots & |x_n|^2 \end{bmatrix}.$$

Note that if $y = x\lambda$ for $|\lambda| = 1$, then x and y have the same image under the map in (9.1). Further, if $xx^* = yy^*$, then multiplication by x on the right gives

$$x = y\, y^* x = y\, \lambda, \qquad \text{for } |\lambda| = |y^* x| = 1.$$

Thus, the map in (9.1) induces an injective map ϕ from FP^n to $H(n+1,F)$. The image of ϕ consists precisely of those matrices in $M(n+1,F)$ satisfying

$$A = A^* = A^2, \quad \text{rank } A = 1.$$

In fact, $\phi(x)$ is just the matrix representation of orthogonal projection of F^{n+1} onto the F-line spanned by the vector x. One can check that ϕ is a smooth immersion on FP^n by direct calculation, or else deduce this fact as a consequence of the equivariance given in Lemma 9.1. Thus, ϕ is a smooth embedding of FP^n into $H(n+1,F)$. We can, and often will, consider $x_0,\ldots,x_n$ as homogeneous coordinates on FP^n.

The condition $|x| = 1$ is equivalent to the condition trace $\phi(x) = 1$. Hence, the image of ϕ lies in the hyperplane in $H(n+1,F)$ given by the linear equation trace $A = 1$. We now show that the image of ϕ does not lie in any lower-dimensional plane.

Let p be an arbitrary point in $S^{(n+1)d-1}$ and let X be a unit vector tangent to the sphere at p. Consider the curve

$$\alpha(t) = \cos t\; p + \sin t\; X.$$

Then

$$(9.2) \qquad \phi_* X = \frac{d}{dt}\,[\alpha(t)\,\alpha^*(t)]\big|_{t=0} = p\,X^* + X\,p^*.$$

Taking $p = e_i$, $X = e_j u$, $j \neq i$, u a unit quaternion, yields a matrix which is zero except for u in the (j,i) position and $\bar{u}$ in the (i,j) position. This shows that all off-diagonal elements of H(n+1,F) occur as tangent vectors to ϕ. Furthermore, if we take $p = e_0$, $X = e_j$ and evaluate at $t = \pi/4$, we get

$$\phi_* X = e_j e_j^* - e_0 e_0^*,$$

showing that all real diagonal matrices with trace = 0 also occur. Thus, ϕ embeds FP^n substantially into the Euclidean space

$$E^N = \{A \in H(n+1,F) \mid \text{trace } A = 1\},$$

$$N = \frac{n(n+1)}{2} d + n.$$

For the remainder of this section, N will always have the value given above. Note also that

$$\langle xx^*, xx^* \rangle = \text{trace}[(xx^*)^2] = \text{trace}(xx^*) = 1,$$

and so $\phi(FP^n)$ lies in the intersection of E^N with the unit sphere in M(n+1,F).

U(n+1,F) acts linearly on M(n+1,F) by

$$U(A) = UAU^*$$

for $U \in U(n+1,F)$, $A \in M(n+1,F)$. An elementary calculation shows that this action preserves the inner product of M(n+1,F). Further, we have

$$\phi(Ux) = (Ux)(Ux)^* = U\,x\,x^* U^* = U(\phi(x)),$$

for $x \in FP^n$, $U \in U(n+1,F)$. Thus, we obtain the following important fact,

<u>Lemma 9.1</u>: <u>The embedding $\phi: FP^n \to H(n+1,F)$ is equivariant with respect to and invariant under the action of U(n+1,F), i.e.</u>

$$\underline{\phi(Ux) = U(\phi(x)) \in \phi(FP^n)}$$

<u>for all $x \in FP^n$, $U \in U(n+1,F)$</u>.

We now prove the main result of this section.

<u>Theorem 9.2</u>: <u>The embedding $\phi: FP^n \to H(n+1,F)$ is taut. Thus, the embedding $\phi: FP^n \to E^N$ is taut and substantial</u>.

<u>Proof</u>: We will show that every non-degenerate linear height function in $H(n+1,F)$ has the minimum number $\beta(FP^n) = n + 1$ critical points. Thus, ϕ is a tight embedding into $H(n+1,F)$ and since ϕ is also spherical, it is taut by Theorem 6.6. Since E^N is a Euclidean subspace of $H(n+1,F)$, every height function in E^N corresponds to a height function in $H(n+1,F)$, and so ϕ is also a taut embedding into E^N.

For $A \in H(n+1, F)$,

$$\ell_A(x) = \langle A, \phi(x)\rangle = \text{Re trace } (Axx^*)$$

$$= \text{Re trace}(x^*Ax) = \langle x, Ax\rangle.$$

Thus ℓ_A has a critical point at x if and only if $\langle X, Ax\rangle = 0$ for all X tangent to the sphere at x. In other words, Ax is normal to the sphere and so must be a (real) multiple of x. This shows that the critical points of ℓ_A correspond to eigenvectors of the matrix A. The usual inductive process (maximizing $\langle x, Ax\rangle$) may be used to produce n+1 real eigenvalues, each with a d-dimensional eigenspace. In fact, if x and λ are found so that $Ax = \lambda x$ (λ real), and u is a unit quaternion, we have $A(xu) = \lambda(xu)$. However, all the points xu for a given x yield the same point of FP^n. Thus ℓ_A has precisely n+1 critical points on FP^n provided that the n+1 eigenvalues are all distinct.

We now compute the Hessian at a critical point x where $Ax = \lambda x$. Here X and Y are tangent to the sphere at x.

$$(9.3) \qquad H(X,Y) = 2\, D_Y < X, Ax>$$

$$= 2(<\nabla_Y X, Ax> - <X,Y> <x, Ax> + <X, AY>)$$

$$= 2 <(A - \lambda I)X, Y>,$$

since $\nabla_Y X$ is tangent at x while Ax is normal. Note we have used the fact that the sphere is umbilic with shape operator $-I$ with respect to the unit normal x. Formula (9.3) shows that x is a non-degenerate critical point if and only if all the eigenvalues of A with eigenspaces orthogonal to the F-line generated by x are distinct from λ. In particular, ℓ_A is a Morse function on FP^n if and only if all n+1 eigenvalues are distinct. As we know (Corollary 3.5), almost every A has this property. Also, there will be one critical point of index k for each of the following values

$$k = 0, d, 2d, \ldots, nd.$$

Thus, ϕ is tight since it is well known that

$$\beta_i(RP^n;Z_2) = 1, \ 0 \leq i \leq n,$$

and since by the lacunary principle of Morse Theory (see Morse-Cairns [1, p. 272]), the argument above actually proves that for F = C or Q,

$$H_i(FP^n;Z) = \begin{cases} Z & \text{for } i=0, d, 2d, \ldots, nd \\ 0 & \text{otherwise} \end{cases}.$$

Q.E.D.

Kuiper [1] presented a variation of the Veronese embedding due to H. Hopf [1] which gives the following result.

Theorem 9.3: There exists a tight substantial embedding of FP^n into E^m for

$$(2n-1)d + 1 \leq m \leq N.$$

Proof: The embeddings are obtained by projecting the Veronese manifold onto an appropriate choice of E^m. We define the following quadratic functions in the homogeneous coordinates $(x_0,\ldots,x_n)$ of FP^n,

$$z_k = \sum_{\substack{i+j=k \\ i\leq j}} x_i \bar{x}_j, \; k = 0,\ldots,2n-1.$$

The values of z_k are real for $k=0$ and in F for $k > 0$. These functions are easily shown to be linearly independent. Thus, the mapping

$$\psi : x \to (z_0,\ldots,z_{2n-1})$$

is a substantial map of FP^n into E^K, $K = (2n-1)d + 1$. Further, the values of all of the homogeneous coordinates $x_0,\ldots,x_n$ can be recovered by knowing $z_0,\ldots,z_{2n-1}$, so the map is injective on FP^n. Finally, one can compute that ψ is an immersion, and hence ψ is a substantial embedding of FP^n into E^K. The embedding ψ may be related to the Veronese embedding as follows. For each k, $0 \leq k \leq 2n-1$, let M_k be the matrix having a 1 in the (i,j) position for $i+j=k$, $i \leq j$ and zero elsewhere. The M_k are mutually orthogonal. Thus we may write

$$\psi(x) = \sum_{k=0}^{2n-1} z_k M_k / |M_k|.$$

Note that $\psi = \sigma \circ \phi$, where ϕ is the Veronese embedding, and σ is orthogonal projection of $H(n+1,F)$ onto the Euclidean space E^K consisting of real multiples of M_0 and F-multiples of the other M_k. By Remark 5.2(c), ψ is tight. To obtain a tight substantial embedding of FP^n into E^m for $K < m < N$, just adjoin appropriate coordinates of the ϕ embedding which are linearly independent from the coordinates of the ψ embedding, i.e. project $\phi(FP^n)$ into an E^m containing E^K. Such an embedding is tight for the same reasons as ψ.

Q.E.D.

The Veronese embeddings can be generalized to produce taut embeddings of Grassmann manifolds over $F = R$, C or Q into E^m (see, for example, Kuiper [13, p. 113]).

For projective planes, even sharper results can be obtained. Theorem 9.3 gives the existence of substantial tight embeddings of FP^2 into E^m for

$$3d + 1 \leq m \leq 3d + 2.$$

In fact, we can obtain taut embeddings in these dimensions. The Veronese embedding $\phi: FP^2 \to E^{3d+2}$ is taut and spherical, so we get a taut embedding of FP^2 into E^{3d+1} by composing ϕ with stereographic projection (see Theorem 6.5). By methods similar to those used in Theorem 9.2, Tai [1] showed that the analogous embedding of the Cayley projective plane OP^2 into E^{26} is taut (also see Kuiper [13, p. 128]). Again, by stereographic projection, one can obtain a substantial taut embedding of OP^2 into E^{25}.

Kuiper [8, pp. 215-217] then established the following.

<u>Theorem 9.4</u>: <u>There exist tight substantial embeddings of $RP^2 \to E^5$ or E^4, $CP^2 \to E^8$ or E^7, $QP^2 \to E^{14}$ or E^{13}, $OP^2 \to E^{26}$ or E^{25} and only in these Euclidean spaces.</u>

<u>Proof</u>: No embeddings into lower dimensional Euclidean spaces exist because the normal Stiefel-Whitney class $\bar{w}_d(FP^2) \neq 0$, where $d = 1,2,4,8$ for $F = R$, C, Q, O, respectively. (See for example, Husemöller [1, p. 263] and Borel-Hirzebruch [1, p. 533] for the case $F = O$.) The upper bound is obtained from Theorem 5.20 (due to Kuiper). We know that the Z_2-Betti numbers of FP^2 are given by

$$(9.4) \qquad \beta_i(FP^2;Z_2) = \begin{cases} 1 & \text{if } i = 0,\ d,\ 2d \\ 0 & \text{otherwise.} \end{cases}$$

By Theorem 5.20, we know that the substantial codimension of a tight smooth immersion is less than or equal to $c(\beta_0,\ldots,\beta_n)$, i.e. the maximum dimension of a linear family of symmetric bilinear forms in n variables which contains a positive definite form and such that no form of the family has index k if $\beta_k = 0$. Hence, the proof will be finished if we show that for the β_i given in (9.4), we have $c(\beta_0,\ldots,\beta_{2d}) = 4,6,10$ for $d = 2,4,8$, respectively. (The case $F = R$ is already proven by Theorem 5.18.) The result we need is contained in Hurwitz [1] (see also Kuiper [8, pp. 232-234]). There it is shown that the desired linear family of symmetric bilinear forms with maximal dimension can be represented by the set of symmetric matrices of the form

$$\begin{bmatrix} \lambda I & B \\ B^t & \mu I \end{bmatrix}$$

where B is the 2×2, 4×4 or 8×8-matrix in the upper left corner of the matrix (9.5), depending on whether F = C, Q, O, respectively. From this, we see that $c(\beta_0,\ldots,\beta_{2d})$ has the desired values.

$$(9.5) \qquad \begin{bmatrix} x_1 & -x_2 & -x_3 & -x_4 & -x_5 & -x_6 & -x_7 & -x_8 \\ x_2 & x_1 & -x_4 & x_3 & -x_6 & x_5 & -x_8 & x_7 \\ x_3 & x_4 & x_1 & -x_2 & -x_7 & x_8 & x_5 & -x_6 \\ x_4 & -x_3 & x_2 & x_1 & x_8 & x_7 & -x_6 & -x_5 \\ x_5 & x_6 & x_7 & -x_8 & x_1 & -x_2 & -x_3 & x_4 \\ x_6 & -x_5 & -x_8 & -x_7 & x_2 & x_1 & x_4 & x_3 \\ x_7 & x_8 & -x_5 & x_6 & x_3 & -x_4 & x_1 & -x_2 \\ x_8 & -x_7 & x_6 & x_5 & -x_4 & -x_3 & x_2 & x_1 \end{bmatrix}.$$

Q.E.D.

Remark 9.5: Tight and taut embeddings of highly connected manifolds. We have seen examples of substantial taut embeddings of FP^2 into E^{3d+1} and E^{3d+2}. Those which lie substantially in E^{3d+2} must be spherical. Otherwise, by composing the embedding with inverse stereographic projection, one could produce a taut substantial embedding of FP^2 into E^{3d+3}, violating Theorem 9.4. Recently, G. Thorbergsson [2] has generalized Theorem 9.4 to show that if M^{2k} is a compact (k-1)-connected but not k-connected taut submanifold of E^m which is substantial and not spherical then either:

(i) $m = 2k+1$, and M^{2k} is a cyclide of Dupin diffeomorphic to $S^k \times S^k$

or

(ii) $m = 3k+1$, and M^{2k} is diffeomorphic to one of the projective planes RP^2, CP^2, QP^2 or OP^2.

The fact (i) that a taut (k-1)-connected but not k-connected compact hypersurface must be a cyclide is due to Banchoff [3] for k=1, and Cecil-Ryan [2, p. 184] and C.S. Chen [3, p. 1101] for $k > 1$. See Theorems 7.2 and 8.4 of Chapter 2.

Kuiper [8, p. 231] and [13, p. 133] showed that a tight immersion of a (k-1)-connected M^{2k} in E^m must satisfy rather stringent conditions. His result has been further refined in a recent paper by Thorbergsson [3]. This paper also contains results on tight immersions of M^{2k} into E^{2k+2} which generalize the earlier work of C.S. Chen [1]-[3] and Chen-Pohl [1]. Hebda [2] has recently shown that the connected sum of arbitrarily many copies of $S^k \times S^k$ can be realized as a tight hypersurface in E^{2k+1}. These are counter examples to a conjecture of Kuiper [13, p. 116] that every smooth (k-1)-connected 2k-dimensional manifold ($2k \geq 4$) which admits a smooth tight immersion into Euclidean space is diffeomorphic to CP^2, QP^2, OP^2 or $S^k \times S^k$. It follows from (i) in the paragraph above that Hebda's examples cannot be made taut, as Hebda himself showed.

A compact manifold with Morse number 3 is called a <u>manifold like a projective plane</u>. Eells and Kuiper [1] gave many examples of such manifolds M^{2k}, all necessarily of dimensions $2k = 2, 4, 8, 16$. They are obtained from E^{2k} under compactification by a k-sphere. Of course, the projective planes FP^2 for F = R, C, Q and O are are examples. Kuiper [13, p. 132] showed that if $f:M^{2k} \to E^m$ is a tight substantial topological embedding of a manifold like a projective plane, then $m \leq 3k + 2$. Further, if $f:M^{2k} \to E^{3k+2}$ is a tight smooth substantial embedding of a manifold like a projective plane, then M^{2k} is embedded as a algebraic submanifold. For k=1, 2, respectively, the manifold M^{2k} must be RP^2 or CP^2, resp., and f must be the Veronese embedding, up to a real projective transformation of E^{3k+2}. It is not known if this is true for k=4,8. Note that even for k=1,2, the hypothesis of smoothness is necessary, as the piece-wise linear embeddings of RP^2 into E^5 of Banchoff [1] and of CP^2 into E^8 by Kühnel and Banchoff [1] demonstrate (see Example 5.21).

In Theorem 5.18, we showed that the substantial codimension k of a tight immersion of an n-manifold must satisfy $1 \leq k \leq n(n+1)/2$. We now show that every value k in this interval can be realized.

Theorem 9.6: For every integer k satisfying $1 \leq k \leq n(n+1)/2$, there exists a tight substantial embedding of an n-dimensional manifold M into E^{n+k}.

Proof: For k=1, take the tight embedding of S^n as a round sphere in E^{n+1}. For k=2, take the standard product embedding of $S^{n-1} \times S^1$ into $E^n \times E^2 = E^{n+2}$ which is tight by Theorem 6.2. More generally, for $2 \leq k \leq n$, take the standard product of S^{n-k+1} with (k-1) copies of S^1,

$$S^{n-k+1} \times S^1 \times \ldots \times S^1 \subset E^{n-k+2} \times E^2 \times \ldots \times E^2 = E^{n+k}.$$

Finally, for codimensions $n+1 \leq k \leq n(n+1)/2$, use tight embeddings of RP^n given in Theorem 9.3.

Q.E.D.

10. TIGHT IMMERSIONS OF MAXIMAL CODIMENSION.

In Theorem 5.18, we saw that if $f:M^n \to E^m$ is a substantial TPP smooth immersion, then $m \leq N = \frac{1}{2} n(n+3)$. In Section 9, it was shown that the Veronese embeddings $\phi:RP^n \to E^N$ are substantial tight embeddings. We now prove the remarkable fact that these Veronese embeddings are the only substantial TPP immersions of an n-manifold into E^N (up to projective transformation, of course). Our treatment follows that of Little-Pohl [1] closely.

We first recall some pertinent details of the Veronese embeddings. In this section, we will only consider the field F = R. Recall that map (9.1) from the unit sphere S^n in E^{n+1} into the space H(n+1,R) of real symmetric matrices, and the induced substantial smooth embedding ϕ of P^n into E^N.

For $X,Y \in T_xS^n$, one easily shows, using (9.2) and the fact that F = R, that

$$\langle \phi_* X, \phi_* Y \rangle = 2 \langle X,Y \rangle. \tag{10.1}$$

This shows that ϕ is an immersion on S^n and that the induced metric has constant sectional curvature $1/\sqrt{2}$.

At times, it is convenient to consider ϕ as an embedding of P^n into the projective space $P^N = H(n+1,R)/\sim$. Then we may regard (9.1) as a parametrization of the Veronese manifold in terms of the homogeneous coordinates on P^n and P^N, where we now disregard the condition trace$(\phi(x)) = 1$. We saw in Remark 5.2(b), that tightness in E^N is invariant under projective transformations of P^N. Thus, if σ is a projective transformation of P^N with $\sigma(\phi(P^n)) \subset E^N$, we will also call $\sigma(\phi(P^n))$ a <u>Veronese manifold</u>. The original embedding ϕ will be called the standard embedding.

The Veronese manifold enters into general differential geometric considerations as a generalization of the curvature ellipse for immersions of surfaces into E^4. Let $f:M \to E^m$ be a smooth immersion of a n-dimensional manifold. For $x \in M$, the <u>osculating space</u> to f at x is defined to be

$$f_*(T_xM) + \{\alpha(X,Y) | X,Y \in T_xM\}. \tag{10.2}$$

Note that if $\gamma(t) = f(x(t))$ is a curve on $f(M)$ with $x(0) = x$ and $x'(0) = X$ for $X \in T_xM$, then

$$\gamma''(0) = D_X X = f_*(\nabla_X X) + \alpha(X,X).$$

Thus, the osculating space contains all first and second derivative vectors of curves on M. The maximum dimension possible for the osculating space is $N = \frac{1}{2} n(n+3)$. If the osculating space at x has dimension N, we say that f is <u>non-degenerate</u> at x. (This should not be confused with the notion of a non-degenerate height function).

Let $T_x^{\perp}M \subset E^m$ denote the normal space to $f(M)$ at $f(x)$. Define a map $\beta:T_xM \to T_x^{\perp}M$ by

$$\beta(X) = \alpha(X,X). \tag{10.3}$$

For $f:M^2 \to E^4$, it is easy to check that as X varies around the unit circle in T_xM, $\beta(X)$ traces out a (possibly degenerate) ellipse in the normal plane centered at the mean curvature vector (see, for example, Little [1, pp. 264-265]). When the ellipse is not degenerate, β covers the ellipse twice as X varies around the unit circle S^1 in T_xM. Hence, β is an embedding on $P^1 = S^1/\sim$. The ellipse is called the <u>curvature ellipse</u> of f, and it is useful in the study of tight surfaces in E^4. (See, for example, C.S. Chen

[3], Chen-Pohl [1], Hempstead [1]).

In his paper, Little noted that the Veronese manifold plays the role of the ellipse in the general case of an immersion $f:M \to E^m$. Let $X_1,\ldots,X_n$ be an orthonormal basis for T_xM with respect to the metric induced by f, and let

$$X = \sum_{i=1}^{n} x_i X_i.$$

Then

$$\beta(X) = \alpha\Big(\sum_{i=1}^{n} x_i X_i, \sum_{j=1}^{n} x_j X_j\Big) = \sum_{i=1}^{n}\sum_{j=1}^{n} x_i x_j \,\alpha(X_i,X_j). \tag{10.4}$$

If f is non-degenerate at x, then $\alpha(X_1,X_1)$, $\alpha(X_1,X_2)$, ..., $\alpha(X_n,X_n)$ form a basis for a subspace $E^{n(n+1)/2}$ in $T_x^{\perp}M$ with origin at f(x).
Let e_{ij}, $1 \leq i \leq j \leq n$ be an orthonormal basis for $E^{n(n+1)/2}$, and let L be the non-singular linear transformation of $E^{n(n+1)/2}$ defined by

$$L(e_{ij}) = \alpha(X_i,X_j), \quad 1 \leq i \leq j \leq n.$$

Let S_x denote the unit sphere in T_xM. Then equation (10.4) shows that the restriction $\beta:S_x \to E^{n(n+1)/2}$ is given by $\beta = L \circ \phi$, where ϕ is a standard (n-1)-dimensional Veronese immersion of S_x into $E^{n(n+1)/2}$. Of course, $\beta(X) = \beta(-X)$ and so β is a Veronese embedding of $P^{n-1} = S_x/\sim$ into $E^{n(n+1)/2}$. The map $\beta:S_x \to E^{n(n+1)/2}$ is called the <u>curvature indicatrix at x</u>, and its image lies in a hyperplane, called the <u>indicatrix plane</u>, which is the image under L of the hyperplane given by the equation trace $\phi(x) = 1$.

We now show that a Veronese manifold is non-degenerate at every point. Let x be any point in P^n. Then $\phi(P^n)$ lies in the osculating space at $\phi(x)$. To see this, take any other point y in P^n and let Γ be the unique line P^n through x and y. Then $\phi(\Gamma)$ is a circle in E^N whose tangent and principal normal at $\phi(x)$ lie in the osculating space. Thus, all of $\phi(\Gamma)$, and in particular $\phi(y)$, lies in the osculating space. Since ϕ is substantial, the osculating space has dimension N, and ϕ is non-degenerate.

If ϕ is followed by a projective transformation σ, then $\sigma\circ\phi(\Gamma)$ is still a plane curve, so the same argument applies.

We next develop some consequences of the two-piece property which will enable us to obtain the desired main result.

Definition 10.1: A point $x \in M$ is called an extreme point of a TPP immersion $f:M \to E^m$ if there exists a height function ℓ_p with a non-degenerate local (and hence absolute) maximum at x.

Lemma 10.2: Let $f:M \to E^m$ be a TPP immersion.

(a) The set of extreme points is open and non-empty.

(b) Every extreme point is a singleton; i.e. if x is an extreme point and $f(x) = f(y)$, then $y = x$.

Proof: (a) This follows immediately from consideration of the Gauss map, exactly as in the proof of Corollary 3.5.

(b) Suppose x is an extreme point and $f(y) = f(x)$ for $y \neq x$. Let ℓ_p be the height function given in Definition 10.1 and suppose that $\ell_p(x) = r$. Then for the half-space h given by $\ell_p \geq r$, $f^{-1}h$ has at least two components, one consisting of $\{x\}$ and one containing y.

Q.E.D.

The following lemma is rather crucial as it brings the curvature indicatrix into play.

Lemma 10.3: Let $f:M \to E^m$ be a TPP immersion. Let $x \in M$ be an extreme point, and let π be a hyperplane containing the tangent space at $f(x)$. Then π supports $f(M)$ at x if and only if $\beta(S_x)$ lies in one of the closed half-spaces determined by π.

Proof: In order to correctly interpret β as a map from T_xM to E^m using (10.3), we assume that $f(x)$ is the origin of E^m. Let p be a unit normal to π. Since p is normal to $f(M)$ at x, ℓ_p has a critical point at x with value $\ell_p(x) = 0$. By Theorem 3.2, the Hessian of ℓ_p is given by

$$(10.5) \qquad H_x(\ell_p)(X,Y) = \langle p, \alpha(X,Y)\rangle, \; X,Y \in T_xM.$$

Let $x(t)$ be a curve on M with $x(0) = x$, $x'(0) = X \in T_xM$. Let $\ell_p(t) = \ell_p(x(t))$. Then $\ell_p'(0) = 0$ and $\ell_p''(0) = \langle p, \alpha(X,X)\rangle = \langle p, \beta(X)\rangle$. Thus, if $\beta(S_x)$ lies on both sides of π, then ℓ_p will have a strict local minimum at x along some curves through x and a strict local maximum at x along others. Hence, π is not a support plane of f at x. Note that the TPP was not used in this argument.

To prove the converse, first suppose that π does not intersect $\beta(S_x)$. Since $\beta(S_x)$ is connected, it lies on one side of π, and thus by (10.5), the Hessian of ℓ_p is definite at x. Thus, ℓ_p has a non-degenerate local extremum at x and by Theorem 5.16, ℓ_p has an absolute extremum at x. Thus, π supports f(M).

Finally, suppose that π meets and supports $\beta(S_x)$, say $\ell_p \leq 0$ on $\beta(S_x)$. Again, we must show that π supports f(M). Since x is an extreme point by assumption, there exists a linear height function ℓ_q which has a non-degenerate absolute maximum at x. Let π' be the plane $\ell_q \equiv 0$. We claim claim that π' does not meet $\beta(S_x)$. If it did, then there would exist a non-zero $X \in T_xM$ with $\langle q, \alpha(X,X)\rangle = 0$. This contradicts the fact that the Hessian of ℓ_q is negative-definite at x. Thus, $\ell_q < 0$ on $\beta(S_x)$ while $\ell_p \leq 0$ on $\beta(S_x)$. Let $p(t) = (1-t)q + tp$, $0 \leq t \leq 1$. Then,

$$\ell_{p(t)} = (1-t)\,\ell_q + t\,\ell_p\,,$$

and so $\ell_{p(t)} < 0$ on $\beta(S_x)$ for $0 \leq t < 1$. Let π_t be the plane $\ell_{p(t)} \equiv 0$. Then for $0 \leq t < 1$, π_t does not intersect $\beta(S_x)$, and by the arguments in the paragraph above, π_t is a support plane of f(M). Since, $\pi = \pi_1$ is the limit of π_t as t approaches 1, it too is a support plane of f(M).

Q.E.D.

Of course, $\beta(T_xM)$ is just the cone with vertex at f(x) through $\beta(S_x)$. Thus if π is a hyperplane through f(x), then π supports $\beta(T_xM)$ if and only if $\beta(S_x)$ lies in one of the closed half-spaces determined by π.

Corollary 10.4: Let $f:M \to E^m$ be a TPP immersion. Then any hyperplane which supports f(M) to the second order at an extreme point x supports f(M) to the third order. That is: if π is a hyperplane in E^m containing the tangent space at $f(x)$ which supports $\beta(T_xM)$, and if $\gamma(t) = f(x(t))$ is a curve with $x(0) = x$ and $\gamma''(0)$ in π, then $\gamma'''(0)$ lies in π. (We are again assuming $f(x) = 0$ for convenience.)

Proof: Suppose there exists such a curve with $\gamma'''(0)$ not in π. Expand $\gamma(t)$ as a finite Taylor series:

$$\gamma(t) = \gamma(0) + t\,\gamma'(0) + \frac{t^2}{2}\gamma''(0) + \frac{t^3}{6}(\gamma'''(0) + t\,R(t)),$$

where $R(t)$ is continuous in t. Let p be a unit normal to π. Then $\langle p, \gamma'(0)\rangle = 0$ and $\langle p, \gamma''(0)\rangle = 0$. Hence,

$$\langle p, \gamma(t) - \gamma(0)\rangle = \frac{t^3}{6}(\langle p, \gamma'''(0)\rangle + t\,\langle p, R(t)\rangle).$$

Since $\gamma'''(0)$ does not lie in π, the function ℓ_p takes on both positive and negative values on $\gamma(t)$ near 0. Thus, π does not support $f(M)$, contradicting Lemma 10.3.

Q.E.D.

Lemma 10.5: Let $f:M \to E^m$ be a TPP immersion and let x be an extreme point of f. Then $f(M)$ is contained in the osculating space to f at x.

Proof: If the osculating space to f at x is all of E^m, then there is nothing to prove. So suppose that the osculating space to f at x is a Euclidean subspace E^r of dimension $r < m$. Since x is an extreme point, there exists a linear height function ℓ_p with a non-degenerate absolute maximum at x. Let π be the hyperplane through $f(x)$ orthogonal to p. As in the proof of Lemma 10.3, π does not meet $\beta(S_x)$, since the Hessian of ℓ_p is negative-definite at x. Since $\beta(S_x)$ lies in the osculating space E^r, it follows that E^r is not contained in π, and thus $\pi \cap E^r$ is a Euclidean subspace of dimension $r-1$, call it E^{r-1}. Since $\beta(S_x)$ is connected and does not meet E^{r-1}, it lies in one of the closed half-spaces of E^r determined by E^{r-1}, call it E^r_+. Thus, any hyperplane in E^m which contains E^{r-1} but not E^r contains $\beta(S_x)$ in one of its closed half-spaces, and hence it supports $f(M)$ by Lemma 10.3. So $f(M)$ lies in the intersection of all of these closed half-spaces which is just $E^r_+ \subset E^r$, as desired.

Q.E.D.

We get immediately

<u>Corollary 10.6</u>: <u>If $f:M^n \to E^N$, $N = \frac{1}{2}n(n+3)$ is a substantial TPP immersion, then f is non-degenerate at every extreme point.</u>

<u>Proof</u>: If f is not non-degenerate at an extreme point x, then the osculating space to f at x has dimension $r < N$. But $f(M)$ is contained in this osculating space by Lemma 10.5, and so f is not substantial into E^N. Contradiction.

Q.E.D.

<u>Lemma 10.7</u>: <u>Let $f:M^n \to E^N$ be a TPP immersion. Then every non-degenerate point is an extreme point.</u>

<u>Proof</u>: Let f be non-degenerate at x, and let π be the hyperplane through $f(x)$ parallel to the indicatrix plane which contains $\beta(S_x)$. Let p be the unit normal to π which points from $f(x)$ away from the indicatrix plane. Then,

$$\langle p, \beta(X)\rangle = \langle p, \alpha(X,X)\rangle < 0$$

for all unit vectors $X \in T_xM$, and so the Hessian of ℓ_p is negative-definite at x. Thus, ℓ_p has a non-degenerate maximum at x, and x is an extreme point.

Q.E.D.

The main theorem now follows fairly quickly from the propositions above and the following characterization of the Veronese manifolds. This was proven for n = 2 by C. Segre [1]-[2], also see Lane [1, p. 426]. The condition that every hyperplane which supports to the second order supports to the third order is equivalent to the condition that "the characteristic curves are degenerate" in Lane [1]. This characterization of the Veronese surface is quite important and many other classical characterizations follow from it. It was used as the final step by Kuiper [6] in his proof of Theorem 10.9 for $n = 2$. The proof of this characterization for $n > 2$ is a major part of the paper of Little and Pohl. We will not reproduce their proof here but will only state their result. An immersion is called <u>non-degenerate</u> if it is non-degenerate at every point.

<u>Theorem 10.8</u>: <u>(C. Segre, Little-Pohl) Let $f:M^n \to E^N$, $n \geq 2$, be a non-degenerate immersion which has the property that every hyperplane which supports f(M) to the second order supports f(M) to the third order. Then f(M) is contained in a Veronese n-manifold.</u>

We now prove the main result of this section.

<u>Theorem 10.9</u>: <u>Let M be a compact n-manifold, $n \geq 2$, and let $f:M \to E^N$, $N = \frac{1}{2} n(n+3)$, be a substantial smooth TPP immersion. Then M is diffeomorphic to RP^n and f embeds M as a Veronese manifold.</u>

<u>Proof</u>: Since f has the TPP, M is connected. By Lemma 10.2, f is injective on the set of extreme points which is open and non-empty. Let M_0 be a connected component of the set of extreme points. By Corollary 10.6, f is non-degenerate at every point of M_0. By Lemmas 10.3 and 10.4, the restriction of f to M_0 has the property that every hyperplane which supports $f(M_0)$ to the second order supports $f(M_0)$ to the third order. By Theorem 10.8, $f(M_0)$ is embedded as an open subset of a Veronese n-manifold V^n in E^N. Suppose M_0 is not all of M. If x is a boundary point of M_0, then x is not an extreme point by Lemma 10.2. Hence, f is degenerate at x by Lemma 10.7. On the other hand, by continuity, f(x) is in V^n, and f and V^n have the same tangent planes and second fundamental forms at f(x). But V^n is not degenerate at any point, so we have a contradiction. Thus, M_0 has no boundary points, i.e. $M_0 = M$, and f embeds M onto the Veronese manifold V^n.

Q.E.D.

We will conclude this section with some similar results on taut embeddings of maximal codimension. Of course, for a compact M, taut implies tight, and so if $f:M \to E^m$ is taut and substantial, then $m \leq N = \frac{1}{2} n(n+3)$. Further, if $f:M \to E^m$ is a taut substantial non-spherical embedding, then $P^{-1} \circ f:M \to S^m \subset E^{m+1}$ is a substantial taut spherical embedding, where $P:S^m-\{q\} \to E^m$ is stereographic projection. Hence, as a consequence of this argument and Theorem 10.9, we have:

Theorem 10.10: Let M^n be a compact manifold, $n \geq 2$, and let $f:M \to E^N$, $N = \frac{1}{2} n(n+3)$ be a substantial taut embedding. Then M is a diffeomorphic to RP^n and f embeds M as a spherical Veronese manifold.

Recall that a map $f:X \to E^m$ of a topological space X is called proper if $f^{-1}K$ is compact for every compact subset K in E^m. If an immersion $f:M \to E^m$ of a non-compact manifold M is proper, then the Morse inequalities for distance functions L_p can still be applied on the compact sets $f^{-1}B$, where B is a closed ball in E^m. This leads to the following definition which is discussed in more detail in Section 1 of Chapter 2.

Definition 10.11: A proper immersion of non-compact manifold M into E^m is called taut if for every closed ball B in E^m, the induced homomorphism

$$H_*(f^{-1}B) \to H_*(M)$$

in Cech homology with Z_2-coefficients is injective.

It is then easy to prove (see Theorem 1.28 of Chapter 2) that if $f:M \to E^m$ is a substantial taut embedding of a non-compact n-manifold, then

$$m \leq \frac{1}{2} n(n+1) - 1$$

We now prove the analogue of Theorem 10.10 for non-compact manifolds. This was first noted by Carter and West [1, p. 711].

Theorem 10.12: Let M^n be a non-compact manifold, $n \geq 2$, and let $f: M \to E^{N-1}$, $N = \frac{1}{2} n(n+3)$ be a substantial taut embedding. Then M is diffeomorphic to $RP^n - \{q\}$, and $f(M)$ is the image under stereographic projection of a Veronese manifold V^n where the pole q of the projection lies on V^n.

Proof: Let $P: S^{N-1} - \{q\} \to E^{N-1}$ be stereographic projection with respect to some pole q. Then $P^{-1} \circ f$ is a spherical map of the one point compactification $X = M \cup \{\infty\}$ into S^{N-1} with $P^{-1} \circ f(\infty) = q$. At all other points, $P^{-1} \circ f$ is a smooth immersion. One shows rather easily that $P^{-1} \circ f$ is a TPP map on the compact space X. For if h is a closed half-space in E^N, then $h \cap S^{N-1}$ is a closed ball D in S^{N-1}. Then $B = P(D)$ is either a closed ball, a closed half-space, or the complement of an open ball in E^{N-1}. If B is a closed ball and $f^{-1}B = (P^{-1} \circ f)^{-1}h$ is not connected, then the homomorphism

$$H_0(f^{-1}B) \to H_0(M)$$

is not injective, contradicting the tautness of f. If B is a closed half-space and $f^{-1}B$ is not connected, then there exists a large closed ball B' closely approximating B such that $f^{-1}B'$ is not connected, again violating the tautness of f. Finally, if B is the complement of an open ball in E^{N-1} centered at a point p, and $f^{-1}B$ is not connected, then only one of the components of $f^{-1}B$ is unbounded. The distance function L_p must have a local maximum value on the bounded component of $f^{-1}B$, and by a result similar to Lemma 5.14 for distance functions, there exists a non-degenerate $L_{p'}$ which has a local maximum on this component. This again contradicts the tautness of f, since $\beta_n(M;Z_2) = 0$ for a non-compact manifold (see Morse-Cairns [1, p. 271]). Thus $P^{-1} \circ f: X \to E^N$ has the TPP.

The proof is completed by adapting the arguments of Theorem 10.9 to the non-compact case. First note that $P^{-1} \circ f$ is smooth at all the points of M. If a height function ℓ_p, p a unit vector in E^N, has a non-degenerate local maximum on $P^{-1}f(M)$ at x, then the same argument used earlier shows that ℓ_p has its absolute maximum on $P^{-1}f(X)$ at x. If, as before, we call such points extreme points, then the proof of Lemma 3.2 still works to show that the set of extreme points is open and non-empty in M, and that each extreme point is a singleton. Further, the lemmas and corollaries 10.3–10.7 still hold, as they essentially involve only local arguments and the fact that every non-degenerate local maximum of a height function is an absolute maximum. We now complete the proof in the same way as in Theorem 10.9. Let $M_0 \subset M$ be a component of the set of extreme points of $P^{-1} \circ f$ in M. The restriction to M_0 of $P^{-1} \circ f$ is a non-degenerate immersion with the property that every hyperplane in E^N which supports $P^{-1}f(M_0)$ to the second order supports $P^{-1}f(M_0)$ to the third order. By Theorem 10.8, $P^{-1}f(M_0)$ is embedded as an open subset of a Veronese manifold V^n in E^N. The argument used in Theorem 10.9 shows that no point of M can be a boundary point of M_0, and so $M = M_0$. Hence $P^{-1} \circ f$ embeds M onto an open subset of V^n. Since $q = P^{-1}(f(\infty))$ is a limit point of V^n, it also lies in V^n. Hence $P^{-1}f(X)$ is a compact subset of V^n, and thus $P^{-1}f(M)$ is a closed subset in the relative topology of $V^n - \{q\}$. But we have already shown that $M = M_0$ which is embedded by $P^{-1} \circ f$ as an open subset of $V^n - \{q\}$. Since, $V^n - \{q\}$ is connected, $P^{-1}f(M) = V^n - \{q\}$, and $f(M) = P(V^n - \{q\})$, as desired.

Q.E.D.

Taut immersions

0. INTRODUCTION.

Banchoff [3] began the study of taut immersions by attempting to find all tight surfaces which lie in a Euclidean sphere S^m in E^{m+1}. Known examples were great spheres, small spheres and the standard product torus in S^3. Since a hyperplane in E^{m+1} intersects S^m in a great or small $(m-1)$-sphere, the usual two-piece property is equivalent to the two-piece property with respect to hyperspheres in S^m for a spherical immersion. Via stereographic projection, this problem is equivalent to the study of surfaces in E^m which have the two-piece property with respect to hyperspheres and hyperplanes, i.e., the spherical two-piece property (STPP). As in the relationship between the usual two-piece property and linear height functions, the spherical two-piece property is equivalent to the requirement that every non-degenerate Euclidean distance function $L_p(x) = |p-x|^2$ have exactly one maximum and one minimum on the surface M.

Carter and West [1] generalized the STPP and defined a smooth immersion f of a compact manifold M to be taut if every non-degenerate distance function L_p has the minimum number of critical points. Further, they noted that if M is non-compact but the immersion f is proper (i.e., f^{-1} of a compact set is compact), then the Morse inequalities still hold on compact sets of the form

$$M_r(L_p) = \{x \in M \mid L_p(x) < r\},$$

and the notion of a taut immersion extends to this case. In Section 1, we further generalize this to define a taut map in terms of an injectivity condition on homology, as was done for tightness. We then prove several basic results and discuss the equivalent formulations of the definition. In particular, a taut map is tight, and a taut smooth immersion of a manifold must be an embedding, as was first shown by Carter-West [1].

If $M \subset E^m$ is a substantial non-spherical taut embedding of a compact manifold, then one can obtain a substantial spherical embedding of M into E^{m+1} via inverse stereographic projection. Using this and the fact that a taut embedding is tight, one obtains from the Chern-Lashof theorem that a taut embedding of a sphere in E^m must be a Euclidean metric sphere. This was observed by Banchoff [3] for surfaces and by Carter-West [1] for higher dimensional spheres. Nomizu and Rodriguez [1] gave a different proof that a taut sphere must be round based on the characterization of round spheres as umbilic submanifolds. Their approach has been generalized to obtain similar results for submanifolds of hyperbolic space and complex projective space, and it is presented in Section 2.

Much of the remainder of the chapter involves the cyclides of Dupin and their generalizations. Classicially, these were characterized as the only surfaces in E^3 for which both sheets of their focal set degenerate to curves. Equivalently, they are the only surfaces such that all lines of curvature in both families are circles or straight lines. Examples are tori of revolution and circular cylinders and cones. It turns out that all others can be obtained from these three by inversion in a sphere in E^3.

In Section 3, we find the formula for the shape operators of a tube of constant radius over a submanifold M. This is useful for our purposes, because it enables one to show that the focal set of a tube over a submanifold M of codimension greater than one is just the union of the focal set of M with M itself. This allows one to answer many questions concerning focal sets, e.g., when are they manifolds, by restricting one's attention to hypersurfaces.

In the fourth section we find conditions under which a sheet of the focal set is a submanifold of codimension greater than one. The classical version of this result is what is needed to show that the two characterizations of the cyclides mentioned above are equivalent. This result is then applied in the next section to give a rather thorough description of the classical cyclides. Noteworthy is the fact that the focal curves must be a pair of so-called "focal conics". These are either an ellipse and hyperbola in orthogonal planes such that the foci of the ellipse are the vertices of the hyperbola and vice-versa, or a pair of parabolas in orthogonal planes such that the vertex of each is the focus of the other. We also study the full family of parallel surfaces to a given cyclide, all of which must be cyclides, since they have the same focal set. Some of these surfaces have one or two singular or nodal points.

A hypersurface M in E^{n+1} is called a Dupin hypersurface if each of the principal curvatures has constant multiplicity and is constant along the leaves of its principal foliation. Examples are images under stereographic projection of isoparametric hypersurfaces in spheres. In Section 6, we generalize the classical characterization of the cyclides and show that a complete Dupin hypersurface with 2 principal curvatures must either be a standard spherical cylinder $S^k \times E^{n-k}$ or the image under stereographic projection of a standard product of spheres $S^k \times S^{n-k}$ in S^{n+1}. By analogy with the classical terminology, these latter hypersurfaces are called ring cyclides if they are compact, and parabolic ring cyclides if they are not compact, i.e., if the pole of the projection is on $S^k \times S^{n-k}$. As in the classical case, the focal set consists of a pair of focal quadrics, an ellipsoid and a hyperboloid of two sheets for the ring cyclides, and two paraboloids for the parabolic ring cyclides.

There has been a good deal of recent research on Dupin hypersurfaces. Using the Lie geometry of spheres, U. Pinkall [1]-[3] classified all Dupin hypersurfaces in E^4 and made several useful observations about the general case. G. Thorbergsson [1] then proved that every complete embedded Dupin hypersurface is taut. This had been shown for stereographic images of isoparametric hypersurfaces in spheres by Cecil and Ryan [6]. The result of Thorbergsson is not only interesting in itself, but in combination with the work of H.F. Münzner [1], it allows one to deduce some important facts about compact Dupin hypersurfaces. In particular, using tautness, one can show that a compact Dupin hypersurface M in S^{n+1} divides the sphere into two ball bundles over the first focal submanifolds on either side of M. It then follows from Münzner's work that the number of distinct principal curvatures of a compact Dupin hypersurface must be 1, 2, 3, 4, or 6, as with isoparametric hypersurfaces in spheres. This leads to the conjecture that every compact Dupin hypersurface embedded in Euclidean space is Lie equivalent to an isoparametric hypersurface in a sphere. The conjecture is true for Dupin hypersurfaces with one or two distinct principal curvatures by the well-known classification of umbilic hypersurfaces and by the theorem mentioned above concerning the cyclides, respectively. Recently, R. Miyaoka [3] has shown that the conjecture holds for Dupin hypersurfaces with three distinct principal curvatures, i.e. they are all Lie equivalent to isoparametric hypersurfaces with 3 principal curvatures which were classified by E. Cartan [2,3] (see Example 7.3 of Chapter 3). Finally, if a principal curvature λ of a hypersurface M in E^{n+1} has constant multiplicity $\nu > 1$, then λ is automatically constant along the leaves of its principal foliation (see Theorem 4.4). Recently, Pinkall [4] (see Lemma 7.5) has shown that if M is taut and λ has constant multiplicity 1, then λ is constant along its lines of curvature. Hence, the converse of Thorbergsson's theorem holds for a hypersurface M whose principal curvatures have constant multiplicties, i.e. taut implies Dupin. There are, however, many known examples of taut hypersurfaces whose principal curvatures do not have constant multiplicities, e.g. a tube over a torus of revolution $T^2 \subset E^3 \subset E^4$. The known examples of taut hypersurfaces do satisfy a generalization of the Dupin condition introduced by Pinkall [3] which we call semi-Dupin. We discuss these examples and conjecture that the conditions of taut and semi-Dupin are equivalent for complete embedded hypersurfaces in Euclidean space. The details of the results mentioned above on Dupin hypersurfaces are given in Section 6.

The complete determination of all taut surfaces in Euclidean space is given in Section 7. A spherical Veronese surface in S^4 is one obvious example, as are its images under stereographic projection into E^4. All other taut surfaces either lie in a 3-sphere S^3 or in E^3. These two classes are equivalent via stereographic projection for compact M. Banchoff [3] showed that a taut compact surface in E^3 is either a round sphere or a ring cyclide. Later, Cecil [3] found that the collection of taut non-compact surfaces in E^3 consists of planes, circular cylinders and parabolic ring cyclides.

In Section 8, we prove some results involving higher dimensional taut hypersurfaces. First, we give a proof of a theorem of Carter and West [1] which states that if M is a taut non-compact hypersurface with the same Z_2-homology as $S^k \times E^{n-k}$, then M is a standard product embedding of $S^k \times E^{n-k}$. Next we show that a taut hypersurface M with the same Z_2-homology as $S^k \times S^{n-k}$ is a ring cyclide. One proves that such an M must have exactly two distinct principal curvatures which are constant along their principal foliations and then invokes the previous characterization. In the case $k = n-k$, the argument actually yields the stronger statement that a taut, compact, (k-1)-connected M^{2k} in E^{2k+1} must be a Euclidean sphere or a ring cyclide. This has been generalized by Thorbergsson [2] to higher codimension.

There are two other large classes of taut submanifolds. The standard symmetric R-spaces (see Takeuchi and Kobayashi [1]) are tight and spherical, and hence taut. Secondly, all isoparametric hypersurfaces in S^{n+1} are taut, as are their images under stereographic projection. This will be shown in Chapter 3. Of course, the standard product embeddings of $S^k \times S^{n-k}$ belong to this class.

Section 9 deals with the notion of <u>totally focal embeddings</u> due to Carter and West [2]-[4]. A proper embedding $f:M \to E^m$ of a smooth manifold is called totally focal if every distance function L_p is either non-degenerate or has only degenerate critical points. Carter and West [2] showed that a totally focal hypersurface in E^{n+1} must be a round sphere, a hyperplane or a standard product embedding of $S^k \times E^{n-k}$. This is, in fact, the list of complete isoparametric hypersurfaces in E^{n+1}, although Carter and West did not use this characterization. Cecil and Ryan [6] pointed out that isoparametric hypersurfaces in the sphere S^{n+1} are also totally focal (as submanifolds of the sphere). Carter and West [4] then showed, conversely, that a totally focal hypersurface in S^{n+1} is isoparametric.

The chapter concludes with a survey of the known results on tight and taut immersions into hyperbolic space.

1. <u>DEFINITIONS AND FIRST RESULTS: EXAMPLES.</u>

Let us recall that the original problem was to find all tight submanifolds which lie in a sphere S^m in E^{m+1}. Since a closed half-space in E^{m+1} intersects S^m in a closed metric ball, one can transfer the problem to a study of submanifolds of E^m via stereographic projection. Under this projection, closed metric balls in S^m are transformed to closed balls (including those with 0-radius, i.e., points), complements of open balls, or closed half-spaces. This motivates the following definition (which was given previously in Section 6 of Chapter 1).

<u>Definition 1.1</u>: A map f of a compact space X into E^m is called <u>taut</u> if for every closed ball, complement of an open ball, or closed half-space Ω in E^m, the induced homomorphism

$$H_*(f^{-1}\Omega) \to H_*(X) \text{ , } Z_2\text{-Cech homology,}$$

is injective.

As an immediate consequence of the definitions of tightness and tautness, we get

Theorem 1.2: A taut map of a compact space X into E^m is tight.

We next note that a tight spherical map (i.e., image is contained in a Euclidean sphere) is taut.

Theorem 1.3: Let $f: X \to S^m \subset E^{m+1}$ be a tight, spherical map of a compact topological space X. Then f is a taut map of X into E^{m+1}.

Proof: Let Ω be a closed ball or the complement of an open ball in E^{m+1}. Then $\Omega \cap S^m = h \cap S^m$ for some closed half-space in E^{m+1}. Since $f(X) \subset S^m$, we have

$$f^{-1}\Omega = f^{-1}(\Omega \cap S^m) = f^{-1}(h \cap S^m) = f^{-1}h.$$

By tightness, the map $H_*(f^{-1}h) \to H_*(X)$ is injective, and so f is taut.

Q.E.D.

Thus, we have already encountered some examples of taut embeddings. A Euclidean sphere S^m in E^{m+1} is taut. The standard product embedding of spheres with radii r_i,

$$S^{k_1}(r_1) \times \dots \times S^{k_j}(r_j) \subset S^m(1) \subset E^{m+1},$$

where

$$E^{m+1} = E^{k_1+1} \times \dots \times E^{k_j+1},$$

is tight and spherical, and thus taut. The spherical Veronese manifolds defined in Section 9 of Chapter 1 are also taut.

Via stereographic projection, E^m is conformally equivalent to $S^m - \{q\}$, or one may write $S^m = E^m \cup \{\infty\}$ as the one-point compactification of E^m. A conformal transformation of $E^m \cup \{\infty\}$ takes the collection of hyperspheres and hyperplanes onto itself. Hence, tautness is obviously preserved by such a conformal transformation. We formulate this specifically in the following statement. In this way, we see that tautness is equivalent to the combination of tight and spherical via stereographic projection.

Theorem 1.4: (Tautness is a conformal invariant) Let X be a compact topological space.

(a) If $f:X \to E^m$ is a taut map and ϕ is a conformal transformation of $E^m \cup \{\infty\}$ such that $\phi(f(X)) \subset E^m$, then $\phi \circ f$ is a taut map of X into E^m.

(b) If $f:X \to S^m \subset E^{m+1}$ is taut and $P_q:S^m - \{q\} \to E^m$ is stereographic projection with pole q not in f(X), then $P_q \circ f$ is a taut map of X into E^m.

(c) If $f:X \to E^m$ is taut and $P_q^{-1}:E^m \to S^m \subset E^{m+1}$ is inverse stereographic projection with respect to any pole q, then $P_q^{-1} \circ f$ is a taut map of X into S^m.

This immediately gives further examples by taking stereographic projection of the tight spherical examples mentioned above. In particular, the classical cyclides of Dupin (without singularities) are obtained from the product torus in S^3 by stereographic projection. One can also obtain a taut embedding of the real projective plane P^2 in E^4 through stereographic projection of the Veronese surface in S^4.

Corollary 1.5: Let M be a compact manifold which is not a sphere. Then there exists a substantial non-spherical taut embedding $f:M \to E^m$ if and only if there exists a substantial spherical taut embedding $\hat{f}:M \to E^{m+1}$.

Carter and West pointed out that Theorem 1.4, when combined with known results for tight immersions, yields theorems for taut immersions. Specifically, in conjunction with the Chern-Lashof theorem, it implies

Theorem 1.6: Let $f:S^n \to E^m$ be a substantial taut immersion. Then $m = n+1$, and f embeds S^n as a Euclidean sphere.

Proof: Since a taut immersion is tight, $m = n+1$ and f embeds S^n as a convex hypersurface in E^{n+1} by the Chern-Lashof theorem. If $f(S^n)$ were not a Euclidean sphere, then by Theorem 1.4(c), $P_q^{-1} \circ f$ would be a taut substantial embedding of S^n in E^{n+2}, contradicting the Chern-Lashof theorem.

Q.E.D.

Returning to the general discussion, in the case where X is a manifold, we wish to relate tautness to the critical point behavior of Euclidean distance functions, as in the original formulation of tautness.

Suppose $f:M \to E^m$ is a smooth immersion. For $p \in E^m$, the Euclidean distance function L_p is defined on E^m by

$$L_p(q) = |p-q|^2.$$

The restriction of L_p to M is defined by $L_p(x) = |p-f(x)|^2$. As in the case of the linear height functions, Sard's theorem implies that for almost all $p \in E^m$, L_p is a Morse function on M. We simply outline the procedure and refer the reader to Milnor [3, pp. 32-38] for a complete proof.

The normal exponential map F from the normal bundle NM to E^m is defined by

$$F(x,\xi) = f(x) + \xi,$$

where ξ is a normal vector to f(M) at f(x). This map plays the role which was assumed by the Gauss map for the ℓ_p functions.

Definition 1.7: A point $p \in E^m$ is called a focal point of multiplicity ν of (M,x) if $p = F(x,\xi)$, and the Jacobian of F has nullity $\nu > 0$ at (x,ξ). The point p is called a focal point of M if p is a focal point of (M,x) for some $x \in M$. The set of all focal points is called the focal set of M.

Since NM and E^m are both m-dimensional, one concludes from Sard's theorem

Proposition 1.8: Let $f:M \to E^m$ be an immersion. Then the focal set of M has measure zero.

A direct computation of the Jacobian of F shows that the notion of focal point is a generalization of the concept of center of curvature for a curve, that is,

Proposition 1.9: Let $p = F(x,t\xi)$ where $|\xi| = 1$. Then p is a focal point of multiplicity $\nu > 0$ of (M,x) if and only if 1/t is an eigenvalue of multiplicity ν of the shape operator A_ξ.

The critical point behavior of the L_p functions is described by the following theorem.

Theorem 1.10 (Index Theorem for L_p Functions): For $p \in E^m$,

(a) A point $x \in M$ is a critical point of L_p if and only if $p = F(x,\xi)$ for some $\xi \in T_x^\perp M$.

(b) L_p has a degenerate critical point at x if and only if p is a focal point of (M,x).

(c) If L_p has a non-degenerate critical point at x, then the index of L_p at x is equal to the number of focal points of (M,x) (counting multiplicities) on the segment from f(x) to p.

As a corollary, we have the following analogue of Corollary 3.5 of Chapter 1 (see Nomizu-Rodriguez [1, p. 199] for a proof).

Proposition 1.11:

(a) For almost all $p \in E^m$, L_p is a Morse function.

(b) Suppose L_p has a non-degenerate critical point of index j at $x \in M$. Then there is a Morse function L_q having a critical point $y \in M$ of index j (q and y may be chosen as close to p and x, respectively, as desired).

We recall some terminology from Chapter 1, namely,

$$M_r(L_p) = \{x \mid L_p(x) < r\}$$

$\mu_k(L_p,r)$ is the number of critical points of index k which L_p has on $M_r(L_p)$, and $\beta_k(L_p,r,Z_2)$ is the k-th Z_2-Betti number of $M_r(L_p)$. The following is an analogue of Theorem 5.4 of Chapter 1 and can be proven similarly using the continuity property of Cech homology.

Theorem 1.12: *An immersion $f:M \to E^m$ of a compact manifold M is taut if and only if*

$$\mu_k(L_p,r) = \beta_k(L_p,r,Z_2)$$

for every non-degenerate distance function L_p, and for every $r \in R$, $k \in Z$.

Corollary 1.13: *An immersion $f:M \to E^m$ of a compact manifold is taut if and only if every non-degenerate distance function L_p has $\beta(M;Z_2)$ critical points on M.*

As with tightness, there is a two-piece property corresponding to tautness. This was introduced by Banchoff [3].

Definition 1.14: A map $f:X \to E^m$ of a compact, connected topological space X is said to have the *spherical two-piece property (STPP)* if $f^{-1}\Omega$ is connected for every closed ball, complement of an open ball or closed half- space, Ω in E^m.

One proves the following results in a manner entirely analogous to Theorems 5.9, 5.11 and Corollary 5.12 of Chapter 1.

Theorem 1.15: *Let f be a continuous map of a compact, connected topological space X into E^m. If f is taut, then f has the STPP.*

Theorem 1.16: *Let $f:M\to E^m$ be an immersion of a compact, connected manifold. Then f has the STPP if and only if every non-degenerate distance function L_p is polar (i.e., has one maximum, one minimum).*

Corollary 1.17: Let $f:M \to E^m$ be an immersion of a compact, connected surface. Then f is taut if and only if f has the STPP.

Next, we discuss how to generalize the concept of tautness to proper immersions of non-compact manifolds. Recall that a map $f:X \to E^m$ of a topological space X is called proper if $f^{-1}K$ is compact for every compact subset K of E^m. If f is proper, then $f^{-1}B$ is compact for every closed ball in E^m, and the Morse inequalities for a non-degenerate distance function can be applied.

Suppose $f:M \to E^m$ is an immersion of a compact manifold and the map

$$H_*(f^{-1}B) \to H_*(M) \quad (Z_2\text{-homology})$$

is injective for all closed balls B in E^m. Then by Theorem 2.1 of Chapter 1 every non-degenerate L_p has $\beta(M;Z_2)$ critical points and by Corollary 1.13, f is taut. Hence, if the injectivity condition holds for closed balls, then it also holds for complements of open balls and for half-spaces. Therefore, the following formulation of tautness for a proper immersion of a manifold is equivalent to the original definition for a compact M, but is also valid for non-compact manifolds.

Definition 1.18: A proper immersion f of a manifold M into E^m is called taut if for every closed ball B in E^m, the induced homomorphism

$$H_*(f^{-1}B) \to H_*(M)$$

in Cech homology with Z_2-coefficients is injective.

As in Theorem 1.12, one can show that it is sufficient to have the injectivity condition for balls determined by non-degenerate distance functions.

Theorem 1.19: Let $f:M \to E^m$ be a proper immersion of a manifold M. Then f is taut if and only if

$$\underline{\mu_k(L_p,r) = \beta_k(L_p,r,Z_2)}$$

for every non-degenerate L_p, and for every $r \in R$, $k \in Z$.

Example 1.20: Taut cylinders

An example of a taut embedding of a non-compact manifold is a circular cylinder in E^3, or more generally, the product embedding

$$f \times g : S^k \times E^{n-k} \to E^{k+1} \times E^{n-k} = E^{n+1}$$

where f embeds S^k as a Euclidean sphere, and g is the identity map on E^{n-k}. Every non-degenerate L_p has two critical points on $S^k \times E^{n-k}$, one of index 0 and one of index k. In fact, suppose $f:M \to E^k$ is any taut immersion of a manifold and g is the identity on E^{m-k}. Then

$$f \times g : M \times E^{m-k} \to E^k \times E^{m-k} = E^m$$

is taut. To see this, note that if $p = (p_1,p_2)$ is in $E^k \times E^{m-k}$, and $x = (x_1,x_2)$, then

$$L_p(x) = L_{p_1}(x_1) + L_{p_2}(x_2).$$

One easily sees that L_p has a critical point at x if and only if $p_2 = x_2$ and L_{p_1} has a critical point at x_1. The critical point of L_p will be non-degenerate if and only if the critical point of L_{p_1} is non-degenerate, and in that case, they have the same index. Since $M \times E^{n-k}$ has the same Betti numbers as M, the result follows.

Another obvious way to obtain taut embeddings of non-compact manifolds is the following.

Theorem 1.21: Suppose $f:M \to E^m$ is a taut embedding of a compact manifold M and Φ is a conformal transformation of $E^m \cup \{\infty\}$ such that $\Phi(f(x)) = \infty$ for some $x \in M$. Then $\Phi \circ f$ is a taut embedding of $M-\{x\}$ into E^m.

Proof: If B is any closed ball in E^m, then $\Phi^{-1}B$ is a closed ball, the complement of an open ball, or a closed half-space in E^m. Since f is taut, the map

$$H_*(f^{-1}(\Phi^{-1}B)) \to H_*(M)$$

is injective. Since this map factors through the homomorphism

$$H_*(M-\{x\}) \to H_*(M),$$

the map

$$H_*(f^{-1}(\Phi^{-1}B)) \to H_*(M-\{x\})$$

must also be injective, as desired.

Q.E.D.

Interesting examples here are the parabolic cyclides of Dupin obtained from $S^k \times S^{n-k}$ through stereographic projection with respect to a pole on the manifold. These will be studied in detail in Sections 5-6. Another noteworthy example is the taut embedding of the Moebius band in E^4 obtained from the Veronese surface in S^4 through stereographic projection with respect to a pole on the surface.

We next show that a taut immersion of a manifold must be an embedding. The proof is quite trivial under the present formulation of tautness, but was much more involved when first proven by Carter and West [1] using the original definition of tautness in terms of non-degenerate distance functions alone. We need the following terminology.

Definition 1.22: A proper immersion f of a connected manifold M into E^m is called k-taut if for every closed ball B in E^m and for every integer $i \leq k$, the induced homomorphism $H_i(f^{-1}B) \to H_i(M)$, (Z_2-Cech homology) is injective.

Theorem 1.23: A 0-taut immersion $f:M \to E^m$ must be an embedding.

Proof: Suppose $f(x_1) = f(x_2) = p$ for two distinct points x_1, x_2 in M. Since f is an immersion, there exist neighborhoods U_1 and U_2 of x_1 and x_2, respectively, on which f is an embedding. Let $B = \{p\}$ be the closed ball of radius 0 about p. Then $f^{-1}B$ has at least two connected components, $\{x_1\} \subset U_1$ and $\{x_2\} \subset U_2$, contradicting 0-tautness.

Q.E.D.

We now follow the development of Carter-West [1, p. 708-710]. The first lemma is Banchoff's [3] observation that for an STPP embedding, every local support sphere must be a global support sphere.

Lemma 1.24:

(a) Let $f:M \to E^m$ be a 0-taut embedding of a compact connected manifold. Suppose p is the first focal point of (M,x) on a normal ray to f(M) at f(x). If q is any point except p on the segment from f(x) to p, then L_q has a strict absolute minimum on M at x. Further, L_p itself has an absolute minimum at x.

(b) Let $f:M \to E^m$ be an STPP embedding of a compact connected n-dimensional manifold. Suppose that p is a focal point of (M,x) such that the sum of the multiplicities of the focal points of (M,x) on the closed segment from f(x) to p is n. If q is a point beyond p on the normal ray from f(x) through p, then L_q has a strict absolute maximum at x. Further, L_p itself has an absolute maximum at x.

Proof: (a) For any $q \neq p$ on the segment [f(x),p], L_q has a strict local minimum at x by the index theorem. But the intersection of f(M) with the closed ball centered at q through f(x) must be connected by 0-tautness, so this intersection consists of the point f(x) alone. Therefore f(M) lies outside the union of these open balls and hence lies outside the open ball centered at p through f(x). Thus L_p has an absolute minimum at x.

The proof of (b) is entirely analogous using maxima rather than minima.

Q.E.D.

This lemma has some useful corollaries. Recall that ℓ_ξ denotes the linear height function $\ell_\xi(z) = \langle \xi, z \rangle$.

Corollary 1.25: Let $f:M \to E^m$ be a 0-taut embedding of a connected manifold M. Suppose there are no focal points of (M,x) on the normal ray to f(M) in the direction ξ at $f(x)$. Then $f(M)$ lies in the closed half-space given by the inequality $\ell_\xi \leq \ell_\xi(x)$.

Proof: By Lemma 1.24(a), for all q on the normal ray in question, $f(M)$ is disjoint from the open sphere centered at q of radius $|q-f(x)|$. Hence, $f(M)$ is disjoint from the union of such balls, which is the open half-space $\ell_\xi > \ell_\xi(x)$.

Q.E.D.

From Lemma 1.24 and Corollary 1.25, we find that the existence of a normal vector ξ such that $A_\xi = \lambda I$ has strong implications for a taut embedding.

Corollary 1.26: Let $f:M \to E^m$ be a 0-taut embedding of a connected manifold M. If $A_\xi = 0$ for some unit normal ξ to $f(M)$ at the point $f(x)$, then $f(M)$ lies in the hyperplane $\ell_\xi \equiv \ell_\xi(x)$.

Proof: Since $A_\xi = 0$, there do not exist any focal points on the normal line determined by ξ. Apply Corollary 1.26 to each of the normal rays determined by ξ, and the result follows.

Q.E.D.

If some $A_\xi = \lambda I$ for $\lambda \neq 0$, one obtains a similar result.

Corollary 1.27: Let $f:M \to E^m$ be a taut embedding of a connected manifold M. If $A_\xi = \lambda I$, $\lambda \neq 0$, for some unit normal ξ to $f(M)$ at $f(x)$, then $f(M)$ lies on the sphere centered at the focal point $p = f(x) + (1/\lambda)\xi$ with radius $1/|\lambda|$.

Proof: Let q be a point on the open segment from $f(x)$ to p. By Lemma 1.24(a), $f(M)$ does not intersect the open ball centered at q of radius $|q-f(x)|$. Hence $f(M)$ is disjoint from the union of such open balls, i.e., the open ball centered at p of radius $1/|\lambda|$. Similarly, by Lemma 1.24(b), $f(M)$ is disjoint from the complement of the closed ball centered at p with radius $1/|\lambda|$. (Tautness and the existence of a critical point of index n imply that M is compact.)

Q.E.D.

Note that in the case where M is a hypersurface, the existence of a single umbilic point implies that M is embedded as a Euclidean hypersphere, as Banchoff [3] first noted for surfaces.

Using Corollaries 1.26 and 1.27, we get bounds on the codimension for a substantial taut embedding. The proof here is essentially that of Kuiper for tight immersions, but it also includes the possibility of a non-compact M. This result is due to Carter and West [1, p. 710].

Theorem 1.28: Let $f:M \to E^m$ be a substantial taut embedding of a connected n-dimensional manifold.

(a) If M is compact, then $m \leq n(n+3)/2$. If $m = n(n+3)/2$ then f is a spherical Veronese embedding of projective space P^n.

(b) If M is non-compact, then $m \leq \frac{1}{2}\, n(n+3) - 1$. If $m = \frac{1}{2}\, n(n+3) - 1$, then f(M) is the image under stereographic projection of a Veronese manifold, where the pole of the projection is on the Veronese manifold.

Proof: We remark that (a) follows from Corollary 1.5 and Theorem 10.9 of of Chapter 1. However the following proof handles (a) and (b) simultaneously.

Let $T_x^{\perp}M$ be the normal space to f(M) at f(x). Let V be the vector space of symmetric bilinear forms on T_xM. Define a linear map $\Phi: T_x^{\perp}M \to V$ by $\Phi(\xi) = A_\xi$, i.e.,

$$\Phi(\xi)\ (X,Y) = \langle A_\xi X, Y\rangle\ ,\ X,Y \in T_xM.$$

We have dimension $T_x^{\perp}M = m-n$, while $\dim V = n(n+1)/2$. Thus, if

$$m-n > n(n+1)/2, \text{ i.e., } m > n(n+3)/2,$$

kernel Φ is non-trivial. In that case, for $\xi \neq 0$ in kernel Φ, we have $A_\xi = 0$. So, by Corollary 1.27, f(M) lies in a hyperplane, and f is not substantial. Thus, $m \leq n(n+3)/2$.

Now suppose $m = n(n+3)/2$. Since f(M) does not lie in a hyperplane, kernel Φ must be trivial. This implies that Φ is surjective, and so there exists a normal ξ (not necessarily of unit length) with $A_\xi = I$. Hence, by

Corollary 1.28, $f(M)$ lies on the sphere S of radius $|\xi|$ centered at the point $p = f(x) + \xi$. Since a taut embedding is proper, we conclude that M is compact, as $M = f^{-1}S$. Hence f is a spherical Veronese embedding.

For a non-compact M, we have already shown that $m < \frac{1}{2}n(n+3)-1$ and the result follows from Theorem 10.12 of Chapter 1.

Q.E.D.

2. TAUT EMBEDDINGS OF SPHERES.

In Section 1 (Theorem 1.6), we showed that a taut sphere in E^m must be a round sphere. This is based on the observation that tautness is equivalent to the combination of tight and spherical and on the Chern-Lashof theorem, where the hard work was done. While this point of view is very useful, it is by no means the only approach to taut embeddings. In fact, most of the known results are obtained from a different approach emphasizing focal sets, the L_p functions and the behavior of the principal curvatures.

In this section, we present a second proof of the determination of taut spheres due to Nomizu and Rodriguez [1]. This proof is based on the characterization of spheres as umbilical submanifolds and uses the index theorem for L_p functions. One uses a similar argument to characterize umbilic submanifolds of hyperbolic space (Cecil-Ryan [3]) and to characterize totally geodesic CP^n and complex quadrics Q^n in CP^m (Cecil [1]) in terms of the critical point behavior of distance functions.

Since the proof relies on the characterization of the sphere as umbilic, it only works for S^n with $n \geq 2$. For $n = 1$, one can use the approach of Theorem 1.6, or Banchoff's [3] elementary direct proof using the circle two-piece property.

The theorem below implies Theorem 1.6, since if $f:S^n \to E^m$ is taut, then every non-degenerate L_p has exactly one maximum and one minimum, i.e., all of its critical points are of index 0 or n. On the other hand, for odd n, the hypotheses of Theorem 2.1 are weaker than those of Theorem 1.6, a priori. However, for an even-dimensional S^n, the hypotheses are easily seen to be equivalent. For if every critical point of an L_p is of index 0 or n, then L_p can have only two critical points, since

$$\sum_{i=0}^{n} (-1)^i \mu_i(L_p) = \mu_0(L_p) + \mu_n(L_p) = \chi(S^n) = 2.$$

Theorem 2.1 (Nomizu-Rodriguez [1]): Let M^n, $n \geq 2$, be a connected, complete Riemannian manifold isometrically immersed in E^m. If every non-degenerate L_p has index 0 or n at any of its critical points, then M is embedded as an n-plane E^n or a round n-sphere $S^n \subset E^{n+1} \subset E^m$.

Proof: As mentioned earlier, the proof is accomplished by showing that f is an umbilic immersion, i.e. for every normal ξ to f(M) at any point f(x), the shape operator A_ξ is a multiple of the identity endomorphism on T_xM.

Let ξ be a unit normal at a point f(x). If $A_\xi = 0$, then we are done. If not, we may assume that A_ξ has a positive eigenvalue, by considering $A_{-\xi} = -A_\xi$, if necessary.

Let λ be the largest positive eigenvalue of A_ξ. Let t satisfy $1/\lambda < t < 1/\mu$, where μ is the next largest positive eigenvalue of A_ξ (if λ is the only positive eigenvalue just consider $1/\lambda < t$). Then for

$$q = f(x) + t\xi,$$

L_q has a non-degenerate critical point of index k at x, where k is the multiplicity of λ, by Theorem 1.10. While L_q may not be non-degenerate, by Proposition 1.11 there exists a non-degenerate L_p having a critical point y of index k (with p and y as close to q and x, respectively, as desired). By the hypotheses of the theorem, since $k > 0$, we must have $k = n$, i.e. $A_\xi = \lambda I$. Thus f is umbilical, and the proof is completed by invoking the result of Cartan [5, p. 231] that a complete Riemannian n-manifold isometrically and umbilically immersed is embedded as a Euclidean n-space or n-sphere. (See also B.Y. Chen, [1, p. 50] for a proof of Cartan's result and Kobayashi-Nomizu [1, Vol. 2, p. 30] for a proof in the case of codimension one.)

Q.E.D.

Remark 2.2: Approaching the question of taut spheres from a different point of view, J. Hebda [1] showed that a complete simply connected n-manifold which admits a taut embedding (suitably defined) of S^{n-1} is either homeomorphic to S^n, diffeomorphic to E^n or diffeomorphic to $S^{n-1} \times R$.

3. SHAPE OPERATORS OF TUBES.

In this section, we find the formula for the shape operators of a tube of constant radius over a submanifold M of a real space form $\hat{M}$ in terms of the shape operators of M itself. In the case where M is a hypersurface, this reduces to the well-known formula for parallel hypersurfaces. When the codimension is greater than one, the focal set of the tube is equal to the union of the focal set of M with M itself. Hence, one can answer many questions about focal sets of arbitrary codimension by dealing solely with hypersurfaces, i.e., tubes.

We will only do the calculations for submanifolds of Euclidean space, but the computations in the sphere or hyperbolic space are very similar. Likewise, formulas can be found for the shape operators of tubes over certain submanifolds of complex projective space endowed with the Fubini-Study metric (see Cecil-Ryan [7].)

Let $f:M \to \hat{M}$ be a smooth immersion. The normal exponential map $F:NM \to \hat{M}$ is defined as follows. For a normal vector ξ to $f(M)$ at $f(x)$, $F(x,\xi)$ is the point in $\hat{M}$ reached by traversing a distance $|\xi|$ along the geodesic in $\hat{M}$ with initial position $f(x)$ and initial tangent vector ξ. For $\hat{M} = E^m$, F is the map of Section 1, $F(x,\xi) = f(x) + \xi$.

As in the Euclidean case, focal points are defined as critical values of F.

Theorem 3.1: *Let M be a submanifold of a real space form $\hat{M}$. Let ξ be a unit normal to $f(M)$ at $f(x)$. Then $F(x,t\xi)$ is a focal point of (M,x) of multiplicity $\nu > 0$ if and only if there is an eigenvalue λ of the shape operator A_ξ of multiplicity ν such that*

(a) $\lambda = 1/t$ if $\hat{M} = E^m$

(b) $\lambda = \cot t$ if $\hat{M} = S^m$

(c) $\lambda = \coth t$ if $\hat{M} = H^m$.

This is proven by straightforward calculation. For details see Milnor [3, p. 32-38] and Cecil [2, p. 243] for hyperbolic space.

Let BM be the bundle of unit normals to M. The map $\phi_t : BM \to \tilde{M}$ defined by

$$\phi_t(x,\xi) = F(x,t\xi)$$

is called the tube of radius t over M. If $(x,t\xi)$ is not a critical point of F, then ϕ_t is an immersion on a neighborhood of (x,ξ) in BM. In particular, given any point $x \in M$, there exists a neighborhood U of x in M such that for all t sufficiently small, ϕ_t is an immersion on BU, the bundle of unit normals to points in U.

In the case where M has codimension 1, the bundle BM is a double covering of M. For local calculations, one can assume that M is orientable with a field of unit normals ξ. Customarily, one then considers $\phi_t : M \to \tilde{M}$ defined by $\phi_t(x) = F(x,t\xi)$. Then, ϕ_t is called the parallel hypersurface to M at oriented distance t. In the remainder of this section, we will handle the case where the codimension is greater than one and leave the necessary modifications in the codimension one case to the reader.

We wish to compute the shape operator of ϕ_t at a point (x,ξ) in BM such that $(x,t\xi)$ is not a critical point of F. For the sake of this local calculation, we consider M as embedded in E^{n+k}, identify T_xM with $f_*(T_xM)$ and suppress further mention of f.

We recall some notation. Let D be the Euclidean connection on E^{n+k}. Let X,Y be vector fields on some open set in M. The decomposition into tangential and normal components,

$$D_XY = \nabla_XY + \alpha(X,Y)$$

defines the Levi-Civita connection ∇ of the induced metric on M and the second fundamental form α. If ξ is a local field of unit normals to M, the decomposition

$$D_X\xi = -A_\xi X + \nabla_X{}^\perp\xi,$$

into tangential component $-A_\xi X$ and normal component $\nabla_X{}^\perp\xi$, defines the shape operator A_ξ and normal connection $\nabla^\perp$.

Now consider a fixed point (x,ξ) in BM such that $(x,t\xi)$ is not a critical point of F for a fixed value t. Let $\xi_1, \ldots, \xi_k$ be an orthonormal normal frame at x with $\xi_1 = \xi$. Let U be a normal coordinate neighborhood of x in M, as defined in Kobayashi-Nomizu [1, Vol. I, p. 148], and extend $\xi_1, \ldots, \xi_k$ to smooth orthonormal normal vector fields on U by parallel translation with respect to $\nabla^{\perp}$ along geodesics in M through x.

For any $u \in U$ and unit normal η to M at u, we can write

$$\eta = \left(1 - \sum_{j=2}^{k} t_j^2\right)^{1/2} \xi_1 + t_2\xi_2 + \ldots + t_k\xi_k,$$

where $0 < |t_j| < 1$ for all j and $\sum_{j=2}^{k} t_j^2 < 1$, i.e. the t_j are the direction cosines of η. The tangent space to BM at the point (u, η) can be considered as

(3.1) $$T_uM \times \operatorname{span}\left\{\frac{\partial}{\partial t_2}, \ldots, \frac{\partial}{\partial t_k}\right\}.$$

We first find formulas for the Jacobian $(\phi_t)_*$ at (x, ξ). This is essentially the same calculation one makes in locating the focal points of M. The vector $(\phi_t)_*\left(\frac{\partial}{\partial t_j}\right)$ is the initial tangent vector to the curve $\beta(s) = \phi_t(\gamma(s))$, where

$$\gamma(s) = (x, \cos s\ \xi_1 + \sin s\ \xi_j).$$

Thus

$$\beta(s) = x + t\,(\cos s\ \xi_1 + \sin s\ \xi_j)$$

and

(3.2) $$(\phi_t)_*\left(\frac{\partial}{\partial t_j}\right) = \beta'(0) = t\xi_j.$$

On the other hand, using the formulation in (3.1) for $T_{(x,\xi)}BM$, we wish to compute $(\phi_t)_*(X,0)$ for $X \in T_xM$. Let $\gamma(s)$ be a curve in M with $\gamma(0) = x$ and $\gamma'(0) = X$. Then $(\phi_t)_*(X,0)$ is the initial tangent vector to the curve

$$\beta(s) = \phi_t(\gamma(s), \xi_1(\gamma(s))) = \gamma(s) + t\ \xi_1(\gamma(s)).$$

We compute

$$\beta'(0) = \gamma'(0) + t\ \xi_1'(0) = X + t\ D_X\xi_1.$$

Since $D_X \xi_1 = -A_\xi X + \nabla_X^\perp \xi_1$ and ξ_1 was chosen so that $\nabla_X^\perp \xi_1 = 0$, we have

$$(3.3) \qquad (\phi_t)_*(X,0) = X - tA_\xi X = (I - tA_\xi)X,$$

where, of course, we are identifying X with its Euclidean parallel translate to $\phi_t(x)$.

The map ϕ_t is an embedding in a neighborhood W of the point (x,ξ) in BM. At the point $\phi_t(y,\eta)$, the Euclidean parallel translate of η is a unit normal to $\phi_t W$. We now let η denote a field of unit normals to $\phi_t W$ and let A_t denote the shape operator of the hypersurface $\phi_t W$. The shape operator is defined by

$$(3.4) \qquad (\phi_t)_* (A_t(X,V)) = -D_{(\phi_t)_*(X,V)}\eta,$$

for (X,V) in $T_{(x,\xi)}BM$. We first do the computation for the vector $(0,V)$ where $V = \frac{\partial}{\partial t_j}$. We know that $(\phi_t)_* (0,V)$ is the initial tangent vector to the curve

$$\beta(s) = x + \cos s\ \xi_1 + \sin s\ \xi_j.$$

Hence $D_{(\phi_t)_*(0,V)}\eta$ is the initial tangent vector to the curve

$$\eta(\beta(s)) = \cos s\ \xi_1 + \sin s\ \xi_j.$$

Thus,

$$(\phi_t)_* A_t(0,V) = -\eta'(0) = -\xi_j.$$

From (3.2), we see that

$$(3.5) \qquad A_t(0, \frac{\partial}{\partial t_j}) = -\frac{1}{t}(0, \frac{\partial}{\partial t_j}),$$

i.e. the normal circle of radius t contributes a principal curvature $-1/t$.

To find $A_t(X,0)$, let $\gamma(s)$ be a curve in M with $\gamma(0) = x$, $\gamma'(0) = X$. Then $(\phi_t)_*(X,0)$ is the initial tangent vector to the curve

$$\beta(s) = \phi_t(\gamma(s), \xi_1(\gamma(s))) = \gamma(s) + t\ \xi_1(\gamma(s)).$$

Thus

$$D_{(\phi_t)_*(X,0)}\eta = \xi_1'(0) = D_X\ \xi_1 = -A_\xi X, \tag{3.6}$$

where again we are identifying parallel vectors in E^{n+k} and using the fact that $\nabla_X^{\perp}\xi_1 = 0$. From (3.3), (3.4) and (3.6), we get

$$A_t(X,0) = ((I - tA_\xi)^{-1}\ A_\xi X,\ 0). \tag{3.7}$$

If $A_\xi X = \lambda X$, this becomes

$$A_t(X,0) = \frac{\lambda}{1 - t\lambda}\,(X,0). \tag{3.8}$$

We summarize these results as follows. The computations for $\tilde{M} = S^{n+k}$ or H^{n+k} are quite similar to those above.

<u>Theorem 3.2: Let M be a submanifold of a real space for $\tilde{M}^{n+k}$ and ξ a unit normal to M at x such that $(x, t\xi)$ is not a critical point of the normal exponential map F. Suppose $X_1, \ldots, X_n$ are a basis of T_xM of principal vectors of A_ξ with $A_\xi X_i = \lambda_i X_i$, $1 \leq i \leq n$. Then the shape operator A_t of the tube ϕ_t of radius t over M at the point $\phi_t(x,\xi)$ is given in terms of its principal vectors by</u>

(1) <u>for $\tilde{M}^{n+k} = E^{n+k}$</u>

(a) $A_t\ (\frac{\partial}{\partial t_j}) = \frac{-1}{t}\ (\frac{\partial}{\partial t_j})$, $2 \leq j \leq k$

(b) $A_t(X_i,0) = \frac{\lambda_i}{1-t\lambda_i}(X_i,0), \quad 1 \leq i \leq n.$

(2) for $\hat{M}^{n+k} = S^{n+k}$

(a) $A_t\left(\frac{\partial}{\partial t_j}\right) = -\cot t\left(\frac{\partial}{\partial t_j}\right), \; 2 \leq j \leq k$

(b) $A_t(X_i,0) = \cot(\theta_i - t)(X_i,0)$, where $\lambda_i = \cot\theta_i$, $0 < \theta_i < \pi$, for $1 \leq i \leq n$.

(3) for $\hat{M}^{n+k} = H^{n+k}$

(a) $A_t\left(\frac{\partial}{\partial t_j}\right) = -\coth t\left(\frac{\partial}{\partial t_j}\right), \; 2 \leq j \leq k.$

(b) $$A_t(X_i,0) = \begin{cases} \coth(\theta_i - t)(X_i,0), \text{ if } |\lambda_i| > 1 \text{ and } \lambda_i = \coth\theta_i \\ (\pm X_i,0), \text{ if } \lambda_i = \pm 1, \\ \tanh(\theta_i - t)(X_i,0) \text{ if } |\lambda_i| < 1 \text{ and } \lambda_i = \tanh\theta_i, \end{cases}$$

for $1 \leq i \leq n$.

As a consequence of Theorems 3.1 and 3.2, we obtain the following useful result.

Theorem 3.3: Let M be a submanifold of a real space form $\hat{M}$ and t a real number such that $\phi_t M$ is a hypersurface.

(a) If M is a hypersurface, then the focal set of $\phi_t M$ is the focal set of M.

(b) If M has codimension greater than one, then the focal set of $\phi_t M$ consists of the union of the focal set of M with M itself.

4. A MANIFOLD STRUCTURE FOR THE FOCAL SET.

In this section, we find a natural manifold structure for the sheet of the focal set of a hypersurface M in a real space form $\hat{M}$ corresponding to a principal curvature λ of constant multiplicity. By Theorem 3.3, this also provides a manifold structure for a sheet of the focal set of a submanifold of arbitrary codimension. These results are contained in Cecil-Ryan [1], and they were suggested by the work of Nomizu [1], who obtained these results for the sheets of the focal set of an isoparametric hypersurface. See also the related work of Reckziegel [1]-[3].

At times, we will need to refer to the following specific models for the space forms. Consider R^{m+1} with the Lorentzian inner product

$$\langle x,y\rangle = -x_{m+1}y_{m+1} + \sum_{i=1}^{m} x_i y_i$$

for $x = (x_1, \ldots, x_{m+1})$ and $y = (y_1, \ldots, y_{m+1})$. Then hyperbolic space H^m is the hypersurface

$$\{x \in R^{m+1} \mid \langle x,x\rangle = -1,\ x_{m+1} \geq 1\},$$

on which $\langle\ ,\ \rangle$ restricts to a Riemannian metric of constant sectional curvature -1. Of course, S^m is just the unit sphere in Euclidean space E^{m+1} on which the Euclidean metric restricts to metric of constant sectional curvature 1.

Let $f:M \to \tilde{M}$ be a smooth immersion of an n-dimensional manifold into a real space form of dimension n+1. Let ξ be a local field of unit normals to f(M) and let A denote the shape operator corresponding to ξ. If the principal curvatures are ordered

$$\lambda_1 \geq \cdots \geq \lambda_n,$$

then each λ_i is a continuous function (see Ryan [1, p. 371]). Further, a continuous principal curvature function λ which has constant multiplicity on an open set U in M is smooth, as is its corresponding principal distribution T_λ of eigenvectors (see Nomizu [3]). In fact, T_λ is integrable (Theorem 4.4) and thus it will be called a <u>principal foliation</u>. On the other hand, if a continuous principal curvature function λ does not have constant multiplicity, then λ may not be smooth. For example, consider the behavior of the principal curvatures of a surface in E^3 at an isolated umbilic point. In fact, the principal curvatures of the monkey saddle given by the graph

$$z = x^3 - 3y^2x$$

are not smooth at the umbilic point at the origin. A straightforward calculation shows that the two principal curvatures are given in polar coordinates by the formula

$$(1 + r^4)^{3/2}\lambda = -r^5 \cos 3\theta \pm 2r(1 + r^4 + \frac{r^8}{4} \cos^2 3\theta)^{1/2}.$$

As r approaches 0, the curvatures are asymptotically equal to $\pm 2r$. Figure 4.1 is a graph of the two principal curvature functions which was produced by T. Banchoff.

If a smooth principal curvature function λ does have constant multiplicity on some open subset U of M, one can define a smooth focal map $f_\lambda : U \to \tilde{M}$ by the following formulas,

$$\text{(a)} \quad f_\lambda(x) = f(x) + \frac{1}{\lambda}\, \xi(x),$$

$$\text{(4.1)} \quad \text{(b)} \quad f_\lambda(x) = \cos\theta\, f(x) + \sin\theta\, \xi(x), \text{ where } \cot\theta = \lambda,$$

$$\text{(c)} \quad f_\lambda(x) = \cosh\theta\, f(x) + \sinh\theta\, \xi(x), \text{ where } \coth\theta = \lambda,$$

for $\tilde{M}$ equal to E^{n+1}, S^{n+1}, H^{n+1}, respectively. Then $f_\lambda(U)$ is called the <u>sheet of the focal set over U corresponding to λ</u>.

Note that f_λ is not defined if $\lambda(x) = 0$ in the Euclidean case, nor if $|\lambda(x)| \leq 1$ in the hyperbolic case. In the spherical case, each principal curvature λ gives rise to two different (antipodal) focal points determined by substituting $\theta = \cot^{-1}\lambda$ and $\theta = \cot^{-1}\lambda + \pi$ into (4.1b). For a point x in the domain of f_λ, the n-sphere through x centered at $f_\lambda(x)$ is called the <u>curvature sphere</u> determined by λ at x.

It is clear that the requirement that λ have constant multiplicity is important in these considerations. We now note that this condition is

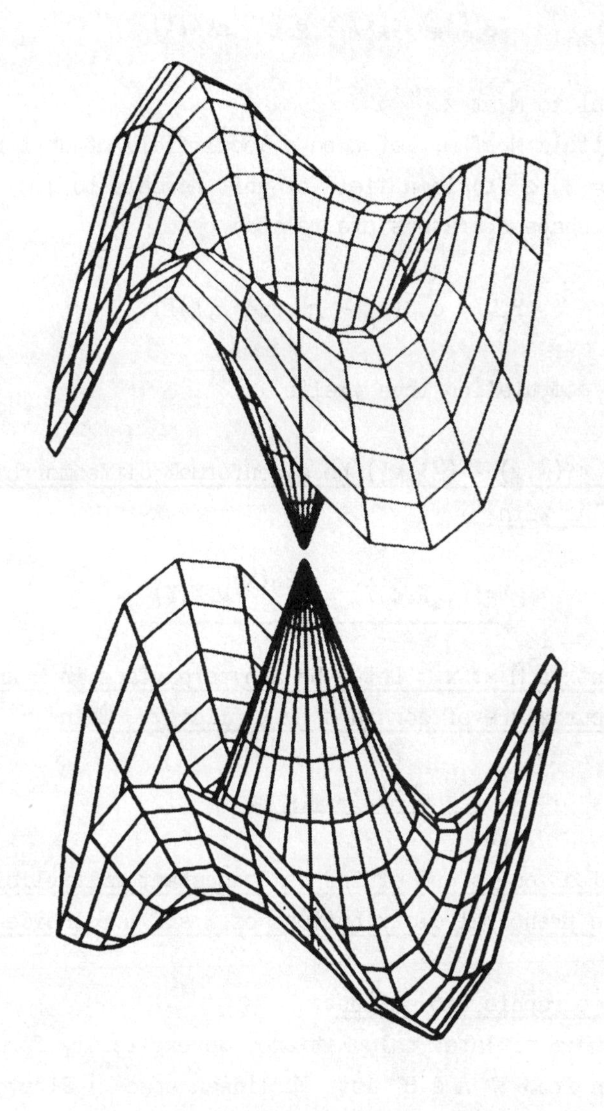

Figure 4.1 Graphs of principal curvatures of the monkey saddle

invariant under conformal transformations of the ambient space.

Let $\psi:(\hat{M},g) \to (\hat{M}',g')$ be a conformal diffeomorphism of Riemannian manifolds with

$$g'(\psi_*X,\psi_*Y) = e^{2\sigma(x)}g(X,Y)$$

for all X,Y tangent to $\hat{M}$ at x.

For a submanifold M of $\hat{M}$, let ξ be a local field of unit normals to M near x. Then $\xi' = \psi_*(e^{-\sigma}\xi)$ is a field of unit normals to $\psi(M)$ near $\psi(x)$ and the corresponding shape operators are related by

$$B_{\xi'} = e^{-\sigma}(A_\xi - g(\text{grad } \sigma,\xi)I).$$

A straightforward computation then yields

Theorem 4.1: *Let $\psi:(\hat{M},g) \to (\hat{M}',g')$ be a conformal diffeomorphism of Riemannian manifolds with*

$$g'(\psi_*X,\psi_*Y) = e^{2\sigma(x)}g(X,Y)$$

for all X,Y tangent to $\hat{M}$ at x. Let M be a hypersurface in $\hat{M}$ and let λ be a smooth principal curvature of constant multiplicity. Then

$$\mu = e^{-\sigma}(\lambda - g(\text{grad } \sigma, \xi))$$

is a smooth principal curvature of the same constant multiplicity on $\psi(M)$, and the respective principal distributions of λ and μ coincide on M.

Remark 4.2: *Stereographic projection*.

For the purposes of later calculations, we explicitly formulate stereographic projection from S^m and H^m into Euclidean space. First, let q be an arbitary point of $S^m \subset E^{m+1}$, and let E^m be defined by

$$E^m = \{x \in E^{m+1} \mid \langle x,q\rangle = 0\},$$

where $\langle\ ,\ \rangle$ denotes the Euclidean inner product. Stereographic projection with pole q is the map P from $S^m - \{q\}$ onto E^m defined by

$$P(x) = q + e^{\sigma(x)}(x-q),$$

where $e^{-\sigma(x)} = 1 - \langle x,q\rangle$.

One can easily show that P is a conformal diffeomorphism with

$$\langle P_*X, P_*Y\rangle = e^{2\sigma(x)}\langle X,Y\rangle$$

for all X,Y tangent to S^m at x.

Similarly, consider hyperbolic space H^m in R^{m+1} with the Lorentz inner product $\langle\ ,\ \rangle$, as defined at the beginning of this section. Let q be a point of R^{m+1} such that $-q \in H^m$. Let D^m be the m-dimensional disk

$$D^m = \{x \in R^{m+1} \mid \langle x,q\rangle = 0,\ \langle x,x\rangle < 1\}$$

on which the metric $\langle\ ,\ \rangle$ restricts to a Euclidean metric which we denote for emphasis by g. Then stereographic projection from the pole q is the map P from H^m onto D^m defined by

$$P(x) = q + e^{\sigma(x)}(x-q),$$

where $e^{-\sigma(x)} = 1 + \langle x,q\rangle$. As in the spherical case, one easily shows that P is a conformal diffeomorphism with

$$g(P_*X, P_*Y) = e^{2\sigma(x)}\langle X,Y\rangle$$

for all X,Y tangent to H^m at x.

If λ is a smooth principal curvature function with constant multiplicity, then the focal map f_λ is smooth. To see if $f_\lambda(M)$ is a manifold, one must compute its rank. As before, we will only prove the result for $\tilde{M} = E^{n+1}$.

The other cases are handled with minor modifications.

<u>Theorem 4.3: Let $f:M^n \to \tilde{M}^{n+1}$ be a smooth immersion into a real space form. Suppose λ is a smooth principal curvature of constant multiplicity $\nu > 0$ near a point $x \in M$. Then at x,</u>

$$\text{rank } f_\lambda = \begin{cases} n+1-\nu, & \text{if there exists } X \in T_\lambda(x) \text{ such that } X\lambda \neq 0 \\ n-\nu, & \text{otherwise.} \end{cases}$$

<u>Proof</u>: Recall that $f_\lambda(y) = f(y) + \rho(y)\xi(y)$ for $y \in M$, where $\rho = 1/\lambda$. Thus, for $X \in T_xM$, one has

$$(f_\lambda)_* X = X + (X\rho)\xi + \rho D_X\xi, \tag{4.2}$$

where we are again identifying vectors which are Euclidean parallel. First suppose that X is in $T_\mu(x)$ for $\mu \neq \lambda$. Since $D_X\xi = -\mu X$, (4.2) becomes

$$(f_\lambda)_* X = (1 - \frac{\mu}{\lambda})X + (X\rho)\xi, \text{ for } X \in T_\mu.$$

Thus, $(f_\lambda)_*$ is injective on $T_\lambda^\perp$, the $(n-\nu)$-dimensional orthogonal complement of T_λ in T_xM. On the other hand, if $X \in T_\lambda$, then $D_X\xi = -\lambda X$, and (4.2) becomes

$$(f_\lambda)_* X = (X\rho)\xi = \frac{-X\lambda}{\lambda^2}\xi, \text{ for } X \in T_\lambda.$$

Thus, if $X\lambda \neq 0$ for some $X \in T_\lambda$, we see that the range of $(f_\lambda)_*$ is the $(n+1-\nu)$-dimensional space spanned by $T_\lambda^\perp$ and ξ, while if $X\lambda = 0$ for all $X \in T_\lambda$, then the range of $(f_\lambda)_*$ is the $(n-\nu)$-dimensional space $(f_\lambda)_* T_\lambda^\perp$.

Q.E.D.

This generalizes the classical result that the normal to a surface in E^3 is tangent its evolute (focal set) when f_λ has maximal rank. (See, for example, Goetz [1, pp. 302-305].) In the situation where λ has constant multiplicity 1 and $X\lambda \neq 0$ for all non-zero X in T_λ on an open set U, then $f_\lambda(U)$ is a hypersurface. In this case, one gets the well-known picture for the sheets of the focal set (see Goetz [1, p. 304]). In general, if λ has constant multiplicity 1, singularities occur at the points where $X\lambda = 0$. For example, the evolute of an ellipse has singularities at the images of the 4 vertices.

We can obtain even more specific information concerning f_λ after we examine the distribution T_λ. If λ has constant multiplicity 1, then T_λ is integrable by the basic existence and uniqueness theorem for ordinary differential equations. However, in general, λ is certainly not constant along its corresponding lines of curvature.

If λ has constant multiplicity $\nu > 1$, then one must prove that T_λ is integrable using the theorem of Frobenius. This does turn out to be true, and moreover, λ is constant along the leaves of the foliation T_λ. The proof here is a generalization of the classical result that if M is a totally umbilic hypersurface in $\tilde{M}$, i.e. $A = \lambda I$ at each $x \in M$, then the function λ is a constant on M.

Theorem 4.4: Let M be an orientable hypersurface of a real space form $\tilde{M}$. Suppose λ is a smooth principal curvature of constant multiplicity $\nu > 1$ on M. Then T_λ is integrable and $X\lambda = 0$ for all $X \in T_\lambda$.

Proof: For a hypersurface in a real space form, the Codazzi equation takes the form (see, for example, Kobayashi-Nomizu [1, vol. II, p. 26])

$$(\nabla_X A)Y = (\nabla_Y A)X,$$

i.e.

$$\nabla_X(AY) - A(\nabla_X Y) = \nabla_Y(AX) - A(\nabla_Y X), \tag{4.3}$$

for X,Y tangent to M. Take X,Y linearly independent (local) vector fields in T_λ. Then (4.3) becomes

$$(X\lambda)Y + \lambda\, \nabla_X Y - A(\nabla_X Y) = (Y\lambda)X + \lambda\, \nabla_Y X - A(\nabla_Y X). \tag{4.4}$$

Since the connection ∇ has no torsion, the Lie bracket is

$$[X,Y] = \nabla_X Y - \nabla_Y X$$

and (4.4) becomes

$$(X\lambda)Y - (Y\lambda)X = (A-\lambda I)[X,Y].$$

The left side of the above equation is in T_λ, while the right side is in $T_\lambda^\perp$. We conclude that both must be zero. Hence $[X,Y]$ is in T_λ, proving that T_λ is integrable, while $X\lambda$ and $Y\lambda$ must both be equal to zero.

Q.E.D.

From Theorems 4.3 and 4.4, one sees that if λ has constant multiplicity $\nu > 1$, then rank $f_\lambda \equiv n-\nu$, thus giving $f_\lambda(M)$ a local structure as an $(n-\nu)$-dimensional manifold. On the other hand, if λ has constant multiplicity 1, one cannot guarantee that $X\lambda$ is identically zero for $X \in T_\lambda$. On account of this, we treat the cases $\nu > 1$ and $\nu = 1$ separately.

We begin with $\nu > 1$. Recall that a submanifold V of any space $\tilde{M}$ is called <u>umbilic</u> if for each $x \in V$, there is a real-valued linear function ω on $T_x^\perp V$ such that the shape operator B_η of V satisfies

$$B_\eta = \omega(\eta)I$$

for every $\eta \in T_x^\perp V$.

<u>Theorem 4.5</u>: <u>Let M be an orientable hypersurface of a real space form $\tilde{M}$. Suppose λ is a smooth principal curvature of constant multiplicity $\nu > 1$ on M. Then the leaves of the foliation T_λ are umbilic ν-dimensional submanifolds of $\tilde{M}$.</u>

<u>Proof</u>: Let V be any leaf of T_λ. The normal space to V in $\tilde{M}$ at a fixed point x is

$$T_x^\perp V = T_x^\perp M \oplus T_\lambda^\perp(x).$$

Choose $\eta \in T_x^{\perp}V$, and let B_η denote the shape operator of V determined by η. If η is normal to M then $B_\eta X = A_\eta X = \lambda X$, for $X \in T_\lambda(x)$, i.e. $B_\eta = \lambda I$.

On the other hand, suppose $\eta \in T_\lambda^{\perp}(x)$ is a principal vector of the shape operator A of M, so that $A\eta = \mu\eta$ with $\mu \neq \lambda$. Extend η to a vector field $Y \in T_\lambda^{\perp}$ on a neighborhood U of x. Then there exists a unique vector field $Z \in T_\lambda^{\perp}$ such that $\langle Z,Y\rangle = 0$ and

$$AY = \mu Y + Z \tag{4.5}$$

for some function μ on W. This is possible since $T_\lambda^{\perp}$ is invariant under A, even though the eigenvalues of A need not be smooth.

Let X be a vector field in T_λ on U. Since Z = 0 at the point x, one easily shows that $\nabla_X Z \in T_\lambda^{\perp}$ at x. Using (4.5), the Codazzi equation becomes

$$(X\mu)Y - (Y\lambda)X + \nabla_X Z = (A - \mu I)\nabla_X Y - (A - \lambda I)\nabla_Y X.$$

Taking the T_λ-component of both sides of the above equation, one finds that the T_λ-component of $\nabla_X Y$ at x is

$$\frac{-(Y\lambda)X}{\lambda - \mu}. \tag{4.6}$$

Recall that if $\tilde{\nabla}$ is the Levi-Civita connection on $\tilde{M}$, then

$$\tilde{\nabla}_X Y = \nabla_X Y + \langle AX,Y\rangle\xi.$$

Since $\langle AX,Y\rangle = 0$, we have that (4.6) is also the T_λ-component of $\tilde{\nabla}_X Y$ at x. But by definition, this is equal to $-B_\eta X$, and thus

$$B_\eta X = \frac{(\eta\lambda)X}{\lambda - \mu}. \tag{4.7}$$

Q.E.D.

Recall that a complete umbilic hypersurface in E^{n+1} is a Euclidean n-sphere or n-plane. In S^{n+1}, the complete umbilic hypersurfaces are the great and small n-spheres. In H^{n+1}, a complete umbilic hypersurface corresponding to a constant principal curvature λ is a metric sphere, a horosphere, or a hypersurface at a fixed oriented distance from a hyperplane (an equidistant hypersurface) depending on whether $\lambda > 0$, $\lambda = 0$, $\lambda < 0$, respectively.

As noted earlier in Theorem 3.1, the domain of f_λ is the set where $\lambda \neq 0$ for $\tilde{M} = E^{n+1}$ and $|\lambda| > 1$ for $\tilde{M} = H^{n+1}$. Thus, at any point x in the domain of f_λ, the leaf of T_λ through x is always part of a ν-dimensional metric sphere.

We have seen that f_λ is constant on the leaves of T_λ and thus it factors through a map of the space of leaves U/T_λ, where U is the domain of f_λ. In fact, we have

<u>Theorem 4.6</u>: <u>Let M be an orientable immersed hypersurface in a real space form $\tilde{M}$, and let λ be a smooth principal curvature of constant multiplicity $\nu > 1$. Then the focal map $f_\lambda: U \to \tilde{M}$ factors through an immersion of the (possibly non-Hausdorff) $(n-\nu)$-dimensional manifold U/T_λ into $\tilde{M}$. If M is complete with respect to the induced metric, then U/T_λ is Hausdorff</u>.

<u>Proof</u>: Since the leaves of T_λ are umbilic, T_λ is a regular foliation as defined by Palais [1, p. 13] (i.e. every point has a coordinate chart distinguished by the foliation, such that each leaf intersects the chart in at most one ν-dimensional slice). This implies that the space of leaves U/T_λ is an $(n-\nu)$-dimensional manifold in the sense of Palais, which may not be Hausdorff. By Theorems 4.3 and 4.4, f_λ factors through a map g_λ from the space of leaves U/T_λ into $\tilde{M}$. Since rank g_λ = rank $f_\lambda = n-\nu$, g_λ is an immersion. Finally, the regularity of T_λ implies that each leaf is a closed subset of M (Palais [1, p. 18]). If M is complete, then each leaf is also complete (see, for example, Kobayashi-Nomizu [1, vol. I, p. 179]). Hence, each leaf which intersects U is a ν-dimensional metric sphere in $\tilde{M}$. Since each leaf of T_λ in U is compact, U/T_λ is Hausdorff, Palais [1, p. 16].

Q.E.D.

The space of leaves of a regular foliation need not be a Hausdorff manifold (see Palais [1, p. 20]). The following example shows that in the present context, the assumption that M is complete is necessary to guarantee that U/T_λ is a Hausdorff manifold.

<u>Remark 4.7</u>: <u>An example with rank f_λ constant but $f_\lambda(M)$ not a Hausdorff manifold.</u>

Let

$$f(t) = \begin{cases} e^{-1/t} & \text{if } t > 0 \\ 0 & \text{if } t \leq 0 \end{cases}$$

Let K be the tube of constant radius 1 in E^3 over the curve,

$$\gamma(t) = (t, 0, f(t)), \ t \in (-1,1).$$

Then $\gamma(t)$ is the sheet of the focal set of K corresponding to the constant principal curvature $\lambda = 1$.

Let N be the intersection of K with the closed upper half-space, $z \geq 0$, with the points satisfying the two conditions $z = 0$, $x \geq 0$ removed (see Figure 4.2). Let M be the union of N with its reflection in the plane $z = 0$. The leaf space M/T_λ is not Hausdorff since the two semi-circular leaves in the plane $x = 0$ cannot be separated in the quotient topology. This is consistent with the fact that the focal set,

$$f_\lambda(M) = \{(x, 0, z) \mid z = |f(x)|, -1 < x < 1\},$$

is not a Hausdorff 1-manifold in a neighborhood of the origin (Figure 4.3). Nevertheless, the rank of f_λ is identically equal to 1 on M.

We remark that while in the above example λ has constant multiplicity $\nu = 1$, one can easily produce similar examples where $\nu > 1$ by imitating the above construction in E^n for $n > 3$.

We now turn to the case where λ has constant multiplicity one, which differs from the previous case, since λ is not automatically constant along

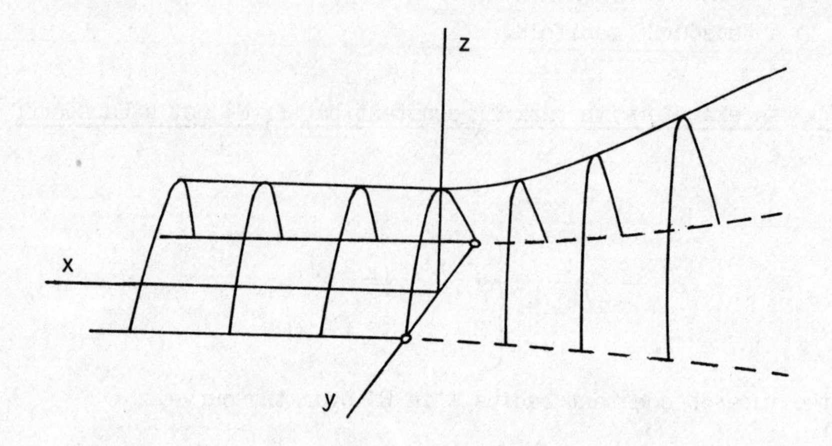

Figure 4.2 The surface N

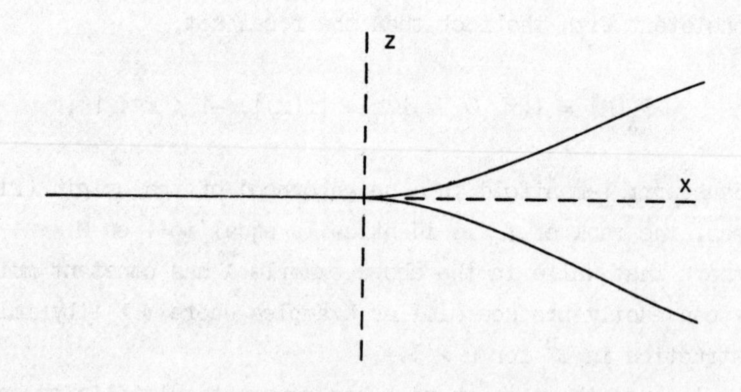

Figure 4.3 The focal set $f_\lambda(M)$

its lines of curvature. By Theorem 4.3, if $X\lambda \neq 0$ for some $X \in T_\lambda(x)$, then f_λ is an immersion of a neighborhood W of x into $\tilde{M}$. This is what one generally expects, and the points at which $X\lambda = 0$ are singularities of the map f_λ.

Thus the main theorem in this case must be formulated, as in the classical case of surfaces in E^3, in terms of conditions which are equivalent to the condition that $f_\lambda(M)$ is a manifold of dimension n-1.

In this case, the fact that for hypersurfaces of H^{n+1}, the domain U of f_λ does not include those $x \in M$ where $|\lambda(x)| < 1$, becomes significant. In fact, the conditions (a), (b), and (c) of Theorem 3.2 are equivalent on U, but not necessarily on all of M. Specifically, (a) and (b) are equivalent on M, and they imply (c). However, one can construct a surface M in H^3 such that the focal set $f_\lambda(M)$ is a curve, and yet not all the lines of curvature of M corresponding to λ are of constant curvature. This is done by beginning with a standard example K on which (a), (b), and (c) are satisfied and modifying K on the region where $|\lambda| < 1$, so as to destroy property (b), but introduce no new focal points and thus preserve (c).

Theorem 4.8: *Let M be an orientable (immersed) hypersurface in $\tilde{M}$. Suppose λ is a smooth principal curvature of constant multiplicity 1 on M. Then the following are equivalent on $\tilde{M}$ if $M = E^{n+1}$ or S^{n+1}, and on U, the domain of f_λ, if $\tilde{M} = H^{n+1}$.*

(a) *λ is constant along each leaf of T_λ (the lines of curvature).*

(b) *The leaves of T_λ are plane curves of constant curvature.*

(c) *The rank of f_λ is identically equal to n-1 on its domain U and f_λ factors through an immersion of the space of leaves U/T_λ.*

Unlike the case $\nu > 1$, one must use a different proof for the different ambient spaces. We first give a proof for the Euclidean case, then handle the others by stereographic projection.

Proof (Euclidean case):

(a) $\longleftrightarrow$ (c). This follows immediately from Theorem 4.3 and the connectedness of the leaves of T_λ.

(b) $\longleftrightarrow$ (a). This is easily shown using the Frenet equations for plane curves.

(a) $\longleftrightarrow$ (b). We imitate the classical proof and show that the domain U of f_λ is the envelope of an (n-1)-dimensional family of spheres, whose centers lie on $f_\lambda(U)$. Further, each λ-line of curvature lies on the intersection of the appropriate sphere with a 2-plane.

Assuming (a), one can find a local chart of the following type in a neighborhood of any point of M. Consider $R^n = R \times R^{n-1}$. The coordinate map $\phi:(-\varepsilon,\varepsilon) \times V \to M$, V open in R^{n-1}, can be chosen so that the leaves of T_λ which intersect the image W of ϕ are precisely the images of the curves v = constant.

First assume that λ is a non-zero constant on each leaf which passes through W and that the field of unit normals ξ has been chosen so that $\lambda > 0$ on W. By Theorem 4.3, the focal map f_λ is constant on the leaves of T_λ, and thus the maps $g_\lambda = f_\lambda \circ \phi$ and $\rho = (\frac{1}{\lambda}) \circ \phi$ are functions of the second coordinate v alone. Moreover, g_λ is an immersion on V.

The point $q = f(\phi(t,v))$ lies on the sphere given by the equation

$$(4.8) \qquad |z - g_\lambda(v)| = \rho(v), \; z \in E^{n+1}.$$

Further, since the normal to f(M) at q is the same as the normal to the sphere (4.8) at q (i.e. f(M) is the envelope of the (n-1)-parameter family of spheres), one can show that for any $\vec{v}$ tangent to V at the point v, the point q also lies in the hyperplane,

$$(4.9) \qquad \langle z - g_\lambda(v), (g_\lambda)_*(\vec{v}) \rangle = -\rho(v)\, \vec{v}(\rho), \; z \in E^{n+1}.$$

Thus, for a particular value of v, the line of curvature in W determined by v lies in the intersection of the n-sphere (4.8) and the 2-plane determined by (4.9), as $\vec{v}$ ranges over the (n-1)-dimensional tangent space to V at v. Hence, each leaf of T_λ in U lies locally on a circle and by connectedness, it is an arc of a circle.

Finally, if γ is a leaf in M on which $\lambda \equiv 0$, then by Theorem 4.1, one can find an inversion I of E^{n+1} in an n-sphere such that $I \circ f(\gamma)$ is a line of curvature of $I \circ f(M)$ on which the associated principal curvature is a non-zero constant. By the above argument, $I \circ f(M)$ lies on a circle, so that $f(\gamma)$ itself lies on a circle or straight line. This completes the proof for $\tilde{M} = E^{n+1}$.

Q.E.D.

Proof of Theorem 4.8 (non-Euclidean case):

(a) <—> (c). As in the Euclidean case, this follows immediately from Theorem 4.3.

(a) <—> (b). Let P be stereographic projection from S^{n+1}, resp. H^{n+1}, into E^{n+1}, resp. D^{n+1}, as formulated in Remark 4.2. Given a principal curvature λ of $f:M \to \tilde{M}$, let $\mu = e^{-\sigma}(\lambda - \langle \text{grad } \sigma, \xi \rangle)$ be the corresponding principal curvature of $P \circ f:M \to E^{n+1}(D^{n+1})$, according to Theorem 4.1 and Remark 4.2. By direct calculation, the leaves of T_λ are plane curves of constant curvature in $\tilde{M}$ if and only if $P \circ f$ embeds the leaves of T_μ as plane curves of constant curvature in $E^{n+1}(D^{n+1})$. Thus, the proof follows from the equivalence of (a) and (b) in the Euclidean case and the following result.

Lemma 4.9: Let $f:M^n \to S^{n+1}$ (resp. H^{n+1}) be a smooth immersion. Suppose λ is a principal curvature of constant multiplicity 1 on M, and let X denote the field of unit principal vectors of λ on M. Let

$$\mu = e^{-\sigma}(\lambda - \langle \text{grad } \sigma, \xi \rangle)$$

be the corresponding principal curvature of the immersion $P \circ f:M \to E^{n+1}$ (resp. D^{n+1}), where P is stereographic projection. Then $X\lambda = 0$ at $x \in M$ if and only if $X\mu = 0$ at x.

Proof: For this local calculation, we consider M as embedded and suppress the mention of f. For M in S^{n+1}, a straightforward computation yields that,

$$\text{grad } \sigma = e^{\sigma}(q - \langle x,q \rangle x).$$

Thus, using the fact that $\langle x, \xi \rangle = 0$, one obtains,

$$\mu = e^{-\sigma}(\lambda - \langle e^{\sigma}q, \xi\rangle) = e^{-\sigma}\lambda - \langle q, \xi\rangle,$$

and

$$X\mu = -e^{-\sigma}(X\sigma)\lambda + e^{-\sigma}(X\lambda) - \langle q, D_X\xi\rangle,$$

where D is the usual Euclidean connection on E^{n+2}. Since $\langle X, \xi\rangle = 0$, it follows that $D_X\xi = \tilde{\nabla}_X\xi$ where $\tilde{\nabla}$ is the Levi-Civita connection in S^{n+1}. However, $\tilde{\nabla}_X\xi = -\lambda X$, so $D_X\xi = -\lambda X$. This and the fact that

$$X\sigma = \langle \text{grad } \sigma, X\rangle = e^{\sigma}\langle q, X\rangle,$$

allow the above expression for $X\mu$ to be written as,

$$X\mu = -\langle q, X\rangle\lambda + e^{-\sigma}(X\lambda) + \langle q, X\rangle\lambda = e^{-\sigma}(X\lambda),$$

and, clearly, $X\mu = 0$ if and only if $X\lambda = 0$.

Similarly, for M in H^{n+1}, grad $\sigma = -e^{\sigma}(q + \langle x, q\rangle\, x)$, and the result follows as in the spherical case.

Q.E.D.

As in the case of Theorem 4.6 for multiplicity $\nu > 1$, if one assumes in addition, that M is complete with respect to the induced metric, then one can produce a natural manifold structure on $f_\lambda(M)$ by introducing the space of leaves U/T_λ, where U is the domain of f_λ.

This case differs slightly from the $\nu > 1$ case. As in the $\nu > 1$ situation, the completeness of M implies that each leaf of T_λ is also complete with respect to the induced metric. This is sufficient to guarantee that each leaf in M_λ is a covering space of the metric circle on which it lies (see, for example, Kobayashi-Nomizu [1, vol. I, p. 176]). Since the circle is not simply connected, however, one cannot conclude that the leaf itself is compact, as in the $\nu > 1$ case. However, using the fact that each leaf is a covering of the circle on which it lies, one can produce a direct argument that U/T_λ is Hausdorff, and we will omit the proof here. Thus, one obtains the following global version of Theorem 4.8.

Theorem 4.10: Let M be an orientable (immersed) hypersurface in a real space form $\tilde{M}$ which is complete with respect to the induced metric. Let λ be a smooth principal curvature of constant multiplicity 1 on M. Suppose the equivalent conditions (a), (b), (c) of Theorem 4.8 are satisfied on the domain U of f_λ. Then the focal map of f_λ factors through an immersion of the (n-1)-dimensional manifold U/T_λ into $\tilde{M}$.

From Theorems 4.5 and 4.8, we see that if λ has constant multiplicity $\nu > 1$ and is constant along the leaves of its principal foliations, then the leaves are umbilic in $\tilde{M}$. The case where λ assumes a critical value along a leaf γ of T_λ has further geometric significance.

Theorem 4.11: Let M be an orientable hypersurface of a real space form $\tilde{M}$. Suppose λ is a smooth principal curvature of constant multiplicity $\nu > 1$ which is constant along the leaves of T_λ. Then λ assumes a critical value along a leaf γ of T_λ if and only if γ is totally geodesic in M.

Proof: The normal space to γ in M is $T_\lambda^\perp$. Take $x \in \gamma$ and $\eta \in T_\mu(x)$, $\mu \neq \lambda$. Then by equation (4.7), the shape operator B_η of γ in M at x has the form

$$B_\eta X = \frac{-(\eta\lambda)X}{\lambda - \mu}, \; X \in T_\lambda(x).$$

The leaf γ is totally geodesic if and only if each $B_\eta = 0$. Thus, γ is totally geodesic precisely when $\eta\lambda = 0$ for all $\eta \in T_\lambda^\perp$. Since λ is assumed constant along γ, this happens precisely when λ assumes a critical value along γ.

Q.E.D.

From Theorems 4.5, 4.8 and 4.11, we immediately obtain the following fact about isoparametric hypersurfaces, which was first noted by Nomizu [1].

Corollary 4.12: Let M be an isoparametric hypersurface of a real space form

$\tilde{M}$. Then for each principal curvature λ, the leaves of T_λ are umbilic in $\tilde{M}$ and totally geodesic in M.

The following result is very similar, but is needed for our characterization of the cyclides of Dupin. Recall that if λ is a principal curvature of a hypersurface M with constant multiplicity $\nu > 1$ and λ is constant along the leaves of T_λ, then the focal map f_λ is constant on each leaf of T_λ in its domain. Thus, each such leaf γ lies on a curvature n-sphere centered at the common focal point $p = f_\lambda(\gamma)$. As before, the case where λ has a critical value along γ has further signficance.

Theorem 4.13: Let M be an orientable hypersurface of a real space form $\tilde{M}$. Suppose λ is a smooth principal curvature of constant multiplicity $\nu > 1$ which is constant along the leaves of T_λ. Then λ assumes a critical value along a leaf γ of T_λ in the domain of f_λ if and only if γ is totally geodesic in the curvature n-sphere S^n determined by γ.

Proof: The leaf γ is a submanifold of S^n and its normal space in S^n is $T_\lambda{}^\perp$, since S^n is tangent to M along γ. The proof is then exactly the same as that of Theorem 4.11.

Q.E.D.

4.14 Examples of hypersurfaces whose focal sets are manifolds

Let M be an isoparametric hypersurface in S^{n+1} and let P be stereographic projection from S^{n+1} to E^{n+1} with respect to any pole. By Theorem 4.1 and Lemma 4.9, the principal curvatures of the hypersurface P(M) in E^{n+1} have constant multiplicities and are constant along the leaves of their corresponding principal foliations (i.e. P(M) is a Dupin hypersurface). Thus, each sheet of the focal set of P(M) is a manifold by Theorems 4.6 and 4.10.

Examples in H^{n+1} are now easily constructed. Let K be a hypersurface of E^{n+1} as constructed above. Let L be the image of K under a contraction of E^{n+1} such that L is contained in the unit disk D^{n+1} with respect to the pole (0, ..., -1) as in Remark 4.2. Then $P^{-1}L$ is a hypersurface in H^{n+1}

with the property that each sheet of its focal set is a manifold.

5. THE CLASSICAL CYCLIDES OF DUPIN.

In 1822, Dupin [1, p. 200] defined a cyclide to be a surface M in E^3 which is the envelope of the family of spheres tangent to three fixed spheres. This was shown to be equivalent to requiring that both sheets of the focal set degenerate into curves. Then M is the envelope of the family of curvature spheres centered along each of the focal curves. The 3 fixed spheres can be chosen to be 3 from either family. All the spheres of the other family are tangent to these three. By Theorem 4.8, one sees that the cyclides are equivalently characterized by requiring that the lines of curvature in both families be arcs of circles or straight lines. Thus, one can obtain three obvious examples: a torus of revolution, a circular cylinder, and a circular cone. It turns out that all cyclides can be obtained from these three by inversion in a sphere in E^3.

From a different point of view, we recall (Theorem 3.3) that parallel surfaces have the same focal set. Thus every parallel surface of a cyclide is also a cyclide, and it is interesting to examine the entire collection of parallel surfaces of a given cyclide, some of which may have one or two singular points.

A detailed treatment of the cyclides can be found in a book published in 1975 by K. Fladt and A. Baur [1, pp. 354-379]. Included there are beautiful illustrations of all the standard forms of the cyclides.

A good classical reference is Lilienthal [1], where the history of the cyclides is traced and various constructions are given. In particular we mention the papers Liouville [1], Cayley [1] and Maxwell [1], the last of which includes stereoscopic figures of the basic models. The books of Eisenhart [1, pp. 312-314], Darboux [1, vol. 2, pp. 267-269], Klein [1, pp. 56-58] and Hilbert and Cohn-Vossen [1, pp. 217-219] also have lengthy discussions of the cyclides. Finally, the physicist Louis Michel has pointed out to us that the focal conics and the cyclides are prominent in the 1923 paper of G. Friedel [1, p. 316 ff] on the structure of crystals.

In recent work, Pinkall [1]-[3] has given a unified approach to the classical cyclides in terms of the so-called Lie geometry of spheres and has investigated the larger class of Dupin hypersurfaces.

A Dupin hypersurface in E^{n+1} with two distinct principal curvatures is called an <u>n-dimensional cyclide</u>. The complete n-dimensional cyclides are the standard products $S^k \times E^{n-k}$ and the images under stereographic projection of a standard product of spheres $S^k \times S^{n-k}$ in S^{n+1} (see Theorem 6.2). Using Lie geometry, Pinkall [3] and Voss [1] have proven a version of this result which does not assume completeness. This will be discussed further in Section 6.

Finally, the classical cyclides were characterized by Banchoff [3] and Cecil [3] as the only taut surfaces in E^3, aside from spheres and planes (see Section 7). This relies on the characterization of the cyclides as those surfaces whose lines are curvature are circles. The higher dimensional complete cyclides can also be characterized in terms of tautness, as will be seen in Section 8.

In this section, we sketch the classical determination of the cyclides following Eisenhart [1, pp. 312-314]. In the next section, we give a different proof for the higher dimensional case which also works for surfaces in E^3. However, it does use the hypothesis that M is complete.

Let $f:M \to E^3$ be a smooth immersion of a surface. Suppose M has two distinct principal curvatures λ and μ at each point and that λ is constant along its lines of curvature. By Theorem 4.8, $f_\lambda(M)$ is a smooth curve, and M is the envelope of the family of curvature spheres centered along $f_\lambda(M)$. Conversely, if $\gamma(s)$ is an arbitrary smooth curve in E^3 and $\rho(s)$ is a smooth positive function, then the envelope of the family of spheres given by

$$|z-\gamma(s)| = \rho(s), \ z \in E^3,$$

is a surface having $\gamma(s)$ as one sheet of its focal set (see, for example, Eisenhart [1, pp. 66-68]). Such a surface was classically called a <u>canal surface</u>. Surfaces of revolution are examples; one sheet of the focal set is the axis of revolution. Suppose now that M is a canal surface and that the other sheet of the focal set $f_\mu(M)$ is a regular surface, i.e., $X\mu \neq 0$ for $X \in T_\mu$ on M.

Let γ be a line of curvature of T_λ. The other focal map f_μ is injective on γ, and by Theorem 4.3, the normal ξ to M at a point $x \in \gamma$ is tangent to the surface $f_\mu(M)$ at $f_\mu(x)$. In fact, the tangent plane to $f_\mu(M)$ at $f_\mu(x)$ is spanned by ξ and X, where X is Euclidean parallel to a non-zero vector in the space $T_\lambda(x)$. This is also the tangent plane to the cone of revolution through γ with vertex at the common focal point $f_\lambda(\gamma)$. Hence, $f_\mu(M)$ is the envelope of a 1-parameter family of cones of revolution with vertices on $f_\lambda(M)$. The characteristics of such an envelope (intersections of $f_\mu(M)$ with each cone) are plane curves, and we find that $f_\mu(M)$ is a locus of conic sections.

Now suppose that M has two distinct principal curvatures at each point, each constant along its own lines of curvature. As in the above discussion, f_μ is still injective on a leaf γ of T_λ, since the focal points $f_\mu(x)$ all lie on different normal lines to M as x varies over γ. These normal lines intersect only at the common focal point $p = f_\lambda(\gamma)$.

The curve $f_\mu(\gamma)$ must be an open subset of the curve $f_\mu(M)$. Further, the curve $f_\mu(\gamma)$ lies on a cone of revolution with vertex p, and as above, $f_\mu(\gamma)$ must be given by a plane intersection with this cone. Hence, $f_\mu(\gamma)$ lies on a conic section. The curve $f_\mu(M)$ is the union of the $f_\mu(\gamma)$, as γ ranges over the leaves of T_λ. By connectedness, we conclude that each component of $f_\mu(M)$ lies on a conic section.

By reversing the roles of λ and μ, we obtain the same conclusion for the curve $f_\lambda(M)$. Further, these conics have the property that the cone through one conic with vertex on the other is always a cone of revolution. This property characterizes a pair of so-called <u>focal conics</u> (see Eisenhart [1, p. 226]). These are, by definition, either an ellipse and hyperbola in mutually orthogonal planes such that the vertices of the ellipse are the foci of the hyperbola and vice-versa or a pair of parabolas in orthogonal planes such that the vertex of each is the focus of the other. Also possible is the degenerate case consisting of a circle and an orthogonal straight line through its center (covered twice). A proof that this property characterizes the focal conics can be found in Salmon [1, p. 186].

We now describe all of the cyclides of Dupin in E^3. First consider a torus of revolution T^2 (Figure 5.1). (The Figures 5.1-5.14 of the cyclides are at the end of this section.) One sheet of the focal set is the core circle determined by the constant principal curvature λ, whose lines of curvature are the meridian circles. The second sheet $f_\mu(M)$ is the axis of revolution covered twice. The principal curvature μ is equal to zero on the top and bottom circles (which are the two top-cycles of this tight torus). These circles separate T^2 into regions, with Gaussian curvature $K > 0$ on one region and $K < 0$ on the other. Each point on the axis of revolution is the center of one sphere which is tangent to the torus along a circle in the region where $K > 0$ and a second sphere which is tangent to T^2 along a circle in the region $K < 0$.

By a result very similar to Lemma 4.9, the property that a principal curvature is constant along its lines of curvature is shown to be invariant under inversion in a 2-sphere in E^3.

Inversion of T^2 in a sphere not centered on T^2, takes it to a ring cyclide (see Figure 5.2). One sheet of the focal set is the core ellipse, while the other is the corresponding focal hyperbola. As with the torus, there are two distingushed latitude circles which separate the region with $K > 0$ from that with $K < 0$. Each of these two regions is the envelope of a family of curvature spheres centered along one branch of the hyperbola.

We next consider the family of parallel surfaces to the ring cyclide in Figure 5.2. We take the ellipse in the xy-plane, with major axis along the x-axis. Thus, the ellipse can be written as

$$x^2 + \frac{y^2}{1-a^2} = C^2,\ z = 0,\ \text{for } 0 < a < 1,\quad C > 0.$$

The corresponding focal hyperbola has the equation

$$\frac{x^2}{a^2} - \frac{z^2}{1-a^2} = C^2,\ y = 0. \tag{5.1}$$

We parametrize the ellipse $\alpha(u)$ and one branch of the hyperbola $\beta(v)$ as

$$\begin{aligned} \alpha(u) &= (C \cos u,\ C\sqrt{1-a^2}\ \sin u,\ 0),\ 0 \leq u \leq 2\pi, \\ \beta(v) &= (C\ a \cosh v,\ 0,\ C\sqrt{1-a^2}\ \sinh v),\ v \in \mathbb{R}. \end{aligned} \tag{5.2}$$

The cyclide M is the envelope of the family of curvature spheres centered along $\alpha(u)$ and also of those centered along $\beta(v)$. Each curvature sphere S inherits an orientation from the orientation of M by choosing the unit normal field for S which agrees with the unit normal field for M along the intersection of S with M. We assign an oriented radius ρ to S by taking $\rho > 0$ if S has the outward normal orientation and $\rho < 0$ for the inward normal. Borrowing a term from Lie Geometry (see Pinkall [1, p. 8]), we say that two oriented spheres S_1 and S_2 are in <u>oriented contact</u> if they are tangent and if their orientations agree at the point of tangency. If p_1 and p_2 are the respective centers of S_1 and S_2 and ρ_1, ρ_2 are the signed radii, then oriented contact is just the equation

$$|p_1 - p_2| = |\rho_1 - \rho_2|. \tag{5.3}$$

For the cyclide in question, suppose that the curvature spheres centered along $\alpha(u)$ have signed radius $\rho(u)$ and those centered along $\beta(v)$ have signed radius $\bar{\rho}(v)$. The two curvature spheres at the point of M corresponding to the parameter values (u,v) are in oriented contact. Substituting $\alpha(u)$ and $\beta(v)$ from (5.2) into (5.3) as p_1, p_2 and squaring, we obtain

$$C^2[(\cos u - a \cosh v)^2 + (1-a^2)(\sin^2 u + \sinh^2 v)] = (\rho(u) - \bar{\rho}(v))^2.$$

As ρ and $\bar{\rho}$ are functions of the single variables u and v, respectively, this implies that up to a sign, the functions must have the form

$$\rho(u) = Ca \cos u - t, \quad \bar{\rho}(v) = C \cosh v - t,$$

where t is an arbitrary constant whose values correspond to parallel surfaces. If one takes the opposite sign for ρ, then one must also take the opposite sign for $\bar{\rho}$. This just amounts to choosing the opposite orientation on the two families of spheres.

Singularities on a parallel surface occur at points where ρ or $\bar{\rho}$ equals zero. Since the original ring cyclide has no singularities, it corresponds to a parameter value satisfying

$$Ca < |t| < C,$$

as one finds by setting the formulas for ρ and $\bar{\rho}$ equal to 0.

Suppose the original ring cyclide M corresponds to a parameter value t_0 with $Ca < t_0 < C$. Then M is oriented with the inward normal since $\rho(u) < 0$

for all u. Begin taking parallel surfaces by moving along the inward normal, i.e. by decreasing t. At $t = Ca$, we have $\rho = 0$ at $u = 0$, and so there is one singularity of the parallel surface which lies on the focal ellipse (see Figure 5.3). Continuing to decrease t, one obtains two singularities by solving $\rho(u) = 0$, i.e.

$$\cos u = \frac{t}{Ca} .$$

These two singularities lie on the focal ellipse and are symmetrically placed with respect to the major axis (see Figure 5.4). In classical terminology, the cyclides in Figures 5.3 and 5.4 are called a <u>limit horn</u> cyclide and a <u>horn</u> cyclide, respectively.

The equation $\rho(u) = 0$ has solutions for $|t| \leqslant Ca$. To get a singularity on the branch of the focal hyperbola $\beta(v)$ in (5.2), one must have $\bar{\rho}(v) = 0$, i.e.

$$\cosh v = \frac{t}{C}.$$

Thus, there is one singularity if $t = C$ and two if $t > C$. These cyclides were classically known as the <u>limit spindle</u> and <u>spindle</u> cyclides, respectively. They are shown in Figures 5.5 and 5.6.

If one parametrizes the other branch of the hyperbola (5.1) as

$$\delta(v) = (-Ca \cosh v, 0, C\sqrt{1-a^2} \sinh v), \; v \in R,$$

one finds the radii of the curvature spheres to be

$$\tilde{\rho}(v) = -C \cosh v - t.$$

Thus, there is one singularity of the parallel surface on this branch of the hyperbola if $t = -C$ and two for $t < -C$. These are also spindle cyclides.

Note that inversion of the horn cyclide with one singularity p (Figure 5.3) in a sphere centered at p, takes all the circles through p into straight lines, which are lines of curvature on the image. Moreover, these lines are all tangent to each curvature sphere of the other principal curvature, and so the surface is a circular cylinder. In other words, the limit horn cyclide in Figure 5.3 is the image of a circular cyclinder under inversion in a sphere, whose center lies outside the cylinder.

Similarly, a horn cyclide with two singularities (Figure 5.4) is obtained from a cone of revolution by inversion in a sphere centered outside the cone. The spindle cyclides in Figures 5.5 and 5.6, respectively, are obtained by inverting a circular cylinder, resp. cone, in a sphere centered inside the cylinder, resp. cone.

Finally, we show that any ring cyclide, i.e. parallel surface with $Ca < |t| < C$, can be inverted into a torus of revolution. Note that the radii of curvature functions $\bar{\rho}(v)$ and $\hat{\rho}(v)$ each have an extreme value at $v = 0$, i.e. at the vertices of the respective branches of the hyperbola. The respective curvature spheres are centered at these vertices, and by Theorem 4.13, the corresponding circles of curvature, call them $\bar{\gamma}$ and $\hat{\gamma}$, are great circles in these curvature spheres. $\bar{\gamma}$ and $\hat{\gamma}$ are the smallest and largest latitude circles, respectively, in Figure 5.2. The plane of $\bar{\gamma}$ is the plane determined by all of the normal lines to M at points of $\bar{\gamma}$. Each such normal line also contains another focal point of M lying on the focal ellipse, and we see that $\bar{\gamma}$ lies in the plane of the ellipse. By the same argument, $\hat{\gamma}$ is also in the plane of the ellipse. Thus, there exists an inversion I, centered at a point in this plane, which takes the two curvature spheres to concentric spheres, and takes $\bar{\gamma}$ and $\hat{\gamma}$ to concentric circles in this same plane. The image I(M) is then seen to be a torus of revolution with $I(\bar{\gamma})$ and $I(\hat{\gamma})$ as inner and outer rims.

We next consider the special case where the original ring cyclide M is a torus of revolution, obtained by revolving a circle Γ about an axis L. In this case, one does not get any horn cyclides as parallel surfaces, but rather the core circle is at a constant distance ρ = radius Γ from M. However, one does obtain spindle cyclides with one or two singularities on the axis of revolution L as parallel surfaces. These are obtained by revolving a circle concentric to Γ about L. One singularity (Figure 5.7) occurs at a point of tangency of the circle of revolution with L, and two singularities (Figure 5.8) occur when the circle intersects L in two points. These surfaces are called a _limit torus_ and _spindle torus_, respectively.

Finally, suppose the focal set consists of the two focal parabolas

$$z - a = \frac{-x^2}{8a},\ y = 0 \quad \text{and} \quad z + a = \frac{y^2}{8a},\ x = 0 \text{ for } a > 0.$$

We parameterize the parabolas as

$$(5.4) \qquad \begin{aligned} \alpha(u) &= (u,\ 0,\ \frac{-u^2}{8a} + a) \\ \beta(v) &= (0,\ v,\ \frac{v^2}{8a} - a) \end{aligned}$$

As before, let $\rho(u)$, $\overline{\rho}(v)$ be the signed radii of the curvature spheres centered along $\alpha(u)$ and $\beta(v)$, respectively. The two curvature spheres at the point of M corresponding to the parameter values (u,v) are in oriented contact. Substituting $\alpha(u)$ and $\beta(v)$ from (5.4) into (5.3), we obtain

$$|\alpha(u)-\beta(v)| = \frac{1}{8a}(u^2+v^2+16a^2) = |\rho(u)-\overline{\rho}(v)|.$$

As ρ and $\overline{\rho}$ are functions of the single variables u and v, respectively, this implies that up to a sign, the functions must have the form

$$\rho(u) = \frac{1}{8a}(u^2+8a^2-t),\ \overline{\rho}(v) = \frac{-1}{8a}(v^2+8a^2+t),$$

where again t is an arbitrary constant whose values correspond to parallel surfaces, and the opposite choice of sign corresponds to the opposite choice of orientations of the two families of spheres.

Singularities on a parallel surface occur at points where ρ or $\overline{\rho}$ equals zero. Thus, there are singularities on the parabola $\alpha(u)$ if $t \geq 8a^2$ and on $\beta(v)$ if $t \leq -8a^2$. There are no singularities for $|t| < 8a^2$. Those surfaces with no singularities are called <u>parabolic ring</u> cyclides, (see Figures 5.9 and 5.10 for two perspectives). Those with one singularity are called <u>limit parabolic horn</u> cyclides (Figure 5.11 and 5.12) and those with two singularities are called <u>parabolic horn</u> cyclides (Figures 5.13 and 5.14). These surfaces are non-compact, and there is one straight line in each family of lines of curvature. Further, they are algebraic of degree 3, while those previously discussed are of degree 4.

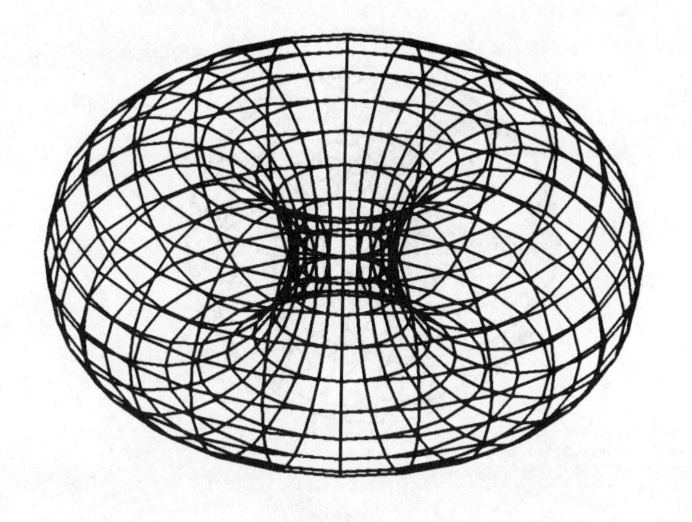

Figure 5.1 Torus of revolution

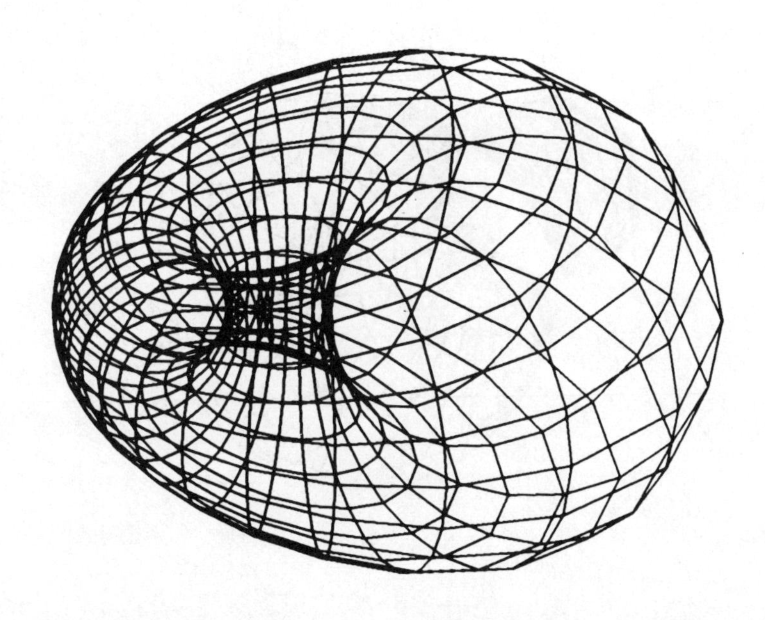

Figure 5.2 Ring cyclide

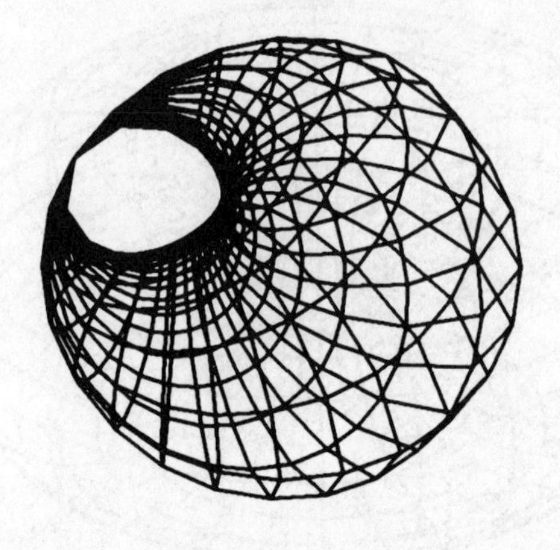

Figure 5.3 Limit horn cyclide

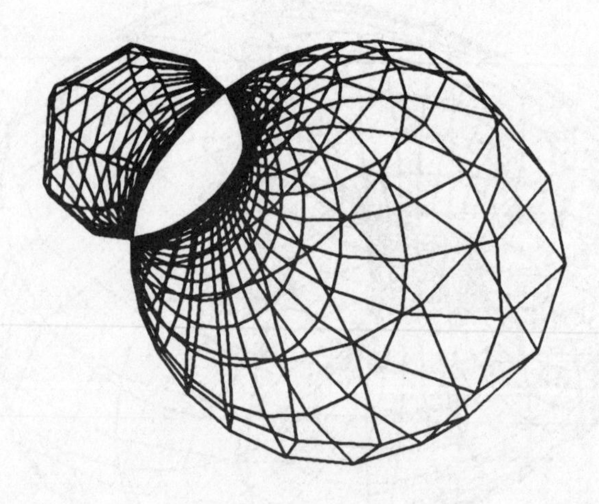

Figure 5.4 Horn cyclide

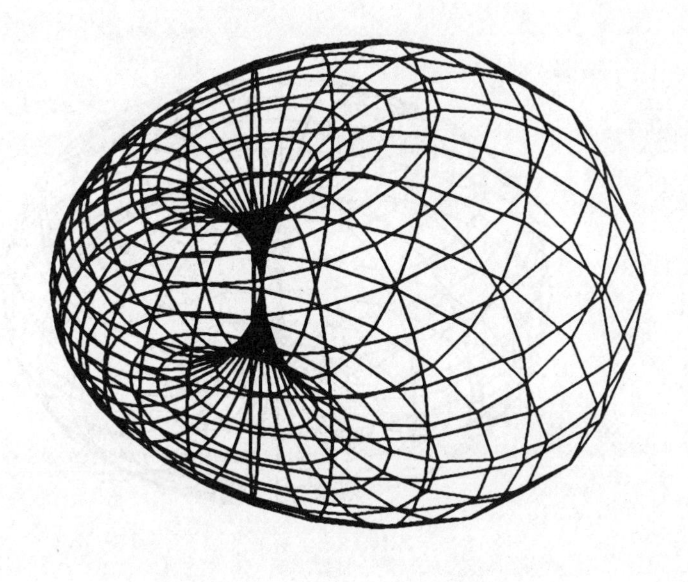

Figure 5.5 Limit spindle cyclide

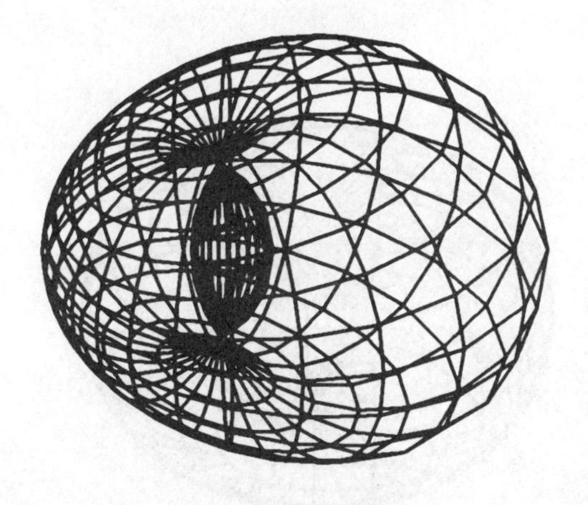

Figure 5.6 Spindle cyclide

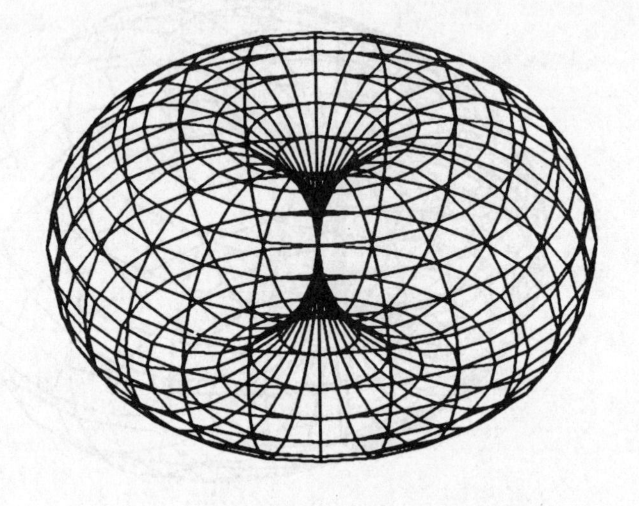

Figure 5.7 Limit torus

Figure 5.8 Spindle torus

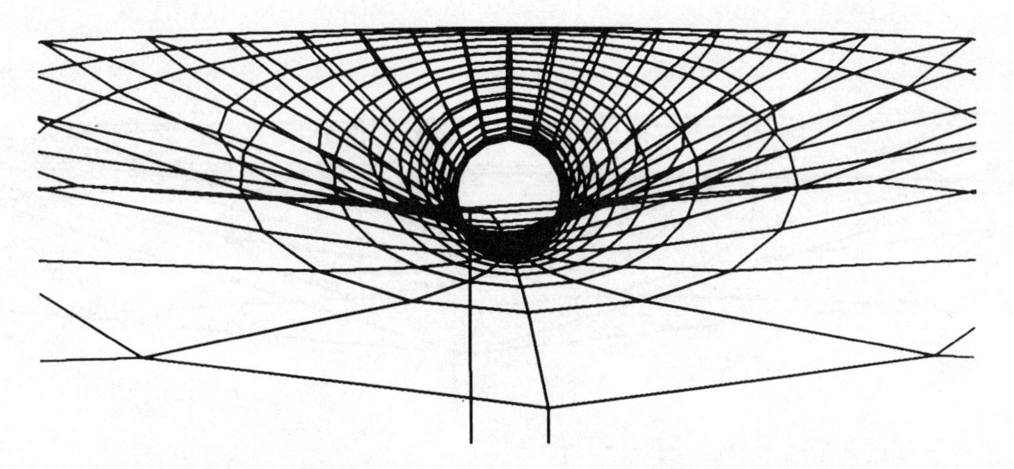

Figure 5.9 Parabolic ring cyclide (first perspective)

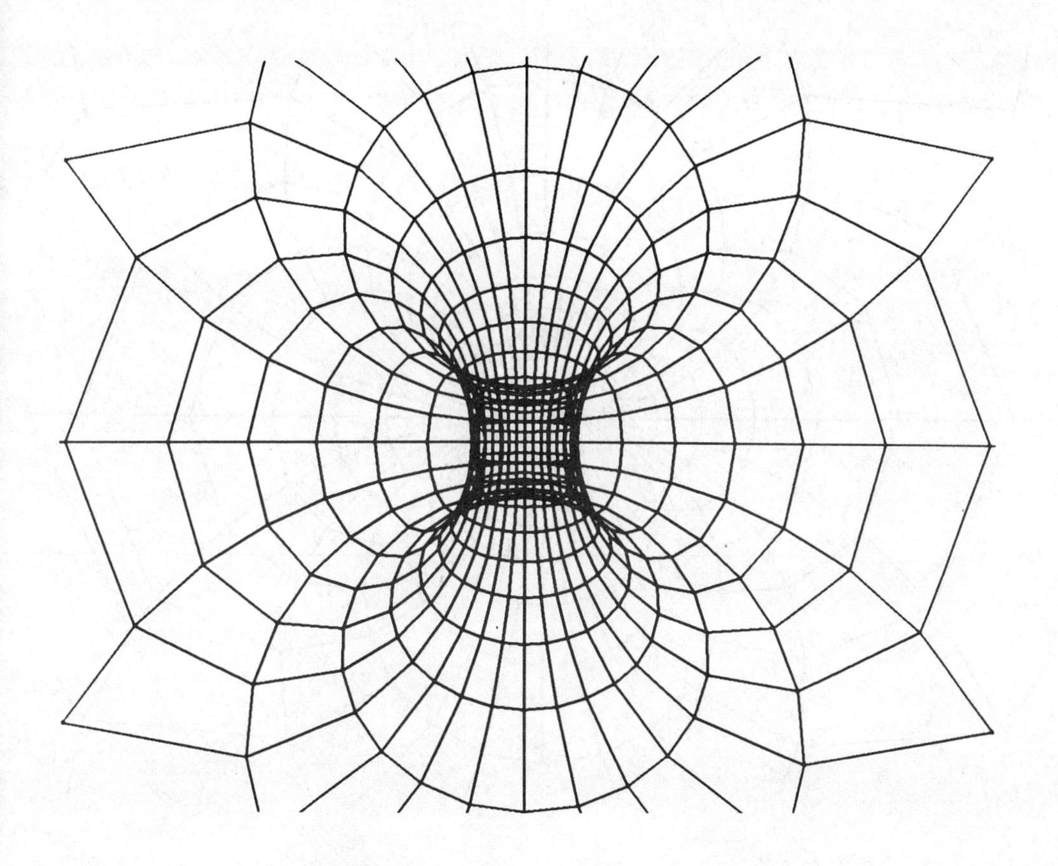

Figure 5.10 Parabolic ring cyclide (second perspective)

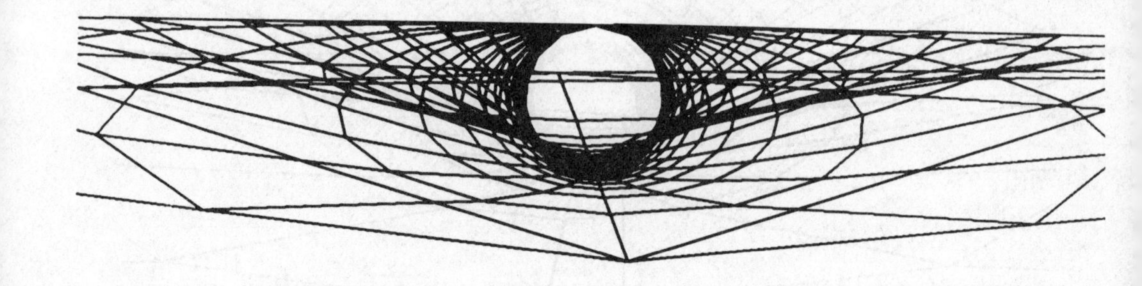

Figure 5.11 Limit parabolic horn cyclide (first perspective)

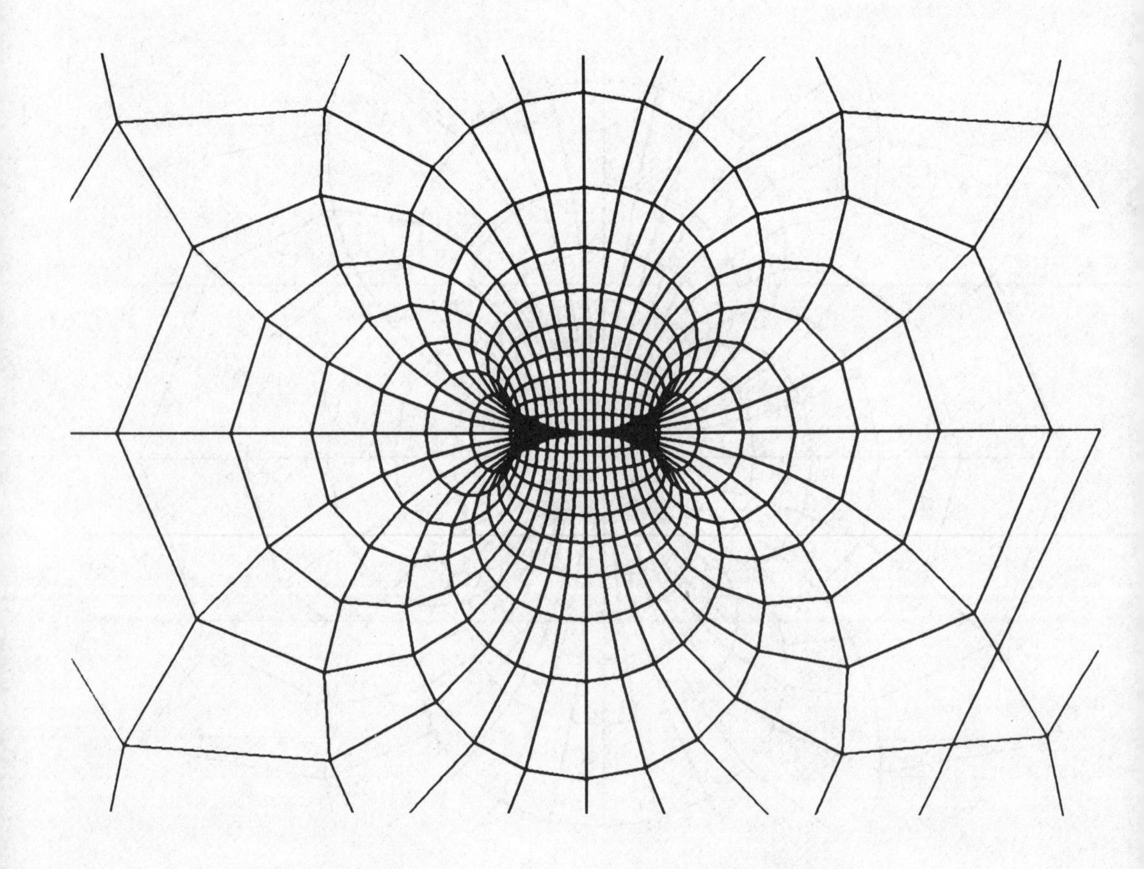

Figure 5.12 Limit parabolic horn cyclide (second perspective)

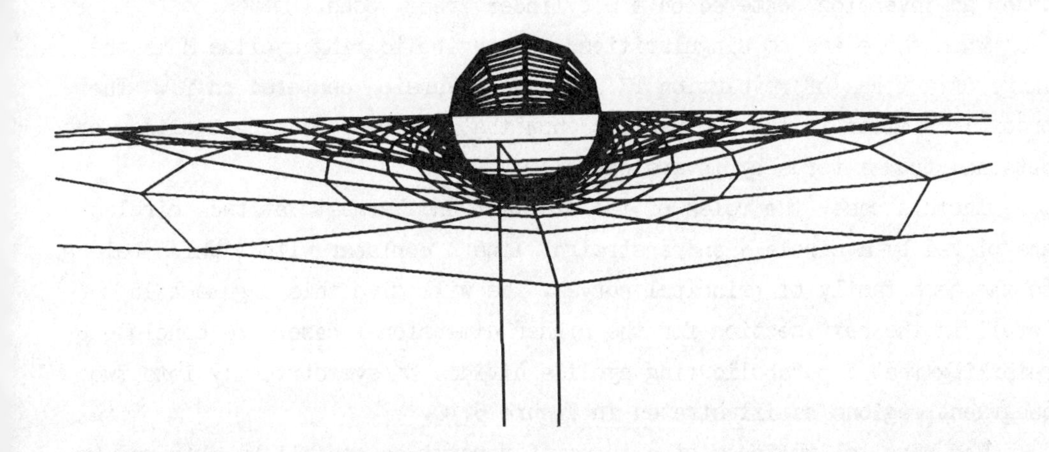

Figure 5.13 Parabolic horn cyclide (first perspective)

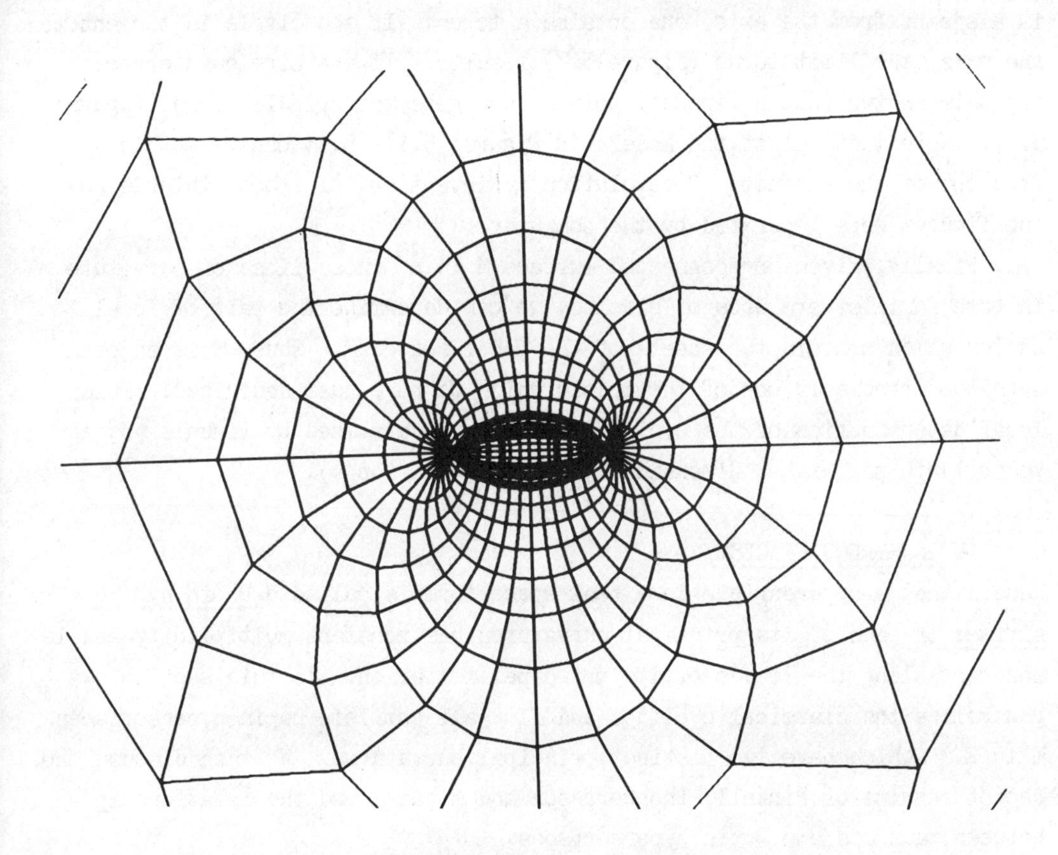

Figure 5.14 Parabolic horn cyclide (second perspective)

As in the previous case, if a parabolic ring cyclide has one, resp. two, singularities, then it is the image of a circular cylinder, resp. cone, under an inversion centered on the cylinder, resp. cone.

When there are no singularities, the parabolic ring cyclide M is the image of a torus of revolution T^2 under an inversion centered on T^2. The proof is similar to the one used to show that the compact ring cyclides are obtained from a torus by inversion.

In this case, the roles of the smallest and largest latitude circles are played by a circle γ and a straight line L coplanar with γ which belong to the same family of principal curves. We will give this argument in detail in the next section for the higher dimensional case. We conclude by remarking that a parabolic ring cyclide divides E^3 symmetrically into two congruent regions as illustrated in Figure 5.10.

The parallel surfaces of a torus of revolution can all be obtained by revolving a circle about an axis in the plane of the circle. If the circle is disjoint from the axis, one obtains a torus. If the circle is tangent to the axis, the limit torus (Figure 5.7) results. If the circle intersects the axis in two (not antipodal) points, one gets the spindle torus (Figure 5.8). Note that all of the models in Figures 5.1 - 5.14 can be obtained from one of the surfaces of revolution by inversion. In fact, this is how the figures were generated by the computer.

Finally, given any connected surface M in E^3 whose lines of curvature in both families are arcs of circles, we can determine the pair of focal conics which contain the sheets of the focal set of M. Thus, M is an open subset of precisely one of the models which we have just described. This local determination of the cyclides has been generalized by Pinkall [3] and Voss [1] to the higher dimensional case (see Section 6).

6. DUPIN HYPERSURFACES.

Recall that a hypersurface in a real space form is called a Dupin hypersurface if each of its principal curvatures has constant multiplicity and is constant along the leaves of its principal foliation. In this section, we generalize the classical cyclides and find all complete Dupin hypersurfaces M in E^{n+1} which have two distinct principal curvatures. We then discuss the recent results of Pinkall, Thorbergsson and Miyaoka and the relationship between tautness and Dupin hypersurfaces.

Suppose M is a complete Dupin hypersurface in E^{n+1} with two distinct principal curvatures. If one of the principal curvatures is identically zero, we find that $M = S^k(r) \times E^{n-k}$, where $S^k(r)$ is a Euclidean sphere in a Euclidean subspace E^{k+1} orthogonal to E^{n-k}. Otherwise, we show that M is the image under stereographic projection of a standard product of two spheres

$$S^{n-k}(r) \times S^k(s) \subset S^{n+1}(\rho), \quad r^2 + s^2 = \rho^2.$$

As in the classical case, if the pole of the projection does not lie on $S^{n-k} \times S^k$, M is called a ring cyclide. Otherwise, M is non-compact and is called a parabolic ring cyclide. In both cases, the two sheets of the focal set are a pair of focal conics.

Of special importance in our method of proof is the generalization of a torus of revolution which we will call a round cyclide.

Example 6.1: The round cyclides.

A round torus in E^3 is obtained by revolving a circle Γ of radius b about an axis in its plane at a distance a from the center of Γ with $a > b$. We can write Γ as

$$\Gamma = \{(x_1,0,x_3) | (x_1-a)^2 + x_3{}^2 = b^2\}.$$

Analogously, write $E^{n+1} = E^{n-k+1} \times E^k$ and

$$\Gamma = \{(x,y) \in E^{n+1} | \ x = ce_1 \text{ for } c \in R, \ |x-ae_1|^2 + |y|^2 = b^2\},$$

as profile submanifold, where $e_1 = (1, 0, \ldots, 0)$. The hypersurface of revolution

$$C(k,a,b) = \{Ax,y) | (x,y) \in \Gamma, \ A \in SO(n-k+1)\},$$

is called a round cyclide. Each integer k between 0 and n determines a family of round cyclides diffeomorphic to $S^{n-k} \times S^k$. We can obtain a round cyclide via stereographic projection as follows. Set $\rho = (a^2-b^2)^{1/2}$. For $q \in S^{n+1}(\rho)$, let P be stereographic projection with pole q onto

$$E_q^{n+1} = \{x \in E^{n+2} | \langle x,q\rangle = 0\},$$

as formulated in Remark 4.2. Let

$$K = S^{n-k}(\frac{\rho^2}{a}) \times S^k(\frac{\rho b}{a}) \subset S^{n+1}(\rho).$$

If the pole q lies in the second factor of the decomposition $E^{n+2} = E^{n-k+1} \times E^{k+1}$, then $P(K) = C(k,a,b)$. K has two distinct principal curvatures $1/b$ and $-b/\rho^2$ as a hypersurface in $S^{n+1}(\rho)$ (see, for example, Ryan [1, p. 367]). Thus, by Theorem 4.1, P(K) has principal curvatures

$$\lambda_1 = 1/b \quad \text{and} \quad \lambda_2 = -b/\rho^2 + \frac{\langle z,q\rangle a^2}{\rho^4 b}$$

at the point P(z) in E_q^{n+1}. Setting $z = \pm(b/a)q$ yields the maximum and minimum values of λ_2 which are $1/(a+b)$ and $-1/(a-b)$. These values are assumed on λ_2-leaves which are the outer and inner rims of C(k,a,b). They are easily seen to be concentric (n-k)-spheres of radii a+b and a-b, both lying in the first factor E^{n-k+1} and centered at the origin. One sheet of the focal set is the core (n-k)-sphere of radius a centered at the origin in E^{n-k+1}, while the second sheet is E^k doubly covered.

The main goal of this section is to prove the following.

<u>Theorem 6.2</u>: <u>Let M be a connected, complete Dupin hypersurface embedded in E^{n+1} with two distinct principal curvatures,</u>

(a) <u>if M is compact, it is a ring cyclide.</u>

(b) <u>if M is non-compact and one principal curvature is identically zero, then M is a standard product $S^k \times E^{n-k}$. Otherwise, M is a parabolic ring cyclide.</u>

We remark that a complete embedded hypersurface in E^{n+1} is orientable (Samelson [1]). We assume that each principal curvature λ_i has multiplicity ν_i and that the unit normal field ξ has been chosen so that $\lambda_1 > \lambda_2$ on M.

Let T_i denote T_{λ_i}.

<u>Lemma 6.3</u>: <u>Each compact T_i-leaf meets each T_j-leaf, $j \neq i$, in exactly one point.</u>

<u>Proof</u>: Let γ be a compact leaf of T_i. We first show that γ intersects any leaf of T_j in at most one point. First assume $\lambda_i \neq 0$ on γ. For $x \in \gamma$, let η_x be the T_j-leaf through x. The leaf γ lies on the curvature n-sphere S centered at the λ_i-focal point $p = f_i(x)$. If $\lambda_j(x) \neq 0$, then η_x lies on the curvature n-sphere S' centered at $q = f_j(x)$. But S and S' are tangent at x, which is therefore the only point in $S \cap S'$ and hence in $\gamma \cap \eta_x$. On the other hand, if $\lambda_j(x) = 0$, then η_x lies in the tangent plane T_xM which also intersects S only in the point x.

Next suppose $\lambda_i = 0$ on γ. Then $\lambda_j(x) \neq 0$ for all $x \in \gamma$, and the argument above shows that $\gamma \cap \eta_x = \{x\}$ for each x.

We now show that γ does indeed intersect each leaf of T_j. Let N be the union of the η_x as x ranges over γ. Using coordinate systems which arise naturally in the theory of foliations and the continuation theorem for such coordinate systems (see Palais [1, p. 10]), one easily shows that N is open in M. We now show that N is also closed in M. Let $\{x_n\}$ be a sequence of points in N converging to $x \in M$. If an infinite number of x_n lie on the same leaf L of T_j, then x is also on L and thus in N. This follows since T_j is a regular foliation, and hence each leaf is a closed submanifold of M (Palais [1, p. 18]). Otherwise, let L_n be the leaf of T_j containing x_n. The leaf space N/T_j is diffeomorphic to γ. Thus, it is compact and Hausdorff. Therefore, $\{L_n\}$ has an accumulation point $L \in N/T_j$. Clearly, x must belong to L and hence to N.

Q.E.D.

<u>Lemma 6.4</u>: <u>For M as in Theorem 6.2, if one principal curvature is identically zero, then M is a standard product $S^k \times E^{n-k}$.</u>

<u>Proof</u>: Assume $\lambda_2 \equiv 0$. Note that M cannot be compact, because a compact manifold would have to contain points at which non-degenerate height functions assume their maxima. The shape operator must be negative definite at such points, contradicting $\lambda_2 \equiv 0$. Since $\lambda_1 > 0$ on M, the leaves of T_1 are

all ν_1-dimensional Euclidean spheres. Let $k = \nu_1$. Suppose that a leaf η of T_2 were compact. Let N be the union of all T_1-leaves γ_x as x ranges over η. Then N is k-sphere bundle over η, and so N is a compact submanifold of M. Since M is complete, it follows that N = M (see, for example, Kobayashi-Nomizu [1, vol. I, p. 179]). This contradicts the fact that M is not compact, and we conclude that all of the leaves of T_2 are non-compact. Hence, the leaves of T_2 are (n-k)-dimensional Euclidean subspaces by Theorems 4.5 and 4.8.

Let γ be an arbitrary T_1-leaf. Then γ is a great or small k-sphere in the curvature n-sphere S centered at the common λ_1-focal point $p = f_1(\gamma)$. At each $x \in \gamma$, the T_2-leaf η_x is the (n-k)-plane orthogonal to the (k+1)-plane spanned by $T_1(x)$ and the normal ξ_x. If γ is totally geodesic in S, then all of these (k+1)-planes are the same space, call it E^{k+1}, and $\gamma = S \cap E^{k+1}$. Thus each leaf η_x is parallel to the orthogonal complement E^{n-k} of E^{k+1}, and $M = \gamma \times E^{n-k}$, as desired.

We now suppose that none of the leaves of T_1 are totally geodesic in their curvature spheres and get a contradiction. Suppose that a T_1-leaf γ is a small k-sphere in its curvature n-sphere S centered at the focal point $p = f_1(\gamma)$. Then γ is the intersection of S with a (k+1)-plane E^{k+1}, and we take the origin O of E^{k+1} to be the center of γ. Let E^{k+2} be the space spanned by E^{k+1} and the vector from O to the point p. The union of the normal lines to M through points of γ is a cone of revolution in E^{k+2} with vertex p. For each $x \in \gamma$, the T_2-leaf η_x contains one line β_x in E^{k+2} which is orthogonal to the (k+1)-plane spanned by $T_1(x)$ and ξ_x. The line β_x lies in the plane determined by x, O and p, and it is orthogonal to the normal line to M from x to p. The union of these lines β_x as x ranges over γ also forms a cone of revolution. This gives a contradiction, since the leaves of T_2 must be disjoint, and the proof is finished.

Q.E.D.

In Lemmas 6.5-6.9 we assume that M is non-compact, and neither principal curvature is identically zero.

<u>Lemma 6.5</u>: <u>Each principal foliation has a non-compact leaf</u>.

Proof: Since neither principal curvature is identically zero, each T_i has some compact leaves. If all of the leaves of T_i were compact, then M would be a ν_i-sphere bundle over a compact leaf of T_j. Thus, M itself would be compact, a contradiction.

Q.E.D.

Corollary 6.6: (a) $\lambda_1 \geq 0 \geq \lambda_2$ on M.

(b) If $\lambda_i = 0$ on a T_i-leaf η, then η is non-compact.

Proof(a): Let γ be a non-compact T_1-leaf. Then $\lambda_1 = 0$ on γ. By Lemma 6.3, every non-zero value of λ_2 is taken on γ. Since $\lambda_1 > \lambda_2$, we have $\lambda_2 \leq 0$ on M. By considering a non-compact T_2-leaf, one gets $\lambda_1 \geq 0$.

(b) Suppose $\lambda_i = 0$ on a compact leaf η. By Lemma 6.3, η intersects a non-compact leaf γ of T_j on which λ_j must be zero. Thus $\lambda_i = \lambda_j$ at the point of intersection of η and γ, a contradiction.

Q.E.D.

Lemma 6.7: There exists a compact T_1-leaf on which λ_1 assumes its maximum and a compact T_2-leaf on which λ_2 assumes its minimum.

Proof: Let η be a compact leaf of T_2. By Lemma 6.3, every value of λ_1 is taken on η, and thus λ_1 has a maximum value. Since this maximum is positive, it is taken on a compact leaf. The same proof produces a compact T_2-leaf on which λ_2 assumes its minimum value.

Q.E.D.

Let γ^+ be a compact T_1-leaf on which λ_1 assumes its maximum and η^+ a compact T_2-leaf on which λ_2 takes its minimum. Let x be the unique point of intersection of γ^+ and η^+ given by Lemma 6.3. By Theorem 4.13, γ^+ and η^+ are great spheres in their respective curvature n-spheres centered at the focal points $p = f_1(x)$ and $q = f_2(x)$. Since λ_1 and λ_2 have opposite signs at x, the points p and q lie on opposite sides of the tangent hyperplane T_xM. The leaf γ^+ lies in a Euclidean (ν_1+1)-plane spanned by $T_1(x)$ and ξ_x, while η^+ lies in the (ν_2+1)-plane spanned by $T_2(x)$ and ξ_x. The two planes intersect orthogonally in the normal line ℓ to M at x. The next lemma shows that the extreme leaves of M have the same configuration as for a parabolic cyclide (see Figure 6.1).

Lemma 6.8: (a) The T_2-leaf η^- through the point y antipodal to x in γ^+ is non-compact. Further, η^- lies in the (ν_2+1)-plane determined by η^+, and the tangent hyperplane T_yM is disjoint from the curvature n-sphere S^+ determined by η^+.

(b) The T_1-leaf γ^- through the point z antipodal to x in η^+ is non-compact. Further, γ^- lies in the (ν_1+1)-plane determined by γ^+, and the tangent hyperplane T_zM is disjoint from the curvature n-sphere determined by γ^+.

Proof: (a) We wish to show that η^- is non-compact. The proof is divided into two cases, according to whether or not the T_1-leaf γ^- through z is compact. Suppose, first that γ^- is compact. We know that there exists a non-compact leaf η of T_2 through some point v of γ^+. Suppose $v \neq y$. Since γ^- is compact, it must intersect η in a unique point $u \neq v$. The normal line β to M at u is orthogonal to the ν_2-plane η. Hence, it cannot meet the (ν_1+1)-plane π_1 containing γ^+ which is also orthogonal to η. On the other hand, the line β and π_1 both contain the common λ_1-focal point of z and u, whose existence is guaranteed by Corollary 6.6(b) and the compactness of γ^-. This is a contradiction, and we conclude that if γ^- is compact, then the only possibility for a non-compact leaf of T_2 through a point on γ^+ is the leaf η^- through y. Hence, η^- is non-compact in this case.

Now, we suppose that γ^- is non-compact and show that η^- must be non-compact. The argument is similar to the preceding one. If η^- is compact, then η^- and γ^- intersect in a unique point $w \neq z$. Since y and w lie on the same compact leaf η^-, the normal lines to M at y and w must intersect at the λ_2-focal point determined by η^-. The normal line to M at y is ℓ which lies in (v_2+1)-plane π_2 determined by η^+ which is orthogonal to the ν_1-plane γ^- at z. On the other hand, the normal line at w is orthogonal to γ^-. Conse-

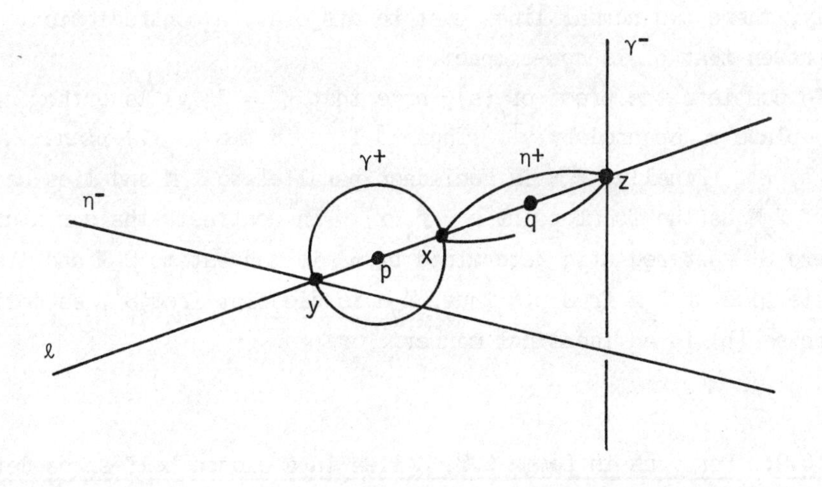

Figure 6.1 The extreme leaves of a parabolic ring cyclide

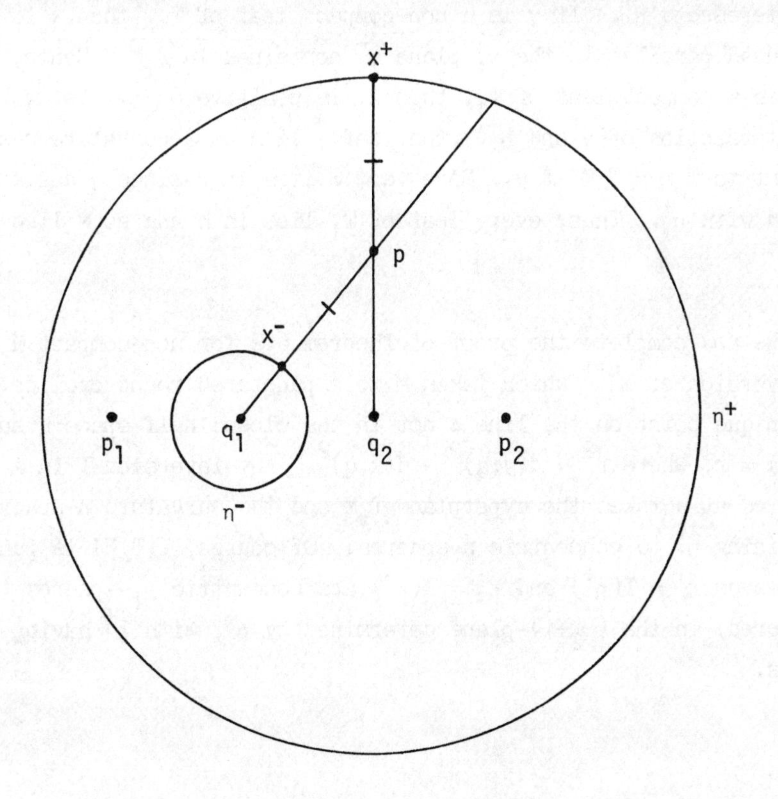

Figure 6.2 Extreme leaves of a ring cyclide

quently, these two normal lines must be disjoint, a contradiction. Thus, we have proven that η^- is non-compact.

To complete the proof of (a), note that $\eta^- = T_2(y)$ is orthogonal to the (ν_1+1)-plane π_1 spanned by γ^+. Thus η^- lies in the (ν_2+1)-plane π_2 determined by η^+. Finally, T_yM is Euclidean parallel to T_xM and lies on the same side of T_xM as the focal point $p = f_1(x)$. In contrast, the curvature n-sphere S^+ centered at q determined by η^+ is tangent to T_xM and lies on the opposite side of T_xM from p. Thus, T_yM is disjoint from S^+, as desired. One proves (b) in an identical manner.

Q.E.D.

Lemma 6.9: For y as in Lemma 6.8, M lies in a closed half-space determined by the hyperplane T_yM.

Proof: Let h be the half-space containing η^+. By Lemma 6.3, every leaf of T_1 interesects η^+. If γ is a non-compact leaf of T_1, then γ is a ν_1-plane Euclidean parallel to the ν_1-plane γ^- contained in T_yM. Hence, γ lies in h. If γ is a compact leaf of T_1, then λ_1 is positive on γ. Let w be the point of intersection of γ and η^-. The leaf γ lies on a curvature n-sphere S tangent to $T_wM = T_yM$ at w. Moreover, S lies in h since γ has a point in common with η^+. Thus, every leaf of T_1 lies in h and so M lies in h.

Q.E.D.

We now complete the proof of Theorem 6.2 for non-compact M by producing an inversion of E^{n+1} which takes M to a punctured round cyclide. Let w be the unique point on the line ℓ not in the closed half-space h such that $d(w,y) = r$, where $r^2 = d(y,q)^2 - d(x,q)^2$. Any inversion I in a sphere centered at w takes the hyperplane T_yM and the curvature n-sphere S^+ containing η^+ to concentric n-spheres. Of course, $I(T_yM)$ is punctured at w. The leaves $\zeta^+ = I(\eta^+)$ and $\zeta^- = I(\eta^-)$ are concentric ν_1-spheres (ζ^- is punctured) in the (ν_2+1)-plane determined by η^+, with ζ^+ having smaller radius.

The hypersurface I(M) has two distinct principal curvatures at each point which are constant along the leaves of their principal foliations by Lemma 4.9. Let μ_2 be the principal curvature on I(M) whose principal foliation contains ζ^+ and ζ^-. Then every leaf of the other principal foliation determined by μ_1 intersects ζ^+, since every leaf of T_1 intersects η^+ on M. Let x^- be an arbitrary point on ζ^-. The μ_1-leaf through x^- intersects ζ^+ in a point x^+. All the points on ζ^+ and ζ^- have the same μ_2-focal point q, the common center of ζ^+ and ζ^-. The points x^- and x^+ also have the same μ_1-focal point p. The normal lines to I(M) at x^- and x^+ both contain the points p and q, and thus they coincide. The points x^- and x^+ are equidistant from p, and therefore p is the mid-point of the segment from x^- to x^+. Hence $1/\mu_1$ is equal to one-half the difference of the radii of ζ^+ and ζ^- at x^+, and this value is clearly independent of the original choice of x^-. We have thus shown that μ_1 has a constant value at every point of ζ^+, except the point whose normal line intersects the sphere containing ζ^- at w. By continuity, μ_1 has the same value at this point also, and we conclude that μ_1 is constant on I(M).

Finally, let $k = \nu_1$ and let C(k,a,b) be the round cyclide whose inner and outer rims coincide with ζ^+ and ζ^-, respectively. We show that $I(M) \cup \{w\}$ coincides with C(k,a,b). For each point $x \in \zeta^+$, the normal lines to both I(M) and C(k,a,b) pass through q, and hence I(M) and C(k,a,b) have the same tangent plane at x. They have the same $T_2 = T_x\zeta^+$, and thus they have the same T_1. They have the same constant value for μ_1, and so the T_1-leaves through x for the two hypersurfaces coincide. Since both hypersurfaces are the union of these T_1-leaves as x varies over ζ^+, they are the same. We conclude that M is the image of C(k,a,b) under the inversion I, i.e. M is a parabolic ring cyclide.

Finally, we handle the case when M is compact. This can be done directly in a manner quite similar to the proof above (see Cecil-Ryan [2]). However, we can also get a short proof by using the results for the non-compact case. Let M be a compact hypersurface satisfying the hypotheses of Theorem 6.2, and let J be an inversion of E^{n+1} whose center is a point x in M. Let $M' = J(M-\{x\})$. Then M' is a connected, complete, non-compact hypersurface which still satisfies the hypotheses of Theorem 6.2 by Lemma 4.9. By the proof above, there is an inversion I which takes M' to a punctured round cyclide. Thus $I \circ J$ is a diffeomorphism of $M-\{x\}$ onto a punctured round cyclide. Up to a similarity of E^{n+1}, $I \circ J$ is another inversion K or the identity. Since similarities preserve round cyclides, $M-\{x\}$ is related to some punctured round cyclide by an inversion K. The center of this inversion cannot be x (since the image of $M-\{x\}$ under K is bounded). By continuity, K must take x to the puncture of the round cyclide, and thus the original manifold M is a ring cyclide. This completes the proof of Theorem 6.2.

Remark 6.10: The focal sets of the cyclides.

As noted earlier, for the round cyclide $C(k,a,b)$ in Example 6.1, one sheet is the core $(n-k)$-sphere of radius a centered at the origin in E^{n-k+1}. The second sheet is the k-plane E^k orthogonal to E^{n-k+1} at the origin covered twice. We next consider a (compact) ring cyclide M in E^{n+1}. All of the leaves in each principal foliation are compact. Thus, each leaf of one foliation intersects each leaf of the other in exactly one point by Lemma 6.3. Consequently, one principal curvature λ_1 is positive on M, while λ_2 assumes both positive and negative values. Let k be the multiplicity of λ_1. There are two distinguished leaves in each family corresponding to the extrema of the associated principal curvatures. These leaves are great spheres in the corresponding curvature n-spheres whose centers lie along a line. Let p_1 and p_2 be the centers of the extreme λ_1-leaves and q_1, q_2 the centers of the extreme λ_2-leaves η^- and η^+ (the inner and outer rims). See Figure 6.2. We first show that the sheet of the focal set $f_1(M)$ coming from λ_1 is the ellipsoid of revolution consisting of all points p in the $(n-k+1)$-plane E^{n-k+1} determined by η^+ and η^- satisfying

$$d(p,q_1) + d(p,q_2) = d(p_1,p_2) = \rho^+ + \rho^-,$$

where ρ^+ and ρ^- are the respective radii of η^+ and η^-. Of course, q_1 and q_2 are the foci of the ellipsoid, and p_1 and p_2 are the vertices along the major axis. First, since every T_1-leaf intersects η^+ and the focal map f_1 is constant on each leaf, we see that $f_1(M) = f_1(\eta^+)$. Since all the normal lines to M at points of η^+ are in E^{n-k+1}, $f_1(M)$ lies in E^{n-k+1}. Now let p be an arbitrary point in $f_1(M)$ (see Figure 6.2). The T_1-leaf which lies on the curvature n-sphere centered at p intersects η^+ in a point x^+ and η^- in a point x^-, and

$$d(p,x^+) = d(p,x^-). \tag{6.1}$$

Now,

$$d(p,q_1) = d(p,x^-) + d(x^-,q_1) = d(p,x^-) + \rho^-,$$

$$d(p,q_2) = d(x^+,q_2) - d(p,x^+) = \rho^+ - d(p,x^+).$$

Adding these two quantities and using (6.1), we obtain

$$d(p,q_1) + d(p,q_2) = \rho^+ + \rho^-.$$

The second focal submanifold is a hyperboloid of revolution of two sheets in the (k+1)-plane E^{k+1} which intersects E^{n-k+1} orthogonally along the line containing p_1 and p_2. The two extreme λ_1-leaves lie in E^{k+1}. The hyperboloid consists of all points p in E^{k+1} satisfying

$$d(p,p_1) - d(p,p_2) = \pm\, d(q_1,q_2) = \pm\,(\sigma^+ - \sigma^-),$$

where σ^+ is the maximum value of $1/\lambda_2$ and σ^- is the minimum value. Of course when n = 2, these focal submanifolds are a pair of focal conics, as in the previous section.

Finally, we consider a parabolic ring cyclide M. Assume $\lambda_1 > 0 > \lambda_2$ and that λ_1 has multiplicity k. The position of the extreme leaves is shown in Figure 6.1. Each principal foliation has one non-compact leaf which we denote by γ^- for T_1 and η^- for T_2. There is one other compact extreme leaf

γ^+ for T_1 and η^+ for T_2, which are great spheres in their corresponding curvature n-spheres centered at the respective focal points p and q. The leaves γ^+ and γ^- lie in the same E^{k+1}, and η^+ and η^- lie in the orthogonal E^{n-k+1} intersecting E^{k+1} along the line determined by p and q. The focal submanifold $f_1(M)$ is an (n-k)-dimensional paraboloid of revolution defined as follows. The focus of the paraboloid is the point q and the directrix is the (n-k)-plane V parallel to η^- in E^{n-k+1} such that $d(p,q) = d(p,V)$. The paraboloid is the set of points z in E^{n-k+1} satisfying

$$d(z,q) = d(z,V).$$

Note that p is the vertex. Similarly, $f_2(M)$ is a k-dimensional paraboloid in E^{k+1}. The focus is p, the directrix is the k-plane W parallel to γ^- in E^{k+1} such that $d(q,p) = d(q,W)$, and the vertex is q. In the case $n = 2$, these are just a pair of focal parabolas.

We remark that our proof of Theorem 6.2 definitely uses the completeness assumption. In particular, completeness is needed in Lemma 6.7 to produce extreme leave of the foliations. Lemmas 6.3 and 6.5 also involve global considerations.

As we mentioned in Section 5, U. Pinkall [3] and K. Voss [1] have proven local versions of Theorem 6.2 by using Lie geometry. We refer the reader to Pinkall's work and to Blaschke [1] for the basic Lie geometric framework. The notion of a hypersurface in Lie geometry is more general than in Euclidean geometry, and Euclidean realizations of Lie-geometric hypersurfaces may have singularities, as do the horn and spindle cyclides of the previous section. A Lie transformation is a transformation of the space of all oriented hypersurfaces in E^m (or S^m) which preserves oriented contact of spheres. The group of Lie transformation is larger than the group of conformal transformations. Pinkall showed that the class of Dupin hypersurfaces is invariant under Lie transformations. Two hypersurfaces which can be obtained from one another by a Lie transformation are said to be <u>Lie equivalent</u>. A cyclide is said to have <u>characteristic (p,q)</u> if the two distinct principal curvatures have respective multiplicities p and q. Pinkall [3] proves that every connected cyclide is contained in a unique compact, connected cyclide (up to Lie equivalence). Further, any two compact connected cyclides with the same characteristic are Lie-equivalent. Of course they need not be conformally equivalent.

From Pinkall's Lie-geometric theorem, one can easily derive a conformal version, as stated by Voss [1]. Here, the standard models are the hypersurfaces of revolution obtained by rotation of a sphere $S^p \subset E^{p+1} \subset E^{n+1}$ about a plane E^p in E^{p+1}. If S^p and E^p are disjoint, one obtains the round cyclides of Example 6.1. Otherwise, one gets hypersurfaces with singularities, analogous to the limit torus and spindle torus of Section 5. Pinkall and Voss show that a connected cyclide of characteristic (p,q) is related to an open subset of one of these models by a conformal transformation of $E^{n+1} \cup \{\infty\}$. This generalizes the classical result for cyclides in E^3.

We now begin a general discussion of Dupin hypersurfaces. Our first step is to show that any choice of multiplicities for the principal curvatures is possible. Note that this construction (due to Pinkall [3]) is only local and does not give complete hypersurfaces.

<u>Theorem 6.11</u>: <u>Given positive integers</u> $\nu_1, \ldots, \nu_g$ <u>with</u> $\nu_1 + \ldots + \nu_g = n$, <u>there exists a Dupin hypersurface whose principal curvatures have respective multiplicities</u> $(\nu_1, \ldots, \nu_g)$.

<u>Proof</u>: The proof is by an inductive construction which will be clear once the first few examples are done. To get a Dupin hypersurface M^3 in E^4 with principal curvatures having multiplicities (1,1,1), one begins with an open subset U of a torus of revolution on which neither principal curvature vanishes. Let M^3 be the cylinder $U \times E^1$ in $E^3 \times E^1 = E^4$. Then, M^3 has 3 distinct principal curvatures at each point one of which is zero. These are clearly constant along their corresponding lines of curvature.

To get a Dupin hypersurface in E^5 with multiplicities (1,1,2), one simply takes $U \times E^2$ in $E^3 \times E^2 = E^5$. To obtain a Dupin hypersurface M^4 with multiplicities (1,1,1,1) in E^5, invert M^3 in a 3-sphere chosen so that the image contains an open set W^3 on which no principal curvature vanishes. Then $M^4 = W^3 \times E^1$ has the desired properties.

Q.E.D.

It is not so easy to produce compact Dupin hypersurfaces, however. In In fact, Pinkall [1], [2] has classified all Dupin hypersurfaces M^3 in E^4 with 3 distinct principal curvatures at each point. He found that the only compact ones are those obtained by a Lie transformation from a stereographic image M^3 in E^4 of an isoparametric hypersurface in S^4. E. Cartan [2] showed that there is only one family of isoparametric hypersurfaces with 3 distinct distinct principal curvatures, the tubes over a Veronese surface in S^4. This will be discussed in detail in example 7.3 of Chapter 3. The hypersurfaces in this Lie-geometric family are distinguished from all of the other (1,1,1) Dupin hypersurfaces in E^4 as the only ones on which the lines of curvature cannot serve as coordinates of a local parametrization (see Pinkall [1, p. 76 ff]). To be precise, we make the following definition.

Definition 6.12: Let M be a hypersurface in a real space from $\tilde{M}^{n+1}$ with g distinct principal curvatures $\lambda_1, \ldots, \lambda_g$ at each point, with respective constant multiplicities $\nu_1, \ldots, \nu_g$. Let T_i be the principal distribution of λ_i. A principal coordinate system is a set of coordinates

$$(x_{11}, \ldots, x_{1\nu_1}, x_{21}, \ldots, x_{2\nu_2}, \ldots, x_{g_1}, \ldots, x_{g\nu_g})$$

on an open set U in M such that

$$T_i = \text{span}\ \{\frac{\partial}{\partial x_{i1}}, \ldots, \frac{\partial}{\partial x_{i\nu_i}}\}, \quad 1 \leq i \leq g,$$

on U. That is, the set of equations

$$x_{jk} = c_{jk}, \quad j \neq i, \quad 1 \leq j \leq g, 1 \leq k \leq \nu_j,$$

for any set of constants c_{jk}, determines a leaf of T_i.

Theorem 6.13: Let M be an isoparametric hypersurface in S^{n+1} with more than two distinct principal curvatures. Then there cannot exist a local principal coordinate system on M.

Proof: Let $X \in T_\lambda$ and $Y \in T_\mu$ be coordinate vector fields in a principal coordinate system. (See Definition 6.12). Then the Lie bracket $[X,Y] = \nabla_X Y - \nabla_Y X = 0$. The Codazzi equation

$$(\nabla_X A)Y = (\nabla_Y A)X$$

yields

$$\nabla_X(\mu Y) - \nabla_Y(\lambda X) = A(\nabla_X Y - \nabla_Y X)$$

$$= A([X,Y]) = 0.$$

Thus

$$0 = \mu \nabla_X Y - \lambda \nabla_Y X = (\mu - \lambda)\nabla_X Y,$$

so that $\nabla_X Y = 0$ for $\lambda \neq \mu$. On the other hand, since $\langle X,Y \rangle = 0$, we have

$$\langle \nabla_X X, Y \rangle = - \langle X, \nabla_X Y \rangle = 0. \tag{6.2}$$

Thus $\nabla_X X$ lies in $T_\mu^\perp$ for all $\mu \neq \lambda$ and hence

$$\nabla_X X \in T_\lambda. \tag{6.3}$$

Using the Gauss equation (see Ryan [1, p. 366]), we find

$$R(X,Y)Y = (\lambda\mu + 1)\ (\langle Y, Y \rangle X - \langle X, Y \rangle Y) \tag{6.4}$$

$$= (\lambda\mu + 1) \langle Y, Y \rangle X.$$

But

$$R(X,Y)Y = \nabla_X \nabla_Y Y - \nabla_Y \nabla_X Y - \nabla_{[X,Y]} Y$$

which, since $\nabla_X Y = 0$ and $[X,Y] = 0$, becomes

$$R(X,Y)Y = \nabla_X \nabla_Y Y.$$

Hence

$$\langle R(X,Y)Y, X\rangle = \langle \nabla_X \nabla_Y Y, X\rangle$$

$$= X \langle \nabla_Y Y, X\rangle - \langle \nabla_Y Y, \nabla_X X\rangle = 0$$

since (6.3) also applies to Y and T_μ. In view of (6.4) we have

$$\lambda \mu + 1 = 0.$$

Since this identity holds for any two distinct principal curvatures, we must have $g \leq 2$, a contradiction.

Q.E.D.

Examples of Dupin hypersurfaces have also arisen in the work of T. Otsuki [1]-[2] and R. Miyaoka [1]-[2] on minimal hypersurfaces in the sphere. For instance, Otsuki [2, p. 17] gives examples of minimal hypersurfaces in the sphere with 3 distinct non-simple (having constant multiplicity greater than one) principal curvatures. By Theorem 4.4, these are Dupin hypersurfaces. Otsuki's examples have the property that each orthogonal complement $T_i^\perp$ of a principal distribution is integrable. The proposition below shows that this is equivalent to the assumption that each point of M has a principal coordinate neighborhood. Otsuki showed that his examples are not isoparametric and cannot be complete. In fact, Miyaoka [2] shows that if M is a complete hypersurface with constant mean curvature in a real space form $\tilde{M}^{n+1}(c)$ with curvature $c \geq 0$ such that M has 3 non-simple principal curvatures, then M is isoparametric and $c > 0$.

<u>Proposition 6.14</u>: <u>Let M be a hypersurface in a real space form $\tilde{M}^{n+1}$ with g distinct principal curvatures $\lambda_1, \ldots, \lambda_g$ at each point. Then each point of M has a principal coordinate neighborhood if and only if each $T_i^\perp$ is integrable on M.</u>

<u>Proof</u>: Suppose there exists a principal coordinate system on an open set U. Then, for a fixed i, the equations

$$x_{ik} = c_{ik}, \quad 1 \leqslant k \leqslant \nu_i.$$

determine a manifold whose tangent space is the direct sum of the T_j for $j \neq i$. But this space is just $T_i^{\perp}$, and so $T_i^{\perp}$ is integrable on U. Since M is covered by such neighborhoods, $T_i^{\perp}$ is integrable on M. Conversely, suppose each $T_i^{\perp}$ is integrable and let $p \in M$. For each i, choose a system of coordinates (see Kobayashi-Nomizu [1, Vol. 1, p. 182])

$$(x_{i1}, \ldots, x_{i\nu_i}, y_{i1}, \ldots, y_{i(n-\nu_i)})$$

with origin at p such that the leaves of $T_i^{\perp}$ are given by the equations

$$x_{ik} = c_{ik}, \quad 1 \leqslant k \leqslant \nu_i.$$

It can now be shown by induction that

$$(x_{11}, \ldots, x_{1\nu_1}, x_{21}, \ldots, x_{2\nu_2}, \ldots, x_{g_1}, \ldots, x_{g\nu_g})$$

is a coordinate system by verifying that for each m, $1 \leqslant m \leqslant g$,

$$(x_{11},\ldots,x_{1\nu_1}, x_{21},\ldots,x_{m\nu_m}, y_{mj_1},\ldots,y_{mj_r}), \qquad r = n - \sum_{1=i}^{m} \nu_i$$

is a coordinate system for a suitable subset $\{j_1,\ldots,j_r\}$ of $\{1,2,\ldots,n-\nu_m\}$. If L is the manifold given by the equations,

$$x_{jk} = c_{jk}, \quad j \neq i, \quad 1 \leqslant j \leqslant g, \quad 1 \leqslant k \leqslant \nu_j,$$

then at each point $q \in L$,

$$T_qL \subset \bigcap_{j \neq i} T_j^{\perp}(q) = T_i(q).$$

But T_qL has the same dimension as $T_i(q)$, so the spaces are equal, and this is a principal coordinate system about p.

Q.E.D.

Every known complete Dupin hypersurface in E^{n+1} is Lie equivalent to an isoparametric hypersurface in the sphere S^{n+1}. Cecil and Ryan [6] proved that every isoparametric hypersurface in a sphere is taut (see Theorem 6.9 of Chapter 3). The proof rests on the work of Münzner [1] who, among other things, used the fact that an isoparametric hypersurface M in S^{n+1} divides the sphere into two ball bundles over the focal submanifolds to compute the homology of M (see Theorem 6.8 of Chapter 3).

Recently, Thorbergsson [1] has shown that any complete embedded Dupin hypersurface is taut. His method is quite different from that mentioned above in that he does not assume knowledge of the homology of the Dupin hypersurface M. Rather, he uses the principal foliations in a general construction which shows that each critical point of a non-degenerate distance function contributes positively to the homology of M (i.e. it is of "linking type" in the terminology of Morse-Cairns [1, p. 258]). Once tautness has been established, it can be used in conjunction with Münzner's results to deduce some important facts about compact Dupin hypersurfaces. In particular, the number g of distinct principal curvatures must be 1, 2, 3, 4 or 6, as for isoparametric hypersurfaces in spheres (see Theorem 6.11 of Chapter 3). All of this leads one to the following conjecture.

<u>Conjecture 6.15</u>: <u>Every compact embedded Dupin hypersurface in Euclidean space is Lie equivalent to an isoparametric hypersurface in a sphere</u>.

The conjecture is true for Dupin hypersurfaces with one or two distinct principal curvatures by the well-known classification of umbilic hypersurfaces and Theorem 6.2. Further support for this conjecture has recently been provided by Miyaoka [3] who has shown that the conjecture holds for Dupin hypersurfaces with 3 distinct principal curvatures.

There is one noteworthy distinction between the two conditions, Dupin and isoparametric. As we shall see in Chapter 3, any local piece of an isoparametric hypersurface can be extended to a compact isoparametric hypersurface. Theorem 6.11 and the restriction on g for compact Dupin hypersurfaces show that this is not true for Dupin hypersurfaces.

Turning now to tautness, suppose that M is a taut hypersurface in E^{n+1} whose principal curvatures have constant mutliplicities. If a principal curvature λ has constant multiplicity $\nu > 1$, then λ is constant along the leaves of its principal foliation by Theorem 4.4. (tautness is not needed for this.) On the other hand, if λ has multiplicity $\nu = 1$, then tautness implies that λ is constant along its lines of curvature (see Lemma 7.5 in the next section). Hence, we have the converse of Thorbergsson's theorem for hypersurfaces whose principal curvatures have constant multiplicities.

Theorem 6.16: If M is a taut hypersurface E^{n+1} whose principal curvatures have constant multiplicities, then M is a Dupin hypersurface.

There are, however, many known examples of taut hypersurfaces whose principal curvatures do not have constant multiplicities. These are obtained from Dupin hypersurfaces via three common constructions which preserve tautness. The first is to take a cylinder $M \times E^1$ in $E^{n+1} \times E^1 = E^{n+2}$, where M is a taut hypersurface in E^{n+1} (see Example 1.20). We now discuss two other standard methods of obtaining taut hypersurfaces from taut submanifolds: tubes over taut submanifolds and rotation hypersurfaces.

Example 6.17: Parallel hypersurfaces and tubes over taut submanifolds.

Here we use an observation of Carter and West [1, p. 717] to produce more examples of taut hypersurfaces. First, suppose $f:M \to E^{n+1}$ is an embedding of a compact n-manifold M with global unit normal field ξ. Identifying $(x,t\xi)$ with (x,t), the normal bundle is diffeomorphic to $M \times R$. Suppose τ is a positive number such that all points of the form (x,t) for $x \in M$, $0 \leq t \leq \tau$, are not critical points of the normal exponential map $F:NM \to E^{n+1}$. Such a τ always exists for a compact manifold. Then the parallel map $\phi_\tau:M \to E^{n+1}$ given by

$$\phi_\tau(x) = f(x) + \tau\, \xi(x)$$

is an immersion, a parallel hypersurface to the original embedding $\phi_0 = f$.

Moreover, ϕ_τ and ϕ_0 have the same focal set by Theorem 3.3. Suppose

$$p = f(x) + t\xi = \phi_\tau(x) + (t - \tau)\xi$$

is not a focal point of these hypersurfaces. Let L_p denote the restriction of the distance function determined by p to the original embedding f, and let $\tilde{L}_p$ denote its restriction to the immersion ϕ_τ. Then L_p has a critical point at x if and only if $\tilde{L}_p$ does, and further, the two critical points have the same index, since there are no focal points of (M,x) on the segment from f(x) to $\phi_\tau(x)$. Obviously, ϕ_τ is taut if and only if f is taut.

Carter and West pointed out that this idea can be generalized to the case where $f:M \to E^{n+k}$ is a taut embedding of a compact manifold with codimension $k > 1$. For suitably small $\tau > 0$, we consider the tube ϕ_τ over M. By Theorem 3.3, the focal set of ϕ_τ is the union of the focal set $F(\Gamma)$ of f and f(M) itself. Let p be a point in the complement of $F(\Gamma) \cup f(M)$. Let L_p denote the restriction of the distance function to f, and let $\tilde{L}_p$ denote its restriction to the hypersurface ϕ_τ. Each critical point $x \in M$ of L_p gives rise to two critical points of $\tilde{L}_p$ on ϕ_τ,

$$z_1 = f(x) + \tau\xi \quad \text{and} \quad z_2 = f(x) - \tau\xi,$$

where the line from p to f(x) intersects the tube. Here ξ is the unit normal in the direction p-f(x). Suppose L_p has a critical point of index j at x. From the index theorem and the formula for the shape operator of ϕ_τ in Theorem 3.2, we see that $\tilde{L}_p$ has a critical point of index j at z_1 and a critical point of index j+k-1 at z_2 (f(x) is a focal point of multiplicity k-1 of z_2). Thus, if the unit normal bundle BM satisfies

$$H_*(BM;Z_2) = H_*(M \times S^{k-1};Z_2),$$

then each non-degenerate $\tilde{L}_p$ has the minimum number of critical points required by the Morse inequalities, and ϕ_τ is taut. Note that essentially the same proof shows that a tube over a tight immersion is tight if the condition above on the homology of BM is satisifed.

This happens, for example, if the normal bundle NM is trivial. In particular, if M is a standard product of spheres

$$M = S^{k_1} \times \dots \times S^{k_r} \subset E^{k_1+\dots+k_r+r},$$

then a tube of sufficiently small radius over M is a taut embedding of $M \times S^{r-1}$ as a hypersurface. Of course, the round cyclide C(k,a,b) of Example 6.1 is a tube over its core (n-k)-sphere.

Another example (which we will not verify now) is the case where M is the standard Veronese embedding of FP^2 into S^m described in Section 9 of Chapter 1. A tube in S^m over M is an isoparametric hypersurface in S^m with three constant principal curvatures. In fact, Cartan showed that these are the only examples of isoparametric hypersurfaces with 3 principal curvatures (see Example 7.3 of Chapter 3). There are, however, many interesting examples of isoparametric hypersurfaces in S^{n+1} with 4 or 6 principal curvatures (see Ferus, Karcher, Münzner [1]). These are all tubes over each of their two (taut) focal submanifolds, as will be seen in Chapter 3.

Example 6.18: Rotation hypersurfaces

A second method of constructing taut hypersurfaces from given taut hypersurfaces is to generalize the construction of the round cyclide in Example 6.1 as a rotation hypersurface. Suppose M is a taut compact hypersurface in E^{k+1} which is disjoint from the hyperplane E^k through the origin. Let e_{k+1} be a unit normal to E^k in E^{k+1}. Embed E^{k+1} in E^{n+1}, and let E^{n-k+1} be the orthogonal complement of E^k in E^{n+1}. Let SO(n-k+1) denote the isometries in SO(n+1) which keep E^k pointwise fixed. If we consider E^{n+1} as $E^k \times E^{n-k+1}$, then each point of M has the form (x,y) where $y = c\, e_{k+1}$ for some $c > 0$. Let

$$W = \{(x,Ay) \mid (x,y) \in M,\ A \in SO(n-k+1)\}$$

be the hypersurface in E^{n+1} obtained by rotating M about E^k. Then W is clearly diffeomorphic to $M \times S^{n-k}$, and the sum of the Z_2-Betti numbers $\beta(W) = 2\beta(M)$. If p is in E^k, then L_p has a critical point (an absolute minimum) on M at a point $z \in M$. Hence, L_p has critical points at all the points of W in the orbit of z under SO(n-k+1). Since these critical points are not isolated, they are degenerate. Thus, each point p in E^k is a focal point of W. Consider any (k+1)-plane of the form

$$V = E^k + \text{span } \{Ae_{k+1}\},$$

for a fixed $A \in SO(n-k+1)$. Then, $W \cap V$ consists of two disjoint congruent copies of M. Further, if $z \in W \cap V$, then the normal line to W through z lies in V. Now suppose L_p is a non-degenerate function on W. Then p does not lie in E^k, so p lies in the V spanned by E^k and p itself. All of the critical points of L_p on W must lie in $W \cap V$. Since M is taut, L_p has exactly $\beta(M)$ critical points on each of the two copies of M in $W \cap V$. Thus, each non-degenerate L_p has $\beta(W) = 2\beta(M)$ critical points on W, and W is taut.

We briefly consider two examples which show that the application of these standard constructions to a Dupin hypersurface does not, in general, lead to a Dupin hypersurface. Let T^2 be a torus of revolution in $E^3 \subset E^3 \times E^1 = E^4$, and let ξ be a field of unit normals to T^2 in E^3. Let M^3 be a tube of sufficiently small radius $\varepsilon > 0$ over T^2 such that M^3 is a compact hypersurface. The normal space to T^2 in E^4 at a point x is spanned by the orthogonal vectors $\xi(x)$ and $e_4 = (0,0,0,1)$. We know that A_ξ has two distinct eigenvalues at each point while the shape operator for e_4 is identically zero on T^2. The shape operator A_θ for the normal

$$\eta(\theta) = \cos\theta\, \xi(x) + \sin\theta\, e_4$$

at x is given by $A_\theta = \cos\theta\, A_{\xi(x)}$. Thus, at all points of M^3 where $x_4 \neq \pm\varepsilon$, there are 3 distinct principal curvatures, which are constant along their corresponding lines of curvature (see Theorem 3.2). So, this part of M is a Dupin hypersurface. However, the principal curvatures do not have constant multiplicities on all of M. On the two tori $T^2 \times \{\pm\varepsilon\}$, the principal curvature 0 has constant multiplicity two.

Similarly, if one considers the cylinder $T^2 \times E^1$ in E^4, there are 3 distinct principal curvatures which are constant along their lines of curvature on the set $U \times E^1$, where U is the subset of T^2 on which the Gaussian curvature $K \neq 0$, i.e. U is T^2 without the top and bottom circles γ_1 and γ_2. On the circular cylinders $\gamma_1 \times E^1$ and $\gamma_2 \times E^1$, the principal curvature 0 has constant multiplicity two.

Pinkall [3] defines a <u>curvature surface</u> of a hypersurface M to be a smooth submanifold N in M such that for each point p of N, the tangent space T_pN is equal to a principal space $T_\lambda(p)$. Thus, the two tori in the first example and the two cylinders in the second are curvature surfaces. A hypersurface M in a real space form $\tilde{M}$ is called a <u>semi-Dupin hypersurface</u> if each principal space T_λ at each point p of M is the tangent space to a curvature surface, and if along each curvature surface, the corresponding principal curvature is constant. Both the tube and the cylinder over T^2 given above satisfy this condition. This is nearly the same condition which Pinkall called "Dupin" in [3]. He does not insist that every principal space be tangent to a curvature surface, but only that corresponding principal curvature be constant along any curvature surfaces which do exist.

For a Dupin hypersurface, the focal set must be a union of submanifolds of codimension greater than 1 (Theorems 4.6 and 4.10). The following example shows that this need not be true for a taut semi-Dupin hypersurface.

Take a rotation hypersurface V^3 in E^4 obtained from a non-round cyclide M in E^3. In the terminology of Example 6.18, we may arrange that the plane of revolution $E^k = E^2$ is perpendicular to the common axis of the focal conics. Then the focal set of V^3 consists of the rotation surfaces of the focal conics plus the plane of rotation. One branch of the focal hyperbola will intersect the plane of rotation at 2 points. At these points, the sheet of the focal set obtained by revolving that branch of the hyperbola has singularities. For each of these two points, there is a toroidal curvature surface which is mapped by the corresponding focal map onto the point. The rank of the focal map is 1 on these tori and 2 elsewhere. (The focal map can be defined because there are three smooth principal curvature functions on V even though they do not have constant multiplicities. Two of these functions coincide on the tori).

As we have seen, the notions of taut and Dupin are equivalent for hypersurfaces whose principal curvatures have constant multiplicities. The standard constructions, when applied to Dupin hypersurfaces, lead to taut hypersurfaces which, in general, are not Dupin but are semi-Dupin. (One should also consider the result of applying the standard constructions to taut submanifolds of higher codimension.) However, every taut hypersurface which we have checked is semi-Dupin and vice-versa, every semi-Dupin hypersurface which we have checked is taut. This leads to:

Conjecture 6.19: For complete embedded hypersurfaces in E^{n+1}, the notions of taut and semi-Dupin are equivalent.

If the conjecture is true, it would lead to the rather startling conclusion that tautness is equivalent to a property which can be verified by purely local considerations. This would be additional evidence in support of an idea of Kuiper that all taut hypersurfaces are algebraic. Finally, we note that all known STPP hypersurfaces are actually taut, and we ask:

Question 6.20: Does the STPP imply tautness

(a) for hypersurfaces?
(b) for submanifolds of arbitrary codimension?

The answer to the corresponding question for tightness is negative, i.e. the TPP does not imply tightness for smooth immersions (see Example 5.23 of Chapter 1). But, of course, TPP immersions can have large flat portions (and thus need not be analytic), and they have proven to be far less rigid than STPP embeddings.

7. TAUT SURFACES.

In this section, we give the complete determination of all taut surfaces in Euclidean space E^m. In the course of the proof, we develop some important facts (Lemmas 7.1 and 7.5) which are useful in the study of higher-dimensional taut hypersurfaces.

Recall by Theorem 1.28 that for a taut surface M substantially embedded in E^m, we must have $m \leq 5$. Further if $m = 5$, then M is a spherical Veronese surface in E^5. If M is substantial and non-spherical in E^4, then M must be the image of a (possibly punctured) Veronese surface under stereographic projection. On the other hand, every taut spherical surface $M \subset S^3 \subset E^4$ is related to a taut surface in E^3 via stereographic projection. Thus, the problem is reduced to finding all taut surfaces in E^3.

Recall that by Corollaries 1.26 and 1.27, if a taut hypersurface M has one umbilic point, then it is totally umbilic. Thus, M is a sphere or plane.

Suppose now that M is a taut surface in E^3 with no umbilic points. We show that M is a Dupin hypersurface in E^3 and apply Theorem 6.2.

We formulate the main lemma for hypersurfaces, since it is needed in the next section. Since a taut hypersurface M in E^{n+1} is properly embedded, it is a closed subset of E^{n+1}. Therefore, it is orientable, with unit normal field ξ. Thus, there are globally defined principal curvature functions

$$\lambda_1 \geq \dots \geq \lambda_n,$$

determined by ξ. If $\lambda_1(x) > 0$, then the focal point p determined by $\lambda_1(x)$ is the first focal point on the normal ray to M at x in the direction ξ, and L_p has an absolute minimum at x by Lemma 1.24(a). If $\lambda_1(x) < 0$, (and hence M is compact by tautness) then the corresponding focal point p is the last focal point on the normal ray at x in the direction $-\xi$, and L_p has an absolute maximum at x by Lemma 1.24(b). Obviously, similar statements can be made about the smallest principal curvature λ_n, which is just the largest principal curvature for the unit normal field $-\xi$. The key point here is that if p is the focal point of (M,x) corresponding to the largest or smallest principal curvature at x, then L_p has an absolute extremum at x. We now prove the result needed to obtain the classification of taut surfaces in E^3. Note that the hypotheses are implied by tautness.

<u>Lemma 7.1:</u> <u>Let M be a properly embedded hypersurface in E^{n+1}. Assume that for every non-degenerate L_p, $\mu_0 = 1$ and $\mu_n = 1$ or 0 depending on whether M is compact or not. Suppose λ is either the largest or smallest principal curvature function on M. If λ has constant multiplicity 1 on some open set</u>

U, then λ is constant along its lines of curvature in U.

Proof: Let γ be a line of curvature of the principal curvature λ in U. If $\lambda \equiv 0$ on γ, then the result holds on γ. Suppose $\lambda(x_0) \neq 0$ for some point x_0 on γ. The set $B = \{x \in \gamma | \lambda(x) = \lambda(x_0)\}$ is closed in γ, since λ is continuous. We now complete the proof by showing that B is also open in γ.

Let x be an arbitrary point in B, and $W \subset U$ a neighborhood of x on which $\lambda \neq 0$. Let X be a unit vector field in T_λ. We will show that $X\lambda \equiv 0$ on W. In particular, $\lambda \equiv \lambda(x_0)$ on $W \cap \gamma$, as desired. For $y \in W$, let β be the normal section of M at y obtained by intersecting W with the plane spanned by X(y) and the surface normal $\xi(y)$. Parametrize β by arc-length so that $\beta(0) = y$ and $\beta'(0) = X(y)$, where the prime denotes differentiation with respect to s. Let $\kappa(s)$ denote the curvature function of β, and let $\lambda(s) = \lambda(\beta(s))$. We know that $\kappa(0) = \lambda(0)$, and we now show that $\kappa'(0) = \lambda'(0)$ also. The normal curvature $k_n(s)$ at $\beta(s)$ in the direction $\beta'(s)$ is given by the formula

$$k_n(s) = \langle A(\beta'(s)), \beta'(s)\rangle,$$

where A is the shape operator of M. An elementary calculation, classically known as Meusnier's theorem, shows that

$$k_n(s) = \kappa(s) \cos \phi(s),$$

where ϕ is the angle between the principal normal to β and the surface normal $\xi(\beta(s))$. Since $\xi(y)$ is the principal normal to β at y, we have $\phi(0) = 0$. Differentiating the equation above and evaluating at $s = 0$, we obtain $k_n'(0) = \kappa'(0)$. On the other hand, suppose we decompose $\beta'(s)$ into its components in T_λ and $T_\lambda^\perp$, i.e.

$$\beta'(s) = a(s)\, X(\beta(s)) + b(s)\, Y(\beta(s))$$

where X is a unit field in T_λ and Y is a unit field along β in $T_\lambda^\perp$. Then X and Y are smooth along β, as are a and b. Further, $a(0) = 1$, $b(0) = 0$, and $a'(0) = 0$, since 1 is the maximum value which the function a can attain.

Then

$$A\beta' = a\ \lambda X + b\ AY,$$

and AY is also orthogonal to T_λ, since $T_\lambda{}^\perp$ is invariant under A. Thus,

$$k_n(s) = \langle A\beta', \beta'\rangle = a^2\lambda + b^2\langle Y, AY\rangle.$$

Differentiating and evaluating at s = 0, we obtain $k_n'(0) = \lambda'(0)$. Thus $\kappa'(0) = \lambda'(0)$, as claimed.

Let p be the focal point $y + (1/\lambda)\xi(y)$. Let C be the osculating circle to the plane curve β at y, i.e. the circle through y centered at p. Using Taylor's formula and the Frenet equations, it is easy to show (see, for example, Goetz [1, p. 84]) that β crosses C at y unless $\kappa'(0) = 0$. Thus, if $\kappa'(0) \neq 0$, the function L_p does not have an extreme value at y, contradicting Lemma 1.24. We conclude that

$$X\lambda = \lambda'(0) = \kappa'(0) = 0.$$

Q.E.D.

From Lemma 7.1 and Theorem 6.2, we obtain immediately the following classification theorem of Banchoff [3] and Cecil [3].

<u>Theorem 7.2</u>: <u>Let M be a taut connected surface in E^3.</u>

(a) <u>if M is compact, then M is a Euclidean sphere or a ring cyclide.</u>

(b) <u>if M is non-compact, then M is a plane, a circular cylinder, or a parabolic ring cyclide.</u>

Combining this with the argument at the beginning of this section, we obtain the following classification of taut surfaces in E^m.

<u>Theorem 7.3</u>: <u>Let M be a taut connected surface substantially embedded in a Euclidean space E^m.</u>

(a) <u>if M is compact, then M is a Euclidean sphere or a ring cyclide in E^3, a spherical Veronese surface in E^5 or a compact surface in E^4 related</u>

to one of these by stereographic projection.

(b) if M is non-compact, it is a plane, a circular cylinder or a parabolic ring cyclide in E^3, or it is the image in E^4 of a punctured spherical Veronese surface under stereographic projection.

Conversely, all of the surfaces listed in (a) and (b) are taut.

In our proof of Theorem 6.16, we needed the fact that for a taut hypersurface, every principal curvature of multiplicity 1 is constant along its lines of curvature. We will require, however, the stronger hypothesis of tautness. The proof is a combination of the arguments given in Cecil-Ryan [2, pp. 187-188] and Pinkall [4].

The first step is to prove the following proposition which is of independent interest. It generalizes the classical result that if the curvature of a plane curve has non-vanishing derivative on a parameter interval, then the corresponding one-parameter family of osculating circles is nested one within another (see Stoker [1, p. 31]).

Proposition 7.4: Let M be a hypersurface in E^{n+1} and assume that a principal curvature λ has constant multiplicity 1. Suppose that λ and λ' are non-zero along a line of curvature γ. Then along γ, the family of curvature spheres determined by λ is nested.

Proof: By appropriate choice of unit normal and direction of arc-length parametrization, we may assume that $\lambda < 0$ and $\lambda' > 0$. Let s_1 and s_2 be any two parameter values with $s_1 < s_2$ and let p_1 and p_2 be the λ-focal points of $x_1 = \gamma(s_1)$ and $x_2 = \gamma(s_2)$, respectively. One easily computes that the arc-length of the evolute curve

$$\alpha(s) = \gamma(s) + \frac{1}{\lambda(s)}\,\xi(s)$$

from p_1 to p_2 is

$$\frac{1}{\lambda(s_1)} - \frac{1}{\lambda(s_2)} = \sigma - \rho,$$

where $\rho = d(x_1,p_1)$ and $\sigma = d(x_2,p_2)$, since $\lambda(s_1)$ and $\lambda(s_2)$ are both negative. We know that α is not a straight line segment because ξ is not constant along γ. Thus, $d(p_1,p_2) < \sigma - \rho$, and by the triangle inequality, the closed ball $B_\rho(p_1)$ is contained in the interior of the ball $B_\sigma(p_2)$.

Q.E.D.

<u>Lemma 7.5</u>: <u>Let M be a taut hypersurface in E^{n+1} and suppose that a principal curvature λ has constant multiplicity 1 in some open set U. Then λ is constant along its lines of curvature in U.</u>

<u>Proof</u>: Suppose that γ is a line of curvature on which the result does not hold. Then there exists a parametrized interval γ with $\lambda > 0$ and $\lambda' < 0$ as in Proposition 7.4. For each s in the parameter interval, let B_s be the closed ball of radius $1/|\lambda(s)|$ centered at the λ-focal point of $\gamma(s)$ and let

$$\beta(s) = \dim H_*(M \cap B_s;\ Z_2).$$

By tautness, $\beta(s)$ is finite. We will obtain a contradiction by showing that the integer-valued function $\beta(s)$ is strictly increasing. To show this, let s_1 and s_2 be any two parameter values with $s_1 < s_2$, and let B_1 and B_2 denote the corresponding closed balls centered at the λ-focal points of $\gamma(s_1)$ and $\gamma(s_2)$, respectively. We prove that the homomorphism

$$j:\ H_*(M \cap B_1) \to H_*(M \cap B_2)$$

induced by the inclusion $B_1 \subset B_2$ is injective but not surjective, and thus $\beta(s_1) < \beta(s_2)$.

The injectivity of j follows immediately from tautness, since the injective map

$$H_*(M \cap B_1) \to H_*(M)$$

factors through the sequence

$$H_*(M \cap B_1) \xrightarrow{j} H_*(M \cap B_2) \to H_*(M).$$

To show that j is not surjective, consider any parameter value s_0 with $s_1 < s_0 < s_2$. Let p_0 denote the λ-focal point of $\gamma(s_0)$ and let B_0 be the closed ball centered at p_0 of radius $1/|\lambda(s_0)|$. Let q be a point on the normal segment from $\gamma(s_0)$ to p_0 such that q is between p_0 and the next focal point (if any) of $(M, \gamma(s_0))$ on the segment. The point q can be chosen arbitrarily close to p_0. By Proposition 1.11, there exists a point p arbitrarily near to q (and hence to p_0) such that L_p is a Morse function having a non-degenerate critical point x arbitrarily near $\gamma(s_0)$ and no other critical points at the same level. Let $r = d(p,x)$. By Proposition 7.4, we have

$$B_1 \subset \text{int } B_0, \qquad B_0 \subset \text{int } B_2.$$

Since p and x can be taken arbitrarily near to p_0 and $\gamma(s_0)$, respectively, there exists $\delta > 0$ such that

$$B_1 \subset \text{int } B_{r-\delta}(p), \qquad B_{r+\delta}(p) \subset \text{int } B_2.$$

Let k be the index of L_p at x. Since M is taut, the k-th Betti number increases by one as the critical point x is passed, and thus the homomorphism

$$H_k(M \cap B_{r-\delta}(p)) \to H_k(M \cap B_{r+\delta}(p))$$

is injective but not surjective. Now the map j factors through the sequence of homomorphisms induced by inclusions

$$H_k(M \cap B_1) \to H_k(M \cap B_{r-\delta}(p)) \to H_k(M \cap B_{r+\delta}(p)) \to H_k(M \cap B_2).$$

By tautness, all of the maps are injective, but the middle one is not surjective. Hence, the map j is not surjective.

Q.E.D.

Finally, we recall from Section 1 that one can define taut compact subsets of E^m (or S^m) and not just taut embeddings of manifolds. An example is the set X obtained by deleting $k \leq \infty$ disjoint open round disks from a closed round disk D. Such a set is called a Swiss cheese, and if the round interiors are everywhere dense in D, it is called a limit Swiss cheese. Banchoff [3] showed that a compact connected X in E^2 has the STPP if and only if X is a point, a circle or a Swiss cheese. More recently, Kuiper [14] has shown that X in E^2 is taut if and only if X is a point, a circle or a limit Swiss cheese. This latter result follows from his lemma which states that if X in S^m is taut and has an interior point, then $X = S^m$.

The determination of taut subsets in E^3 is substantially more difficult, but Kuiper [14] has found:

Theorem 7.6: A taut compact connected ANR subset X of E^3 is a point, a circle, a round 2-sphere or a cyclide of Dupin.

8. HIGHER DIMENSIONAL TAUT HYPERSURFACES.

In Theorem 1.6 and in Section 2, we discussed taut embeddings of spheres. We now present some further classification results for taut, codimension one embeddings of manifolds with relatively simple homology. We begin with two results (Theorems 8.1 and 8.3) due to Carter and West [1, p. 714-715].

Theorem 8.1: Let $f:M \to E^{n+1}$ be a taut embedding of a connected, non-compact n-manifold with $H_k(M;Z_2) = Z_2$ for some k, $0 < k < n$, and $H_i(M;Z_2) = 0$ for $i \neq 0, k$. Then M is diffeomorphic to $S^k \times E^{n-k}$ and f is a standard product embedding.

A key step in the proof is to show that the complement of the focal set in E^{n+1} is path-connected. We remark that this is true in all examples of taut submanifolds for which we have found the focal set, and it is also important in the proof of another main result (Theorem 8.6) of this section. For Theorem 8.1, this fact is a fairly easy consequence of the following lemma.

We let Γ denote the set of critical points of the normal exponential map $F:NM \to E^m$. Then, of course, the focal set is $F(\Gamma)$.

Lemma 8.2: Let $f:M \to E^m$ be a taut embedding of a connected manifold. Suppose that for each $x \in M$, there is at most one focal point of (M,x) on each normal ray to $f(M)$ at $f(x)$. Then $f(M)$ is disjoint from the focal set $F(\Gamma)$ and the complement of $F(\Gamma)$ in E^m is path-connected.

Proof: Suppose that p is in E^m and that L_p has a non-degenerate minimum at $x \in M$. By Lemma 1.24(a), L_p has a strict absolute minimum on M at x. If p is a focal point of M, then $p = F(y,\eta)$ for $(y,\eta) \in \Gamma$. By hypothesis, p is the only focal point of (M,y) on the normal ray to $f(M)$ at $f(y)$ in the direction η. Again by Lemma 1.24(a), L_p has an absolute minimum at y, a contradiction. In particular, this applies when $p = f(x)$, and we conclude that $f(M) \cap F(\Gamma) = \emptyset$.

Now suppose that p and q are two points in the complement of $F(\Gamma)$ in E^m. Let x and y be the respective points in M where L_p and L_q achieve their strict absolute minima. By the work in the paragraph above, the closed line segments from p to $f(x)$ and from q to $f(y)$ are disjoint from $F(\Gamma)$. We now construct a path from p to q in $E^m - F(\Gamma)$ as follows. First travel along the line segment from p to $f(x)$. Then take any path in $f(M)$ from $f(x)$ to $f(y)$, and finally, traverse the line segment from $f(y)$ to q.

Q.E.D.

Proof of Theorem 8.1: By the hypotheses on the homology of M, tautness implies that every non-degenerate L_p function has exactly two critical points on M, one of index 0 and one of index k. Further, if L_p is not a Morse function, but L_p has a non-degenerate critical point of index j, then $j = 0$ or k by Proposition 1.11. Thus, there is at most one focal point along each normal ray, which must have multiplicity k. By Lemma 8.2, $\Sigma = E^{n+1} - F(\Gamma)$ is path-connected. The restriction of F to $F^{-1}\Sigma$ is a double covering, since F is an immersion on $NM-\Gamma$, and since for each $p \in \Sigma$, the function L_p has exactly two critical points. Since a taut embedding is a proper map, $f(M)$ is a closed set in E^{n+1} and is therefore orientable with unit normal field ξ. By identifying $(x,t\xi)$ with (x,t), the normal bundle is diffeomorphic to $M \times R$. Let

$$N^+ = \{(x,t) \mid t > 0\} \text{ and } N^- = \{(x,t) \mid t < 0\}$$

in $M \times R$, and let $M_0 = M \times \{0\}$, the zero-section of NM. For each (x,t) in $F^{-1}\Sigma$, let $p = F(x,t)$, and note that L_p has a critical point at x, whose index will be denoted by Index (x,t). This index function is continuous and is, therefore, constant on any path-component of $F^{-1}\Sigma$. Since Σ is connected, we conclude that $F^{-1}\Sigma$ consists of just two components V_0 and V_k, on which the index function is constantly equal to 0 and k, respectively. Since $M_0 \cap F^{-1}\Sigma$ lies in V_0, the component V_k lies entirely in either N^+ or N^-, say N^+. Now for each $x \in M$, there is at most one focal point on the normal line to $f(M)$ at $f(x)$, which must lie on the ray determined by N^+. On the other hand, there must be at least one focal point on each normal line, or $f(M)$ would lie in a hyperplane by Corollary 1.26. In summary, every normal line contains exactly one focal point of multiplicity k. Therefore, at each point x, there are exactly two principal curvatures, λ of multiplicity k which is never zero, and μ of multiplicity n-k which is identically zero.

We take advantage of Theorem 6.2 to complete the proof in a different way from Carter and West. If $k > 1$, then λ is constant along the leaves of its principal foliation by Theorem 4.4. If $k = 1$, then tautness implies that λ is constant along its lines of curvature by Lemma 7.1. We now invoke Theorem 6.2 and conclude that $f(M)$ is a standard product $S^k \times E^{n-k}$.

Q.E.D.

Theorem 8.3: Let $f: M^n \to E^{n+q}$ be a substantial taut embedding of a connected non-compact manifold whose Z_2-Betti numbers satisfy $\beta_k(M) = j > 0$ for some k, $\frac{n}{2} < k < n$, and $\beta_i(M) = 0$ for $i \neq 0,k$. Then $q = 1$, $j = 1$, and f embeds M as a standard product $S^k \times E^{n-k}$ in E^{n+1}.

Proof: As in the previous proof, the index of any non-degenerate critical point of any L_p function is 0 or k. Thus, every focal point has multiplicity k. The sum of the multiplicities of the focal points on any normal line cannot exceed n. Thus, since $2k > n$, there is at most one focal point on every normal line. On the other hand, since f is substantial, there exists at least one focal point on every normal line by Corollary 1.26. Again as in the preceding proof, we conclude that for every unit normal ξ, the shape operator A_ξ has exactly two eigenvalues, λ of multiplicity k which is never zero, and μ of multiplicity n-k which is identically zero. If the codimension $q > 1$, then the unit normal bundle BM is connected and the continuous function λ does not change sign on BM. This contradicts the fact that $A_{-\xi} = -A_\xi$. Thus, $q = 1$, and one concludes as in the previous theorem that f(M) is a standard product embedding of $S^k \times E^{n-k}$.

Q.E.D.

We note that the taut substantial embedding of the Moebius band $M^2 = P^2 - \{p\}$ into E^4 obtained from a Veronese surface V^2 in S^4 by stereographic projection with respect to a pole on V^2 shows that the hypothesis $k > \frac{n}{2}$ in Theorem 8.3 and the hypothesis that the codimension is one in Theorem 8.1 are both necessary. The Veronese embeddings of the other projective planes (see Theorem 9.4 of Chapter 1) yield further examples.

Our next theoem generalizes the classification of taut surfaces in E^3 (Theorem 7.2). In 1971, using a rather lengthy argument, Carter and West [1, pp. 712-714] showed that if M^{2k} is a taut compact (k-1)-connected hypersurface in E^{2k+1}, then $H_k(M;Z)$ is either 0 or $Z \oplus Z$. In 1978, Cecil and Ryan [2] proved that if M in E^{n+1} is a taut compact hypersurface with the same integral homology as $S^k \times S^{n-k}$, $1 \leq k \leq \frac{n}{2}$, then M is a ring cyclide. C.S. Chen [3] using tightness arguments and top-sets, independently proved a similar result. His theorem differs, in that he only assumes that M is tight and lies in an ovoid in E^{n+2}. However, he excludes the case where $k/(n-k)$ is equal to 2 or $\frac{1}{2}$.

We first handle the case $k = n-k$. The proof given in Cecil-Ryan [2, p. 184] actually works under the following weaker hypotheses.

<u>Theorem 8.4</u>: <u>Let M^{2k} be a taut connected hypersurface in E^{2k+1} such that $H_i(M;Z_2) = 0$ for i not equal to 0, k, 2k. Then M is a Euclidean sphere or hyperplane, a standard product $S^k \times E^k$, a ring cyclide or a parabolic cyclide</u>.

<u>Proof</u>: For $k = 1$, there is no restriction on the homology of M, and the result is just Theorem 7.2. By Corollaries 1.26 and 1.27, if a taut hyper-

surface has one umbilic point, then it is totally umbilic. Assume now that M has no umbilics. Suppose a function L_p has a non-degenerate critical point of index j at $x \in M$. If L_p is a Morse function, then tautness and the assumption on the homology of M imply that j is equal to 0, k or 2k. The same conclusion holds if L_p is not a Morse function by Proposition 1.11. On the other hand, by the index theorem, j is equal to the number (counting multiplicities) of focal points of (M,x) on the line segment from p to x. From the restriction on the values of j and the fact that there are no umbilic points, we see that the first focal point along either normal ray to M at x must have multiplicity k. If there is a second focal point, it too must have multiplicity k. Finally, there must be a focal point on every normal line, since there are no planar points. We conclude that at every point $x \in M$, there are exactly two distinct principal curvatures, each of multiplicity k. Since $k > 1$, these are constant along the leaves of their corresponding principal foliations by Theorem 4.4, and the result now follows from Theorem 6.2.

Q.E.D.

By standard Morse theory, one sees quickly that the hypotheses of Theorem 8.4 imply that M^{2k} is (k-1)-connected. Specifically, it was rather immediate to show in the proof of Theorem 8.4 that tautness and the assumption on the homology of M imply that each non-degenerate L_p has only critical points of index 0, k, and 2k. Thus applying basic Morse theory to any choice of non-degenerate L_p function, we get that M^{2k} has the homotopy type of a CW-complex with cells of dimension 0, k and 2k only. In particular, the (k-1)-skeleton of M^{2k} is just the 0-skeleton, and M^{2k} is (k-1)-connected (see, for example, Maunder [1, p. 297]). Thus, the assumption on the homology of M^{2k} in Theorem 8.4 is often replaced in this context by the hypothesis that M^{2k} is (k-1)-connected.

Remark 8.5: Thorbergsson [2] has recently generalized Theorem 8.4 to higher codimension. He has shown that if M^{2k} is a compact, (k-1)-connected (not k-connected) taut submanifold of E^m which does not lie in any umbilic hypersurface of E^m then either:

(i) $m = 2k+1$ and M^{2k} is a cyclide of Dupin diffeomorphic to $S^k \times S^k$ (as in

Theorem 8.4) or

(ii) $m = 3k+1$ and M^{2k} is diffeomorphic to one of the projective planes RP^2, CP^2, QP^2 or OP^2.

He has a similar result for non-compact submanifolds of E^m and for taut compact (k-1)-connected hypersurfaces in hyperbolic space H^{2k+1}. See also Remark 9.5 of Chapter 1.

The proof in the case $k \neq n-k$ is substantially more difficult.

Theorem 8.6: A taut compact hypersurface M in E^{n+1} with the same Z_2-homology as $S^k \times S^{n-k}$ is a ring cyclide.

Proof: We assume $k < n-k$, as the case $k = n-k$ was handled in the previous theorem. Every non-degenerate L_p has 4 critical points with respective indices 0, k, n-k, n. Using the index theorem as in the previous result, one can conclude that there are at most three distinct principal curvatures at each point. On the open set U where there are 3 distinct principal curvatures, there are smooth principal curvature functions

$$\lambda_1 > \lambda_2 > \lambda_3,$$

which must have respective multiplicities k, n-2k, k. We now begin a series of lemmas which eventually lead to the conclusion that U is empty, and thus there are only two distinct principal curvatures at each point. The most important lemma is the following.

Lemma 8.7: The number of distinct principal curvatures is a constant which is either 2 or 3.

Proof: Assume that the open set U on which there are 3 distinct principal curvatures is neither empty nor equal to M. Since M is an embedded compact hypersurface, the concept of inner and outer normal ray is well-defined. Pick any x on the boundary of U in M. There must be exactly two distinct principal curvatures at x, since U is open and since there are no umbilic points. By Corollary 1.25, there must be a focal point on every inner normal ray. By taking an inversion of M, if necessary, we can assume that the first focal point on the inner normal ray to M at x has multiplicity n-k. Of course, an inversion preserves tautness and the multiplicities of the principal curvatures. The proof of the lemma will now be accomplished by producing a non-degenerate distance function which has at least five critical points.

Let $p = x + \rho\xi$, where ξ is the globally defined unit inner normal field and $\rho = 1/\lambda$, where λ is the largest principal curvature at x. We have arranged that p has multiplicity n-k. By Lemma 1.24, the function L_p has an absolute minimum value $\alpha = \rho^2$ at x. Hence, p must lie inside M (i.e. in the bounded component of the complement of M), since the segment from x to p cannot intersect M. Let y be a point in M where L_p has an absolute maximum. Since M is not a Euclidean sphere we know that $L_p(x) < L_p(y)$. By the index theorem, p must lie on the inner normal to M at y, since the inner normal contains at least one focal point of (M,y) by Corollary 1.25. Let q be the first focal point of (M,y) on the inner normal ray. Then $q \neq p$ since L_q has an absolute minimum at y. Let $\gamma = L_q(y)$.

We first show that the sets

$$M_\alpha(L_p) = \{z \in M | L_p(z) \leq \alpha\}$$

and $M_\gamma(L_q)$ are disjoint. If z is a point in the intersection, then both L_p and L_q have an absolute minimum at z. Since p and q are inside M, and the segments [z,p] and [z,q] contain no points of M (other than z), these segments must lie on the inner normal ray to M at z. Further, the normal lines at z and y must coincide. We know $z \neq y$ since L_p does not have a minimum at y. If $z = x$, then p is the first focal point on the inner normal ray to M at z. If $z \neq x$, then L_p does not have a strict absolute minimum at z since $L_p(x) = L_p(z)$. By Lemma 1.24, L_p has a degenerate minimum at z, and again we conclude that p is the first focal point on the inner normal to M at z. Thus, q lies beyond the first focal point on the normal ray, and it is impossible for L_q to have a minimum at z, a contradiction.

Since $M_\alpha(L_p)$ and $M_\gamma(L_q)$ are compact, there exists $\varepsilon > 0$ so that $M_{\alpha+\varepsilon}(L_p)$ and $M_{\gamma+\varepsilon}(L_q)$ are also disjoint. Since the focal set has measure zero in E^{n+1}, one can find $p' \in E^{n+1}$ and $r > 0$ such that $L_{p'}$ is non-degenerate and r is a non-critical value of $L_{p'}$ with

$$M_\alpha(L_p) \subset M_r(L_{p'}) \subset M_{\alpha+\varepsilon}(L_p).$$

Recall that x was chosen to have an arbitrarily close neighbor at which there are three distinct principal curvatures. Thus, there are points a, b in E^{n+1} arbitrarily near p such that L_a and L_b are Morse functions with respective critical points u and v with indices k and n-k, arbitrarily close to x. Let $\sigma = L_a(u)$ and $\delta = L_b(v)$. The points can be chosen so that $M_\sigma(L_a)$ and $M_\delta(L_b)$ are both contained in $M_r(L_{p'})$. The function L_a has at least two critical points in $M_\sigma(L_a)$, a minimum and a critical point of index k at u. By tautness and the homology of M, we conclude that the Z_2-Betti numbers

$$\beta_0(M_\sigma(L_a)) = 1, \qquad \beta_k(M_\sigma(L_a)) = 1.$$

Similarly, we have

$$\beta_0(M_\delta(L_b)) = 1, \qquad \beta_{n-k}(M_\delta(L_b)) = 1.$$

We have the inclusion maps

$$M_\sigma(L_a) \xrightarrow{i} M_r(L_{p'}) \xrightarrow{j} M.$$

By tautness, the induced maps on homology j_* and $(j \circ i)_*$ are injective, and so i_* is also injective. Thus $\beta_k(M_r(L_{p'})) = 1$. The same argument applied to $M_\delta(L_b)$ gives that $\beta_{n-k}(M_r(L_{p'})) = 1$. By the Morse inequalities, $L_{p'}$ has at least 3 critical points on $M_r(L_{p'})$.

On the other hand, we know that $M_r(L_{p'})$ and $M_{\gamma+\varepsilon}(L_q)$ are disjoint. Let q' be a point just beyond q on the normal line to M at y and s a real number such that $M_s(L_{q'}) \subset M_{\gamma+\varepsilon}(L_q)$. $L_{q'}$ has a non-degenerate critical point of index h at y, where h is the multiplicity of the focal point q. Thus h is either k or n-k. By tautness, $\beta_h(M_s(L_{q'})) = 1$. For

$$M_r(-L_{p'}) = \{z \mid L_{p'}(z) \geq r\},$$

we have $M_s(L_{q'}) \subset M_r(-L_{p'}) \subset M$. As before, tautness implies that $\beta_h(M_r(-L_{p'})) = 1$, and the Morse inequalities imply that the function $-L_{p'}$ has critical points of index 0 and h on $M_r(-L_{p'})$. Thus, $L_{p'}$ has critical points of index n and n-h on $M_r(-L_{p'})$. These two critical points are distinct from the three in $M_r(L_{p'})$ since r is not a critical value, and so $L_{p'}$ has at least five critical points, contradicting tautness.

Q.E.D.

<u>Lemma 8.8</u>: <u>Each sheet of the focal set of M is an immersed submanifold with codimension greater than 1. Thus, the complement of the focal set is path-connected.</u>

<u>Proof</u>: By Lemma 8.7, a given principal curvature λ has constant multiplicity $\nu \geq 1$. If $\nu > 1$, then the corresponding sheet of the focal set $f_\lambda(M)$ is an immersed submanifold of dimension $n-\nu$ (codimension $\nu+1$) by Theorem 4.6. If $\nu = 1$, then tautness implies that λ is constant along its lines of curvature by Lemma 7.5. By Theorem 4.10, $f_\lambda(M)$ is an immersed (n-1)-dimensional manifold.

Finally, we show that $\Sigma = E^m - F(\Gamma)$ is path-connected as follows. We have shown that $F(\Gamma)$ is the union of the images of at most 3 immersions $g_i : M_i \to E^{n+1}$, where dimension $M_i = n - \nu_i$ for ν_i the multiplicity of λ_i. Let p and q be any two points in Σ. Cover each M_i by a countable number of compact $(n-\nu_i)$-disks D_{ij} such that the restriction of g_i to each D_{ij} is an embedding. Transversality (see, for example, Hirsch [1, p. 74]) implies that there is a path from p to q which is disjoint from $g_i(D_{ij})$ for all i,j, and so Σ is path-connected.

Q.E.D.

We complete the proof of Theorem 8.6 following the approach of Carter and West used in Theorem 8.1.

<u>Lemma 8.9</u>: <u>There exist exactly two principal curvatures on M and they are constant along their principal foliations. Hence M is a ring cyclide.</u>

<u>Proof</u>: In the previous lemma, we showed that the principal curvatures are

constant along their principal foliations. It remains to show that there are two not three principal curvatures. The normal bundle NM is diffeomorphic to $M \times R$ and we write $NM = N^+ \cup N^- \cup M_0$ as in Theorem 8.1. Let Σ be the complement of the focal set in E^{n+1} and $V = F^{-1}\Sigma$. Then $F:V \to \Sigma$ is a four-fold covering map with $F^{-1}(p)$ consisting of four points in V corresponding to the four critical points of L_p. As in Theorem 8.1 the index function is locally constant on V, and induces the decomposition of V into disjoint open sets

$$V = V_0 \cup V_k \cup V_{n-k} \cup V_n.$$

Since Σ is connected by Lemma 8.8, each V_i is connected. Since $M_0 \cap V$ is contained in V_0, the connected set V_k must lie entirely in N^+ or N^-. Now suppose there are three distinct principal curvatures $\lambda_1 > \lambda_2 > \lambda_3$ with respective multiplicities k, n-2k, k. Since M is not a convex hypersurface, there exists a point x with focal points on both inner and outer normal rays. But the first focal point on each normal ray at x must have multiplicity k, and we can construct Morse functions to show that both $V_k \cap N^+$ and $V_k \cap N^-$ are non-empty, a contradiction. Now invoke Theorem 6.2 to get that M is a ring cyclide.

Q.E.D.

This completes the proof of Theorem 8.6. In Cecil-Ryan [5], we found a similar characterization of taut conformally flat hypersurfaces in Euclidean space. Recall that a Riemannian manifold (M,g) is said to be <u>conformally flat</u> if every point has a neighborhood conformal to an open set in Euclidean space. The basic starting point of research on conformally flat hypersurfaces is the following pointwise result of Schouten [1]: Let M be a hypersurface immersed in Euclidean space E^{n+1}, $n \geq 4$. Then M is conformally flat in the induced metric if and only if at least n-1 of the principal curvatures coincide at each point of M. Note that this characterization fails when $n = 3$. An example of a conformally flat hypersurface in E^4 with 3 distinct principal curvatures was given by Lancaster [1, p. 6]. Using the result of Schouten, Theorem 6.2 and some basic results on tautness, one obtains the following result.

Theorem 8.10: Let M^n, $n \geq 4$, be a connected manifold tautly embedded in E^{n+1}. Then M is conformally flat in the induced metric if and only if it is one of the following:

(1) a hyperplane or round sphere;
(2) a cylinder over a circle or an (n-1)-sphere;
(3) a ring cyclide (diffeomorphic to $S^1 \times S^{n-1}$);
(4) a parabolic ring cyclide (diffeomorphic to $S^1 \times S^{n-1} - \{p\}$).

Proof: If M has a umbilic point, them M is totally umbilic by Corollaries 1.26 and 1.27, and hence M is a hyperplane or sphere. Otherwise, M has two distinct principal curvatures λ and μ with respective multiplicities n-1 and 1 at each point. Since $n-1 > 1$, λ is constant along the leaves of its principal foliation by Theorem 4.4. On the other hand, tautness implies that μ is constant along its lines of curvature by Lemma 7.1. The result now follows from Theorem 6.2.

Q.E.D.

By slightly modifying the arguments Lemma 8.8, we can obtain the following result of independent interest.

Theorem 8.11: The complement of the focal set of a Dupin hypersurface is connected.

9. TOTALLY FOCAL EMBEDDINGS.

This section is based on three papers of Carter and West [2]-[4]. A proper embedding $f:M \to E^m$ of a smooth manifold is called totally focal if every distance function L_p is either non-degenerate or has only degenerate critical points. In terms of the normal exponential map $F:NM \to E^m$, this can be expressed by the condition $\Gamma = F^{-1}(F\Gamma)$, where Γ is the set of critical points of F. Immediate examples are Euclidean spheres and planes. One easily shows that a product of totally focal embeddings is totally focal, and thereby obtains the standard products of spheres and planes. In their first paper on the subject, Carter and West [2] found that a totally focal hypersurface M in E^{n+1} must be a Euclidean sphere or plane, or a standard product $S^k \times E^{n-k}$. This is precisely the list of all complete isoparametric hypersurfaces in E^{n+1} (see, for example, Nomizu [2]), although Carter and West do not need to use this characterization in their proof.

In [6], Cecil and Ryan deduced from Münzner's work that all isoparametric hypersurfaces M in S^{n+1} are taut and, moreover, totally focal in S^{n+1}. Carter and West [4] showed conversely that a totally focal hypersurface in S^{n+1} is isoparametric. This result does not depend on Münzner's work.

In their second paper on the subject, Carter and West [3] showed that a totally focal embedding of S^1 in E^m must be a plane circle, while a totally focal embedding of S^2 in E^4 must be a round sphere in some $E^3 \subset E^4$. Even the determination of all totally focal embeddings of S^2 into E^5 is now open. On the other hand, every known totally focal embedding is taut, and in fact, totally focal is a stronger requirement than taut for hypersurfaces. Certainly, a major question to be resolved in this theory is whether a totally focal embedding must be taut. If so, many other questions, e.g. the determination of totally focal embeddings of spheres, would be solved.

The arguments in this section are in many ways similar to those for tautness. In particular, the complement of the focal set of a totally focal embedding is connected. Moreover, the restriction of F to $NM-\Gamma$ is a covering of $E^m-F(\Gamma)$ and covering space arguments form a substantial portion of the proofs in this section. In the determination of totally focal hypersurfaces in E^{n+1} and S^{n+1}, one eventually proves as a key step that such a hypersurface must be 0-taut, i.e. every non-degenerate L_p has exactly one minimum. This again suggests that a totally focal embedding must be taut.

We begin with the formal definition.

Definition 9.1: A smooth proper embedding $f:M \to E^m$ of a connected manifold is called totally focal if $\Gamma = F^{-1}(F\Gamma)$, where Γ is the set of critical points of the normal exponential map F.

Next we note two elementary ways to produce new examples from known examples of totally focal embeddings.

Proposition 9.2: Let $f:M \to E^m$ be totally focal. Then the embedding $\bar{f}:M \to E^m \times E^k = E^{m+k}$ given by $\bar{f}(x) = (f(x),0)$ is totally focal.

Proof: Clearly the normal bundle of $\bar{f}$ is $NM \times E^k$ and the normal exponential map $\bar{F}$ is $F \times I$, where I is the identity on E^k. A simple calculation shows that the critical set $\bar{\Gamma}$ is $\Gamma \times E^k$, and thus since f is totally focal, we have

$$\bar{F}^{-1}(\bar{F}\bar{\Gamma}) = \bar{F}^{-1}(F(\Gamma) \times E^k) = F^{-1}(F\Gamma) \times E^k = \Gamma \times E^k = \bar{\Gamma}.$$

Q.E.D.

Proposition 9.3: Let $f_i:M_i \to E^{m_i}$ $(i = 1,2)$ be proper embeddings. Then the product embedding

$$f_1 \times f_2 : M_1 \times M_2 \to E^{m_1} \times E^{m_2} = E^{m_1+m_2}$$

is totally focal if and only if both f_1 and f_2 are totally focal.

Proof: Write F_1, F_2 and F for the normal exponential maps of f_1, f_2 and $f_1 \times f_2$, respectively, and use a similar notation for the corresponding normal bundles and critical sets. Then $N = N_1 \times N_2$, $F = F_1 \times F_2$, and so $\Gamma = (\Gamma_1 \times N_2) \cup (N_1 \times \Gamma_2)$. Thus

$$F^{-1}(F\Gamma) = (F_1^{-1}(F_1\Gamma_1) \times N_2) \cup (N_1 \times F_2^{-1}(F_2\Gamma_2))$$

and the theorem is clearly true.

Q.E.D.

Theorem 9.4: Let $f:M \to E^m$ be a totally focal embedding. Then $E^m - F(\Gamma)$ is connected and contains $f(M)$.

Proof: First, since F has no critical points on the zero-section $M \times \{0\}$ in NM, the condition $\Gamma = F^{-1}(F\Gamma)$ implies that $f(M)$ is disjoint from $F(\Gamma)$. Now let p be a point in $\Sigma = E^m - F(\Gamma)$. Since f is proper, L_p has an absolute minimum at some point $x \in M$. By the index theorem, if q is a point on the segment from p to $f(x)$, then L_q has a non-degenerate minimum at x. Since f

is totally focal, this means that q is not in the focal set. Now let p' be any other point in Σ, and suppose $L_{p'}$ has an absolute minimum at a point y. We construct a path from p to p' in Σ as follows. First travel along the line segment from p to f(x); then take any path in f(M) from f(x) to f(y), and finally, traverse the segment from f(y) to p'.

Q.E.D.

We now wish to show that the restriction of F to NM−Γ is a covering map. Actually, we only use the hypothesis that f is totally focal in order to get the base space $\Sigma = E^m - F(\Gamma)$ to be connected. Thus, Theorem 9.6 can be stated in a more general setting than totally focal embeddings. First we need the following lemma.

<u>Lemma 9.5</u>: <u>Let $f:M \to E^m$ be a proper immersion of a manifold. Let γ be a smooth path in $\Sigma = E^m - F(\Gamma)$ with initial point</u> p <u>and let $\zeta \in NM$ such that $F(\zeta) = p$. Then there exists a unique smooth path $\tilde{\gamma}$ in $F^{-1}\Sigma$ with initial point ζ such that $F \circ \tilde{\gamma} = \gamma$.</u>

<u>Proof</u>: Let I be the domain of the curve $\gamma(t)$. Then I can be taken to have the form $[0,b]$, $b > 0$, or $[0,\infty)$. Let J be the subinterval of I over which γ can be uniquely lifted. We show that J is both open and closed in I, and is, therefore, I itself. Since F is a local diffeomorphism, J is clearly open and non-empty. Assume now that $\tilde{\gamma}$ has been defined on $[0,\tau)$. We must show that $\tilde{\gamma}(\tau)$ can be uniquely defined.

Write $\tilde{\gamma}(t) = (x_t, v_t)$ for $t \in [0,\tau)$ and let ℓ be the length of the curve $\gamma(t)$, $0 \leq t \leq \tau$. We first show that $|v_t| \leq |v_0| + \ell$ for $0 \leq t < \tau$. Define maps $\tilde{f}$ and ν from NM to E^m by $\tilde{f}(x,v) = f(x)$ and $\nu(x,v) = v$. Then $F = \tilde{f} + \nu$ so $F_* = \tilde{f}_* + \nu_*$. Note that for any value of t, $\tilde{f}_*(\tilde{\gamma}'(t))$ is Euclidean parallel to a tangent vector to f(M) at $f(x_t)$, while $\nu(\tilde{\gamma}(t)) = v_t$ is parallel to a normal vector to f(M) at $f(x_t)$. Thus

$$\langle \tilde{f}_* \tilde{\gamma}', \nu \tilde{\gamma} \rangle = 0$$

for all t in $[0,\tau)$. Therefore, on any interval where $\nu\tilde{\gamma}$ does not vanish, we have

$$\langle F_*\tilde{\gamma}', \nu\tilde{\gamma}\rangle = \langle \hat{f}_*\tilde{\gamma}' + \nu_*\tilde{\gamma}, \ \nu\tilde{\gamma}\rangle = \langle \nu_*\tilde{\gamma}', \nu\tilde{\gamma}\rangle$$

$$= \frac{1}{2}\frac{d}{dt} |\nu\tilde{\gamma}|^2 = |\nu\tilde{\gamma}| \ |\nu\tilde{\gamma}|'.$$

This and the Cauchy-Schwarz inequality for $|\langle F_*\tilde{\gamma}', \nu\tilde{\gamma}\rangle|$ imply that

$$| \ |\nu\tilde{\gamma}|'| \leq |F_*\tilde{\gamma}'| = |\gamma'|.$$

For any fixed t in $[0,\tau)$, let a be the largest number in $[0,t]$ for which $\nu\tilde{\gamma} = 0$. If $\nu\tilde{\gamma}$ is never zero on $[0,t]$, let $a = 0$. In either case, $|v_a| \leq |v_0|$. Using the computations above, we have

$$||v_t| - |v_a|| = |\int_a^t |\nu\tilde{\gamma}|' \ | \leq \int_a^t ||\nu\tilde{\gamma}|' \ | \leq \int_a^t |\gamma'| \leq \ell.$$

Hence, $|v_t| \leq |v_0| + \ell$. Next, since $\gamma([0,\tau])$ is compact, there exists some constant $C > 0$ such that $|\gamma(t)| \leq C$ for all $t \in [0,\tau]$. Hence, since $F(\tilde{\gamma}(t)) = f(x_t) + v_t$,

$$|f(x_t)| = |\gamma(t)-v_t| \leq |\gamma(t)| + |v_t| \leq C + |v_0| + \ell = K.$$

Let B be the closed ball of radius K centered at the origin in E^m. Since f is proper, $f^{-1}B$ is compact, and we have shown that for each $t \in [0,\tau)$, $\tilde{\gamma}(t)$ lies in the set

$$W = \{(x,\eta) | x \in f^{-1}B \text{ and } |\eta| \leq K\},$$

which is compact in NM. Thus, there is a sequence $\{t_i\}$, of points in $[0,\tau)$ converging to τ such that $\{\tilde{\gamma}(t_i)\}$ converges to a point (x_1,v_1) in W. Since F is continuous, $F(x_1,v_1) = \gamma(\tau)$, so $(x_1,v_1) \in F^{-1}\Sigma$. There exists a neighborhood U of (x_1,v_1) on which F restricts to a diffeomorphism. We extend $\tilde{\gamma}$

to $[0,\tau)$ by defining $\tilde{\gamma}$ on U to be equal to $F|_U^{-1}(\gamma)$. This agrees with the previous definition of γ, since if t_i is some point in the above sequence which lies in U, then both definitions of $\tilde{\gamma}$ give the unique lift of γ on $[t,\tau)$. Clearly, we have now extended $\tilde{\gamma}$ to $[0,\tau]$, as required.

Q.E.D.

Theorem 9.6: Let $f:M \to E^m$ be a proper smooth immersion of a connected manifold such that the complement Σ of the focal set is connected. Then the restriction of the normal exponential map F to $F^{-1}\Sigma$ is a covering map.

Proof: Let p be an arbitrary point in Σ and let B be an open ball in E^m centered at p such that B lies in Σ. We will show that each path-component of $F^{-1}B$ is mapped diffeomorphically by F onto B. Each point $u \in F^{-1}(p)$ determines a path-component U, namely the set of points of $F^{-1}B$ which can be reached by starting at u and following the unique (by Lemma 9.5) lift of a line segment in B with initial point p. Clearly $F^{-1}B$ is the disjoint union of the sets U.

We now complete the proof by showing that each U is open in $F^{-1}\Sigma$. Let v be an arbitrary point of U with $F(v) = q$. Let γ be the line segment $[p,q]$ and let $\tilde{\gamma}$ be its unique lift beginning at u and thus ending at v. For each t in the parameter interval, there is a neighborhood V_t of $\tilde{\gamma}(t)$, open in $F^{-1}\Sigma$ on which F restricts to a diffeomorphism onto an open cube in E^m centered at $\gamma(t)$. The curve $\tilde{\gamma}$ is covered by a finite number of these open sets, and we let V denote their union. Since each point in F(V) can have at most one pre-image in each V_t, it follows that the restriction of F to V is a covering map. Further since F(V) is contractible, it is a diffeomorphism. Now choose an open neighborhood Q of q in F(V) such that all line segments $[x,p]$ for $x \in Q$ lie in F(V). Let $W = F^{-1}Q \cap V$. For each $w \in W$, the segment from p to $x = F(w)$ is covered by a path in V. Since u is the only point of $F^{-1}(p)$ in V, we conclude that $w \in U$. Thus, U contains the open neighborhood W of the point q, and U is open.

Q.E.D.

From the theory of covering spaces, we obtain immediately,

Corollary 9.7: *Let $f:M \to E^m$ be a smooth immersion of a compact connected manifold such that the complement Σ of the focal set is connected. Then all non-degenerate L_p functions have the same number of critical points.*

Proof: Let $p \in \Sigma$. Since M is compact, the number of critical points of L_p is finite and equal to the number of points in $F^{-1}(p)$. The result now follows from Theorem 9.6.

Q.E.D.

In the case where $f:M \to E^m$ is totally focal, we have by definition that $F^{-1}\Sigma = NM-\Gamma$. By Theorems 9.4 and 9.6, the restriction of F to $NM-\Gamma$ is a covering map, as is the restriction of F to any connected component U of $NM-\Gamma$. We will use the notation Fr U to denote the frontier or topological boundary of the set U, and reserve the term "boundary" for the manifolds with boundary which appear later in this section.

Lemma 9.8: *Let $f:M \to E^m$ be a totally focal embedding and U a connected component of $NM-\Gamma$. Then $F(\text{Fr } U) = F(\Gamma)$.*

Proof: Let $p \in F(\Gamma)$ and let x be a point in M where L_p attains its absolute minimum value. Then f(M) lies outside the open ball centered at p through x. The same is true for any point q on the segment $\gamma = [f(x),p]$, i.e. L_q also has an absolute minimum at x. If q were a focal point of (M,x), then $L_{q'}$ would not have a minimum at x for q' between q and p, a contradiction. Thus, there are no focal points on the segment other than p. Take a point $u \in U$ with $F(u) = f(x)$; such a point exists since $F|_U$ is a covering of Σ. By Lemma 9.5, we can lift the path $\gamma:[0,1) \to \Sigma$ to a path $\tilde{\gamma}:[0,1) \to U$ beginning at u. Now apply the method of Lemma 9.5 to produce a point $\hat{u}$ in $\bar{U}$ such that $F(\hat{u}) = \gamma(1) = p$. Since $p \in F(\Gamma)$, and $F^{-1}(F\Gamma) = \Gamma$, we have that $\hat{u} \in \bar{U} \cap \Gamma = \text{Fr } U$.

Q.E.D.

Remark 9.9: All that we have done so far holds for totally focal embeddings into the sphere as well as into Euclidean space. We now prove some results which are particular to the Euclidean case, and then handle the appropriate modifications in the spherical case.

If $(x,v) \in NM-\Gamma$ and $F(x,v) = p$, then L_p has a non-degenerate critical point at x, whose index we denote by Index (x,v). This index function is continuous and is thus constant on any connected component of $NM-\Gamma$. There exists a component U of $NM-\Gamma$ on which the index function takes its maximal value k. The value k will be $n = \dim M$ if M is compact, but otherwise it might be less than n. We will assume $k > 0$, for otherwise $f(M)$ is just an n-plane in E^m.

<u>Lemma 9.10: Let $f:M \to E^m$ be a totally focal embedding and let U be a connected component of $NM-\Gamma$ of maximal index. Then $\bar{U}$ is homeomorphic to $\mathrm{Fr}\, U \times [1,\infty)$ and Fr U is connected.</u>

<u>Proof</u>: Let $(x,v) \in U$. Assume the index function has the maximal value $k > 0$ on U. Then k is equal to the sum of the multiplicities of the positive eigenvalues of A_v. Let λ be the smallest positive eigenvalue of A_v. Then $(x,tv) \in U$ if and only if $t > 1/\lambda$. The function λ is clearly continuous on $\bar{U}$, and is identically equal to one on Fr U. The continuous map $(x,v) \to (x,v/\lambda)$ sends $\bar{U}$ onto Fr U. Thus Fr U is connected. The desired homeomorphism $\phi:\bar{U} \to \mathrm{Fr}\, U \times [1, \infty)$ is now given by $\phi(x,v) = ((x,v/\lambda), \lambda)$.

Q.E.D.

From Lemmas 9.8 and 9.10, we obtain immediately,

<u>Corollary 9.11: Let $f:M \to E^m$ be a totally focal embedding. Then the focal set $F(\Gamma)$ is connected.</u>

We now turn to the determination of all totally focal hypersurfaces $f:M \to E^{n+1}$. Since f is proper, $f(M)$ separates E^{n+1} and M is orientable. Let ξ be a global field of unit normals. Then, the map $(x,t\xi) \to (x,t)$ from NM to $M \times R$ is a diffeomorphism. The first step in the classification is to show that $f(M)$ is a convex hypersurface. Using this, one is then able to show that a totally focal hypersurface must be a round sphere, a hyperplane or a standard product $S^k \times E^{n-k}$.

As in Lemma 9.10, we let U denote the component of $NM-\Gamma$ of maximal index $k > 0$. Since the index is zero on the zero-section $M \times \{0\}$, we can assume that if $(x,t) \in U$, then $t > 0$. We let W^+ and W^- denote the two connected components of $E^{n+1}-f(M)$.

Lemma 9.12: *Let $f:M \to E^{n+1}$ be a totally focal hypersurface and U a connected component of NM-Γ of maximal index. Then the focal set lies in one of the components W^+ of $E^{n+1}-f(M)$ and $U^+ = U \cap F^{-1}(W^+)$ is connected.*

Proof: By Corollary 9.11, the focal set $F(\Gamma)$ is connected, and $F(\Gamma) \cap f(M) = \emptyset$ by Theorem 9.4. Thus, $F(\Gamma)$ lies in one of the components of $E^{n+1}-f(M)$, call it W^+. Let $U^+ = U \cap F^{-1}W^+$ and $U^- = U \cap F^{-1}W^-$. Let $\pi:NM \to M$ be the bundle projection and let Ω be the open connected set πU. By Lemma 9.10,

$$U = \{(x,t)|x \in \Omega,\ t > \rho = 1/\lambda\}$$

and

$$\text{Fr } U = \{(x,t)|x \in \Omega,\ t = \rho\}$$

where λ is the smallest positive principal curvature at x. We now construct a smooth function $\alpha:\Omega \to R$ such that $\alpha > \rho$, and such that $(x,t) \in U^+$ if $\rho(x) < t < \alpha(x)$. To do this, let

$$\mu(x) = \inf\{t|(x,t) \in \overline{U}^-\}, \text{ for } x \in \Omega.$$

Since $F(x,\rho) \in F(\Gamma) \subset W^+$, we know that $\mu(x) > \rho(x)$. The function μ is lower semi-continuous and it assumes a minimum value on any compact set. Using this and a partition of unity, one can construct a smooth function α such that $\rho < \alpha < \mu$ on Ω. Since $U^+ = U - \overline{U}^-$, we also have $(x,t) \in U^+$ for $\rho(x) < t < \alpha(x)$. The collar

$$C = \{(x,t)|x \in \Omega,\ \rho(x) < t < \alpha(x)\}$$

is an open connected set in U^+. Further, the set $D = C \cup \text{Fr } U$ is an open connected set in $\overline{U}$ containing Fr U. Thus, any component of U^+ whose closure intersects Fr U must also intersect C. Then, since C is connected, this component must contain C. We now complete the proof by showing that every component of U^+ must contain points of Fr U in its closure, and so there can only be the one component which contains C. Let U_0^+ be an arbitrary component of U^+. Since $F|_U$ is a covering of $E^{n+1}-F(\Gamma)$, the restriction of F to U_0^+ must cover $W^+-F(\Gamma)$. Now, as in Lemma 9.8, let $p \in F(\Gamma)$ and let x be a point in M where L_p attains its absolute minimum. Let $\gamma(t) = (1-t)f(x) + tp$, $0 \leq t \leq 1$. Since L_p has an absolute minimum at x, there are no points of f(M) on $\gamma(t)$, $0 < t \leq 1$. Since $p \in W^+$, the curve $\gamma(t)$ lies in W^+ for $0 < t \leq 1$. Further, if $q = \gamma(t)$ for $t \in (0,1)$, then L_q has a non-degenerate minimum at x. Thus, for $0 < t < 1$, the curve $\gamma(t)$ lies in $W^+-F(\Gamma)$. Since the restriction of F to U_0^+ covers $W^+-F(\Gamma)$, we can lift γ to a path $\tilde{\gamma}:(0,1) \to U_0^+$. Now apply the method of Lemma 9.5 to produce a point (x_1,v_1) in $\overline{U}_0^+$ such that $F(x_1,v_1) = \gamma(1) = p$. Since $p \in F(\Gamma)$, we have $(x_1,v_1) \in \Gamma$, and thus $(x_1,v_1) \in \overline{U}_0^+ \cap \mathrm{Fr}\, U$, as desired.

Q.E.D.

Using Lemma 9.12 and the theory of covering spaces, we obtain the following.

<u>Lemma 9.13</u>: <u>Let V be any connected component of NM-Γ. Then $V \cap F^{-1}(W^+)$ is connected</u>.

<u>Proof</u>: Using the notation of Lemma 9.12, we have the following commutative diagram of fundamental groups and induced homomorphisms, where i and j are inclusions.

$$\begin{array}{ccc} \pi_1(U^+) & \xrightarrow{j_*} & \pi_1(U) \\ F_* \downarrow & & \downarrow F_* \\ \pi_1(W^+-F(\Gamma)) & \xrightarrow{i_*} & \pi_1(E^{n+1}-F(\Gamma)). \end{array}$$

Since $F:U \to E^{n+1}-F(\Gamma)$ is a covering, the fact that U^+ is connected is equivalent to saying that the image of i_* meets every coset of the image of F_* in $\pi_1(E^{n+1}-F(\Gamma))$ (see Massey [1, p. 178]). Further, the collar C in Lemma 9.12 is clearly a deformation retract of U, and hence j_* is surjective. This and the commutativity of the diagram then imply that image (i_*) contains image

(F_*), and we conclude that i_* is surjective. If one now replaces U and U^+ by V and $V^+ = V \cap F^{-1}W^+$ and uses the same argument in reverse, the fact that i_* is surjective implies that V^+ is connected.

Q.E.D.

The following result is a key step in the ultimate classification.

<u>Theorem 9.14</u>: <u>A totally focal hypersurface $f:M \to E^{n+1}$ is the boundary of an open convex set in E^{n+1}.</u>

<u>Proof</u>: Let V be the connected component of $NM-\Gamma$ which contains the zero section M_0. Let $V^+ = V \cap F^{-1}W^+$ and $V^- = V \cap F^{-1}W^-$. We know that the restriction of F to V is a covering and that V^+ is connected by Lemma 9.13. Since $V^+ \cap F^{-1}(f(M))$ is empty by definition, V^+ is disjoint from M_0. With the appropriate choice of unit normal field ξ, we thus have $V^+ \subset M \times (0,\infty)$. Next, since $F(\Gamma) \subset W^+$ and $Fr\, V \subset \Gamma$, we have $Fr\, V^- \cap Fr\, V = \emptyset$. Thus, $Fr\, V \subset Fr\, V^+ \subset M \times [0,\infty)$. Since V is obviously not contained in $M \times [0,\infty)$, we must have $M \times (-\infty,0) \subset V^-$. Thus, $M \times (-\infty,0)$ is a connected component of V^-, since $V^- \cap M_0 = \emptyset$. This further implies that $F^{-1}(f(M))$ lies in $M \times [0,\infty)$, and so $M \times (-\infty,0]$ is a connected component of V in $F^{-1}(W^- \cup f(M))$. Thus, the restriction of F to $M \times (-\infty,0]$ is a covering of $W^- \cup f(M)$ (see, for example, Massey [1, p. 150]). But f(M) is covered just once (by the zero section) in this covering, and so F restricts to a homeomorphism on $M \times (-\infty,0]$.

Now take $x \in M$ and consider a point $p = F(x,t)$ for $t < 0$. We show that L_p has a unique absolute minimum at x. We have shown that $(x,t) \in V^-$ and so $p \in W^-$. Suppose now that L_p has an absolute minimum at some point x', and thus p may be written as $F(x',t')$. Since no point of f(M) can lie on the segment from p to f(x'), we have $F(x',s) \in W^-$, and so $(x',s) \in V^-$ for every s between t' and 0. But for s sufficiently small, this implies that $s < 0$, and therefore $t' < 0$. Now, since F is a homeomorphism on $M \times (-\infty,0]$, we must have $(x',t') = (x,t)$, and L_p has a unique absolute minimum at x. Thus, f(M) lies outside every open ball tangent to f(M) at f(x) and centered at a point $F(x,t)$, $t < 0$. Hence, $f(M)$ lies outside the open half-space which is the union of these open balls, i.e. the tangent plane to $f(M)$ at $f(x)$ is a support plane. This is true for all $x \in M$, and thus $f(M)$ is a convex hypersurface, and the component W^+ is a convex set in E^{n+1}.

Q.E.D.

Convex hypersurfaces in E^{n+1} have been classified (see Busemann [1, p. 3]). There are two basic types given by embeddings of the form

(i) $f:S^n \to E^{n+1}$

(ii) $g:R^n \to E^{n+1}$, where $g(R^n)$ does not contain any straight line.

The others are given as product embeddings

$$f \times I : S^k \times E^{n-k} \to E^{k+1} \times E^{n-k} = E^{n+1},$$

$$g \times I : R^k \times E^{n-k} \to E^{k+1} \times E^{n-k} = E^{n+1},$$

for f and g as above, and I the identity map. Finally, there is the embedding of E^n as a hyperplane in E^{n+1}.

We now determine which convex hypersurfaces are totally focal.

Theorem 9.15: A totally focal embedding $f:S^n \to E^{n+1}$ is a round sphere.

Proof: Let $M = S^n$, and we use the notation of Theorem 9.14. Since $f(M)$ is convex and compact, each normal line to $f(M)$ must intersect $f(M)$ in exactly two points. Thus, $F^{-1}f(M)$ consists of two cross-sections of $M \times R$, the zero-section and some other section in $M \times (0,\infty)$. Therefore, $F^{-1}(W^+)$ is homeomorphic to $M \times (0,1)$ and $F^{-1}(W^-)$ is homeomorphic to the union of $M \times (-\infty,0)$ and $M \times (1,\infty)$.

Every non-degenerate L_p function must have at least one maximum and one minimum on M. On the other hand, $F^{-1}(W^-)$ has only two connected components and $F^{-1}(W^-)$ lies in $NM-\Gamma$. Since $F^{-1}(W^-)$ intersects every component of $NM-\Gamma$, there are exactly two components V_0 and V_n on which the index is equal to 0 and n, respectively. The restriction of F to each component is a covering of $E^{n+1}-F\Gamma$, and $F^{-1}f(M) \cap V_0$ is just the zero-section, while $F^{-1}f(M) \cap V_n$ is the other cross-section lying in $M \times (0,\infty)$. F restricts to a homeomorphism on the zero-section, and thus F is a homeomorphism on the component V_0. Hence, every non-degenerate L_p has exactly one minimum on M. All of the other critical points of L_p have index n. So the Morse relations imply that L_p also has exactly one maximum. Thus, f is a taut embedding of S^n, and $f(S^n)$ is a round sphere by Theorem 1.6.

Q.E.D.

We next show that there are no totally focal embeddings of the second basic type of convex hypersurfaces.

<u>Theorem 9.16</u>: <u>There does not exist a totally focal embedding $f:R^n \to E^{n+1}$ such that $f(R^n)$ does not contain a straight line.</u>

<u>Proof</u>: We again adopt the notation of Theorem 9.14 with $M = R^n$. From the classification of convex hypersurfaces, if $f(R^n)$ does not contain a straight line, then the convex set $\overline{W}^+$ does not contain any straight lines. However, since $\overline{W}^+$ is not compact, it contains at least one half-line.

We use the notion of the characteristic cone C of $\overline{W}^+$. Fix an origin O in E^{n+1}. Take any point x in $\overline{W}^+$. The vector y is in C if $x + ty$ is in $\overline{W}^+$ for all $t \geq 0$. It is known that C does not depend on the choice of $x \in \overline{W}^+$ and that C is a closed convex cone. Further, since $\overline{W}^+$ contains no lines, C has a vertex at O, i.e. there exists a supporting hyperplane for C which intersects C only at the origin.

Let D be the open convex cone consisting of all vectors z such that $\langle y,z\rangle > 0$ for all $y \neq 0$ in C. The cone D is non-empty, since $C-\{0\}$ is contained in an open half-space. Next observe that the closure $\overline{D}$ of D is given as the set of all z such that $\langle y,z\rangle \geq 0$ for all $y \in C$. To see this, note that if $\langle y,z\rangle \geq 0$ for all $y \in C$ and $z' \in D$, then $(1-t)z + tz'$ is in D for $0 < t \leq 1$.

We next show that $C \cap D$ is non-empty. For if C and D are disjoint, then since D is open, there exists a hyperplane which strictly separates C and D (see, for example, Valentine [1, p. 25]), i.e. there exists a unit vector w (normal to the hyperplane) such that $\langle y,w\rangle \geq 0$ for all $y \in C$ and $\langle z,w\rangle < 0$ for all $z \in D$. But then, $w \in \overline{D}$ and so for z in D near w, $\langle z,w\rangle$ is near $\langle w,w\rangle = 1$, contradicting $\langle z,w\rangle < 0$.

The proof is now completed by showing that there exists a point x_0 in M such that $F^{-1}(x_0)$ consists of just one point, i.e. only one normal line passes through $f(x_0)$. This implies that $NM-\Gamma$ has only one component and so there are no focal points and $f(M)$ is a hyperplane, a contradiction.

Let $p \in C \cap D$ with $|p| = 1$. We first show that $p \in D$ implies that there exists a point $x_0 \in M$ where the linear height function ℓ_p has an absolute minimum on M. Consider the hyperplane H given by the equation $\ell_p \equiv \ell_p(q)$, where q is some point in the open set W^+. The intersection $H \cap W^+$ is a convex open subset of H, which contains no half-lines since $p \in D$. Hence, $H \cap \overline{W}^+$ is homeomorphic to a closed $(n-1)$-ball and its boundary in H is an $(n-1)$-sphere S in $f(R^n)$. The $(n-1)$-sphere S bounds a disk in $f(M)$, the set where $\ell_p \leq \ell_p(q)$. Thus, there exists a point x_0 in the disk where ℓ_p has an absolute minimum. Now, we show that no other normal line passes through $f(x_0)$. For suppose that $f(x_0) = F(x,t)$ for some $x \neq x_0$. We know from Theorem 9.14 that $t > 0$, since F takes $M \times (-\infty, 0)$ onto W^-. We have

$$F(x,t) = f(x) + t\,\xi(x) = f(x_0).$$

Thus, $f(x)-f(x_0) = -t\,\xi(x)$. Since ℓ_p has an absolute minimum at x_0, we have

$$0 \leq \ell_p(x)-\ell_p(x_0) = \langle f(x)-f(x_0), p \rangle = -t \langle \xi(x), p \rangle.$$

If $\ell_p(x)$ were equal to $\ell_p(x_0)$, then ℓ_p would have a critical point at x and $\xi(x)$ would be $\pm p$ contradicting $\langle \xi(x), p \rangle = 0$. Thus, the above inequality is strict and $\langle \xi(x), p \rangle$ must be negative. On the other hand, p is in the characteristic cone C, and so $f(x) + p$ is in $\overline{W}^+$, by definition of C. Since the tangent hyperplane at $f(x)$ supports $\overline{W}^+$, we have $\langle \xi(x), z \rangle \geq \langle \xi(x), f(x) \rangle$ for all $z \in \overline{W}^+$, since $\xi(x)$ points towards W^+. In particular,

$$\langle \xi(x), f(x)+p \rangle \geq \langle \xi(x), f(x) \rangle,$$

i.e. $\langle \xi(x), p \rangle \geq 0$, contradicting the previous conclusion.

Q.E.D.

From Theorems 9.14-9.16, Proposition 9.3 and the classification of convex hypersurfaces, we obtain immediately the main theorem of Carter and West [2].

Theorem 9.17: A totally focal hypersurface in E^{n+1} is a round sphere, a hyperplane or a standard product embedding of $S^k \times E^{n-k}$.

Totally focal submanifolds of spheres

A smooth embedding $f:M \to S^m$ of a compact connected manifold is called totally focal if every spherical distance function L_p is either non-degenerate or has only degenerate critical points. While the arguments in this case have much in common with those for submanifolds of E^m, the compactness of S^m does lead to some new twists. In particular, for a hypersurface M in S^{n+1}, a key step is to show that the focal set intersects both components of $S^{n+1}-M$. This contrasts with the Euclidean case where one needed to prove that the focal set was contained inside one of the components. In specific terms, the fact that each principal curvature always produces two antipodal focal points in the sphere contrasts with the situation in Euclidean space, where each principal curvature gives only one focal point, and a principal curvature equal to zero gives no focal point. This affects the possible configurations of the components of the non-critical set $NM-\Gamma$ of F and necessitates some changes in the overall approach to the problem.

For the remainder of this section, we will suppress the mention of the embedding f and consider a compact connected hypersurface M in S^{n+1}. As in the Euclidean case, M separates S^{n+1} and M is orientable. Let ξ be a global field of unit normals to M. We identify NM with $M \times R$ by writing (x,t) for $(x,t\xi)$. Let ζ be the projection of NM onto M.

For the normal exponential map $F:M \times R \to S^{n+1}$, we have $F(x, t + 2\pi k) = F(x,t)$ for all integers k. Let Γ denote the set of critical points of F. Because of the periodicity of F in the t-variable, it suffices to determine $\Gamma \cap (M \times [-\pi,\pi])$. Each principal curvature λ at a point x gives rise to two critical points in $\{x\} \times [-\pi,\pi]$, i.e. $t = \cot^{-1}\lambda$ and $t = \cot^{-1}\lambda - \pi$. Note that 0 and $\pm\pi$ are not possible values of t at critical points.

If $p = F(x,t)$ for $(x,t) \in \Gamma$, then L_p has a non-degenerate critical point at x, whose index is denoted by Index (x,t). This index function is constant on any connected component of $NM-\Gamma$.

Suppose we write the principal curvatures $\lambda_i = \cot \theta_i$, $1 \leq i \leq n$, where

$0 < \theta_1 \leq \ldots \leq \theta_n < \pi$. Then the functions θ_i are continuous on M (see, for example, Ryan [1, p. 371]). If $V_0 \subset M \times [-\pi, \pi]$ is the component of NM−Γ on which the index is zero, then clearly

$$V_0 = \{(x,t) \mid \theta_n(x) - \pi < t < \theta_1(x)\}$$

The frontier of V_0 consists of two disjoint cross-sections determined by $\theta_1(x)$ and $\theta_n(x) - \pi$. The cross-section $M \times \{\pi\}$ is contained in a component V_n on which the index is equal to n. If we consider V_n as a subset of $M \times [0, 2\pi]$, then

$$V_n = \{(x,t) \mid \theta_n(x) < t < \theta_1(x) + \pi\},$$

and Fr V_n consists of two disjoint cross-sections of NM. Aside from V_0 and V_n, any other component of Γ in $M \times [-\pi, \pi]$ lies in $M \times (0, \pi)$ or $M \times (-\pi, 0)$. By taking the opposite choice of unit normal field, if necessary, we can consider any such component as lying in $M \times (0, \pi)$.

Lemma 9.18. *Let M be a compact connected hypersurface in S^{n+1}. Let V be a component of NM−Γ which lies in $M \times (0, \pi)$. If Fr V is disconnected, then Fr V consists of two disjoint cross-sections of $M \times (0, \pi)$.*

Proof: The index function has a constant value, say k, on V. Since V is contained in $M \times (0, \pi)$, we know $0 < k < n$. Thus, if $(x,t) \in V$, then

$$\theta_k(x) < t < \theta_{k+1}(x).$$

Suppose that for the projection $\zeta : NM \to M$, we have $\zeta(V) = M$. Then $\theta_k < \theta_{k+1}$ for all x, and V is clearly equal to the connected set

$$V = \{(x,t) \in M \times (0, \pi) \mid \theta_k(x) < t < \theta_{k+1}(x)\}.$$

Then, the frontier of V is obviously the union of two disjoint sections.

Now suppose $\zeta(V) \neq M$. Then, we will show that Fr V is connected. First, since F is a local diffeomorphism on NM−Γ, Fr V is a subset of Γ. The set $\zeta(V)$ is an open connected subset of M. Let

$$A = \{(x, \theta_k(x)) \mid x \in \zeta(V)\}, \quad B = \{(x, \theta_{k+1}(x)) \mid x \in \zeta(V)\}.$$

The sets A and B are connected subsets of Fr V. Thus $\bar{A} \cup \bar{B}$ lies in Fr V.

We now show that Fr V $\subset \bar{A} \cup \bar{B}$. Suppose (y, τ) is a point in Fr V. If $y \in \zeta(V)$, then clearly y is either in A or B. Now suppose y is not in $\zeta(V)$. There exists a sequence of points $\{(x_j, t_j)\}$ in V converging to (y, τ). We know that

$$\theta_k(x_j) < t_j < \theta_{k+1}(x_j) \quad \text{for all } j,$$

and hence by continuity of θ_k and θ_{k+1}, we get

$$\theta_k(y) \leq \tau \leq \theta_{k+1}(y).$$

We claim $\theta_k(y) = \theta_{k+1}(y)$. For suppose $\theta_k(y) < \theta_{k+1}(y)$. If τ lies strictly between $\theta_k(y)$ and $\theta_{k+1}(y)$, then (y, τ) is not in Γ, so (y, τ) cannot be in Fr V. Next suppose $\tau = \theta_k(y) < \theta_{k+1}(y)$. Then the sequence $(x_j, (t_j + \theta_{k+1}(x_j))/2)$ lies in V and converges to $\eta = (y, (\tau + \theta_{k+1}(x_j))/2)$. The point η lies in $\bar{V}$ but not in Fr V, since η is not in Γ. Hence $\eta \in V$, and $y \in \zeta(V)$, contradicting our original assumption. One obtains a contradiction in a similar way if $\theta_k(y) < \theta_{k+1}(y) = \tau$. The only possibility remaining is that $\theta_k(y) = \tau = \theta_{k+1}(y)$. Note then that (y, τ) is the limit of the sequence $(x_j, \theta_k(x_j))$ lying in the set A, and (y, τ) is the limit of the sequence $(x_j, \theta_{k+1}(x_j))$ lying in B. Thus, in fact (y, τ) lies in $\bar{A} \cap \bar{B}$. We have shown that Fr V $= \bar{A} \cup \bar{B}$. The argument in the previous paragraph shows that $\bar{A} \cap \bar{B} \neq \emptyset$ (and hence Fr V is connected) provided that not all points $(y, \tau) \in$ Fr V have $y \in \zeta(V)$. Since $\zeta(V) \neq M$, we may choose a point $y \in$ Fr $\zeta(V)$ and a sequence $\{(x_i, t_i)\}$ in V with $\{x_i\}$ converging to y. By compactness of $\bar{V}$, this sequence has a limit point (y, τ) in Fr V. This completes the proof.

Q.E.D.

<u>Lemma 9.19</u>: <u>Let M be a compact connected hypersurface in S^{n+1} such that every component of NM$-\Gamma$ has disconnected frontier. Then the principal curvatures of M all have constant multiplicities.</u>

Proof: Let $V_1, \ldots, V_\ell$ be the components of $NM-\Gamma$ which lie entirely in $M \times (0,\pi)$. The index function is constant on each component, and we denote its value on V_i by σ_i. By Lemma 9.18, the frontier of each V_i consists of disjoint cross-sections of $M \times (0,\pi)$. We label the cross-sections as $\Gamma_1, \ldots, \Gamma_{\ell+1}$ in increasing order. Thus, $\mathrm{Fr}\, V_i = \Gamma_i \cup \Gamma_{i+1}$. The value σ_1 is equal to the multiplicity ν_1 of the principal curvature $\lambda_1 = \cot \theta_1$ on M. Next, at each point (x,t) of V_2, the index σ_2 is equal to ν_1 plus the multiplicity ν_2 of the next largest principal curvature at x. Thus, $\nu_2 = \sigma_2 - \nu_1$ is constant. Proceed inductively, taking the components in increasing order. At the last step, as one passes through the section $\Gamma_{\ell+1}$, one enters the region where the index is n. Hence, $\nu_1 + \ldots + \nu_\ell = n$, and all of the principal curvatures are accounted for and have constant multiplicity.

Q.E.D.

We next require several steps which are almost identical with those for hypersurfaces in E^{n+1} done earlier, and we will omit the proofs here. Let M in S^{n+1} be a totally focal hypersurface, i.e. $\Gamma = F^{-1}(F\Gamma)$. Then M separates S^{n+1} into two open components W^+ and W^-. One first shows, exactly as in Theorem 9.4, that $S^{n+1}-F(\Gamma)$ is connected and contains M. Then using a result similar to Lemma 9.5, one shows that the restriction of F to $NM-\Gamma$ is a covering map of $S^{n+1}-F(\Gamma)$. Next, as in Lemma 9.8, one proves that if U is any connected component of $NM-\Gamma$, then $F(\mathrm{Fr}\, U) = F(\Gamma)$. In fact, the same method yields the following stronger result.

Lemma 9.20: Let M be a totally focal hypersurface in S^{n+1}. Let U^+ be a connected component of $F^{-1}(W^+)-\Gamma$. Then $F(\mathrm{Fr}\, U^+) = M \cup (W^+ \cap F(\Gamma))$.

The following result is an important step and is analogous to Corollary 9.11, where it was shown that the focal set of a totally focal embedding into E^m is connected. Here, in fact, we want to show that the focal set is disconnected. This and Lemma 9.19 will then enable us to show that the principal curvatures of a totally focal hypersurface all have constant multiplicities.

Theorem 9.21: Let M be a totally focal hypersurface in S^{n+1}. Then

$F(\Gamma) \cap W^+$ and $F(\Gamma) \cap W^-$ are both non-empty.

Proof: The proof is a rather involved covering space argument very similar to one first given by Carter [1] in a different context. We assume that $F(\Gamma) \cap W^+ = \emptyset$, and eventually obtain a contradiction. Let $W = W^+ \cup M$. Since M is disjoint from $F(\Gamma)$, we have $W \cap F(\Gamma) = \emptyset$, and F restricts to a covering map on $F^{-1}W$. Let Q be the connected component of $F^{-1}W$ which contains the zero-section M_0. Then Q is a manifold with boundary ∂Q, with $M_0 \subset \partial Q$ and $F(\partial Q) = M$. We first show that Q is contained in the component V_0 of NM-Γ on which the index is zero. As we noted prior to Lemma 9.18, the frontier of V_0 consists of two cross-sections which are contained in Γ, i.e.

$$\mathrm{Fr}\, V_0 = \{(x, \theta_n(x)-\pi) | x \in M\} \cup \{(x, \theta_1(x)) | x \in M\}.$$

Any path from a point in NM $- \overline{V}_0$ to a point in M_0 must cross Fr V_0. Since Fr V_0 lies in $F^{-1}(W^-)$, Q must be contained in V_0.

We assume that the unit normal field ξ on M has been chosen to point in the direction of the component W^+. For sufficiently small $\varepsilon > 0$, the cross-section $M \times \{\varepsilon\}$ lies in Q, while the cross-sections $M \times \{-\varepsilon\}$ and

$$\Gamma_1 = \{(x, \theta_1(x)) | x \in M\}$$

are disjoint from Q, since they are mapped into W^-. Thus, for each $x \in M$, the segment

$$\gamma_x = \{x\} \times [-\varepsilon, \theta_1(x)]$$

intersects ∂Q an even number of times, and it intersects M_0 precisely once. So there is some component $\Delta \neq M_0$ of ∂Q whose mod 2 intersection number with γ_x is 1 for all $x \in M$. Suppose (x,t) is not in Δ and $-\varepsilon \leq t \leq \theta_1(x)$. Define $i(x,t)$ to be the mod 2 intersection number of Δ with the interval $\{x\} \times [-\varepsilon, t]$. Define $i(x,t) = 0$ if $(x,t) \in \Delta$. The function $i(x,t)$ is continuous on $Q-\Delta$ and hence constant there. Since it is zero on M_0, we have $i \equiv 0$ on $Q - \Delta$. Since $i = 0$ on Δ by definition, we have

$$Q \subset P = \{(x,t) | i(x,t) = 0\}.$$

We see that the set P is connected as follows. If $(x,t) \in P$, then there is a point $(x,\tau) \in \Delta$ such that $(x,s) \in P$ for $t \leq s \leq \tau$. So every point in P can be joined by a path in P to Δ. Since Δ is connected, so is P.

We next show that $P = Q$. Suppose $(x,t) \in P$. As noted above, there exists a τ, $t < \tau < \theta_1(x)$ such that $(x,\tau) \in \Delta$. Since $\Delta \subset Q \subset V_0$, there are no points of Γ on the segment from $(x,0)$ to (x,τ). In particular, (x,t) is not in Γ. Hence $P \subset NM-\Gamma$ and the restriction of F to P is an immersion of the $(n+1)$-dimensional manifold P with boundary $M_0 \cup \Delta$ into S^{n+1}. Thus, the frontier of $F(P)$ in S^{n+1} is M. We now show that $F(P) \cap W^- = \emptyset$. We have

$$\mathrm{Fr}(F(P) \cap W^-) \subset \mathrm{Fr}(F(P)) \cup \mathrm{Fr}(W^-) = M.$$

Hence, $F(P) \cap W^-$ is a subset of the connected open set W^- with $\mathrm{Fr}(F(P) \cap W^-) \subset \mathrm{Fr}\, W^-$. From this, it is easy to see that $F(P) \cap W^-$ is both open and closed in W^-, and therefore, it equals W^- or $\emptyset$. But $F(P) \cap F(\Gamma) = \emptyset$ and $F(\Gamma) \subset W^-$, so we conclude $F(P) \cap W^- = \emptyset$. Thus, P is a connected subset of $F^{-1}W$ which contains Q and so $P = Q$.

In order to obtain the desired contradiction, it is necessary to duplicate the above procedure on the simply connected covering space of NM. Briefly, let $\widetilde{M}$ be the simply connected covering space of M. Then, $\widetilde{M} \times R$ is the simply connected covering space of $NM = M \times R$. Let $\beta: \widetilde{M} \times R \to M \times R$ be the covering projection, i.e. $\beta(\tilde{x},t) = (x,t)$, and let $\widetilde{Q}$ be the connected component of $\beta^{-1}Q$ which contains $\widetilde{M}_0 = \widetilde{M} \times \{0\}$. Then choose a connected component $\widetilde{\Delta}$ of $\beta^{-1}\Delta$, $\widetilde{\Delta} \subset \partial\widetilde{Q}$, and define $\tilde{i}(\tilde{x},t)$ as the mod 2 intersection of $\widetilde{\Delta}$ with the segment $\{\tilde{x}\} \times [-\varepsilon,t]$ for $0 \leq t \leq \theta_1(x)$ and $(\tilde{x},t)$ not in $\widetilde{\Delta}$. Define $\tilde{i}(\tilde{x},t) = 0$ if $(\tilde{x},t) \in \widetilde{\Delta}$. Then, let $\widetilde{P}$ be the set where $\tilde{i} = 0$. As above, $\widetilde{Q} \subset \widetilde{P}$ and $\partial\widetilde{P} = \widetilde{M}_0 \cup \widetilde{\Delta}$. Then since $\tilde{i} = i \circ \beta$, one has $\widetilde{P} = \beta^{-1}P = \beta^{-1}Q$. One shows that $\widetilde{P}$ is connected as above and thus concludes that $\widetilde{P} = \widetilde{Q}$. Therefore, $\partial\widetilde{Q} = \widetilde{\Delta} \cup \widetilde{M}_0$.

Next, we use a covering space argument to show that $\widetilde{\Delta}$ is simply connected. Since V_0 is homeomorphic to $M \times R$, $\widetilde{V}_0$ is homeomorphic to $\widetilde{M} \times R$ and thus is simply connected. We have $\widetilde{Q}_0 \subset \widetilde{V}_0$, the component of $\widetilde{M} \times R$ containing the zero-section $\hat{M}_0$. Let ϕ be the restriction of $F \circ \beta$ to $\hat{V}_0$. Then $\phi: \hat{V}_0 \to S^{n+1} - F(\Gamma)$ is a covering map. $\hat{M}_0$ is one component of $\phi^{-1}M$. Since $\hat{M}_0$ is simply connected, every component of $\phi^{-1}M$ must be simply connected (see, for example, Massey [1, p. 177]). Thus, $\hat{\Delta}$ is simply connected.

We now use Van Kampen's theorem (see, Massey [1, p. 122]) to show that $\hat{Q}$ is simply connected. The set $\hat{\Delta}$ is an embedded compact submanifold of $\hat{M} \times [0,\infty)$ with codimension 1. Thus, $\hat{\Delta}$ separates $\hat{M} \times [0,\infty)$ into two closed components $\hat{Q}$ and

$$A = ((\hat{M} \times [0,\infty)) - \hat{Q}) \cup \hat{\Delta}$$

whose intersection is the simply connected set $\hat{\Delta}$. If A and $\hat{Q}$ were open in $\hat{M} \times [0,\infty)$, then one could apply Van Kampen's theorem to get

$$0 = \pi_1(\hat{M} \times [0,\infty)) \cong \pi_1(\hat{Q}) * \pi_1(A),$$

where the star denotes free product, and thus $\pi_1(\hat{Q}) = 0$. To get open sets which give the desired conclusion, one enlarges them slightly using a tubular neighborhood of $\hat{\Delta}$, and the intersection remains homotopic to $\hat{\Delta}$.

Thus, $\hat{Q}$ is the simply connected covering space of Q and, therefore, of W. Let α denote the restriction of $F \circ \beta$ to $\hat{Q}$. Then α is a covering map. If $j: \partial W \to W$ is the inclusion map, then the index of $j_*(\pi_1(\partial W))$ in $\pi_1(W)$ is the number of components of $\alpha^{-1}(\partial W) = \hat{\Delta} \cup \hat{M}_0$ (see Massey [1, p. 179]). Thus, this index is 2.

The restriction of F to Δ is a k-fold ($k \geq 1$) covering of f(M), and since F restricts to a homeomorphism on M_0, F restricts to a $(k+1)$-fold covering of Q onto W. We now show that $k = 1$.

Consider the commutative diagram,

$$\begin{array}{ccc} \pi_1(M_0) & \xrightarrow{F_*} & \pi_1(\partial W) \\ \ell_* \downarrow & & \downarrow j_* \\ \pi_1(Q) & \xrightarrow{F_*} & \pi_1(W), \end{array}$$

where $\ell : M_0 \to Q$ is the inclusion map. Since F restricts to a homeomorphism on M_0, the map $F_* : \pi_1(M_0) \to \pi_1(\partial W)$ is an isomorphism. Thus $j_*(\pi_1(\partial W))$ lies in $F_*(\pi_1(Q))$. So the index of $F_*(\pi_1(Q))$ in $\pi_1(W)$ is less than or equal to the index of $j_*(\pi_1(\partial W))$ in $\pi_1(W)$. That is, $(k+1) \leq 2$. We conclude that $k = 1$, and F restricts to a diffeomorphism on Δ.

Now, we obtain a contradiction as follows. The restriction of F to Q is a 2-fold covering of W, and F restricts to a diffeomorphism on M_0 and Δ. We attach two copies of W^- to Q, one along M_0 and one along Δ. This produces a 2-fold covering of S^{n+1}, which is impossible. Hence, the original assumption that $F(\Gamma) \cap W^+ = \emptyset$ is false, and so likewise is the assumption that $F(\Gamma) \cap W^- = \emptyset$.

Q.E.D.

Combining this result with Lemma 9.19, we obtain

<u>Corollary 9.22</u>: <u>Let M be a totally focal hypersurface in S^{n+1}. Then the principal curvatures of M all have constant multiplicities.</u>

<u>Proof</u>: Since $F(\Gamma) \cap W^+$ and $F(\Gamma) \cap W^-$ are non-empty, while $F(\Gamma) \cap M = \emptyset$, $F(\Gamma)$ is not connected. As we noted before Lemma 9.20, if U is any component of NM−Γ, then $F(\text{Fr } U) = F(\Gamma)$. Thus, Fr U is disconnected for every component U, and the principal curvatures have constant multiplicities by Lemma 9.19.

Q.E.D.

As in the Euclidean case, an immersion $f : M \to S^m$ of a compact manifold is called <u>0-taut</u> if every non-degenerate L_p, $p \in S^m$, has exactly one minimum.

<u>Lemma 9.23</u>: <u>Let M be a totally focal hypersurface in S^{n+1}. Then $F(\Gamma)$ has just two components which are separated by M. Further, the embedding $M \subset S^{n+1}$ is 0-taut.</u>

<u>Proof</u>: Each component of NM−Γ, in particular the component V_0 containing

M_O, has a frontier which consists of two disjoint cross-sections of NM. From this and Corollary 9.22, we know that $F(\Gamma)$ has precisely two connected components, one in W^+ and one in W^-. Now consider the connected components of $F^{-1}(W^+)$ and $F^{-1}(W^-)$. Each of these is an open connected set which contains a component of Γ. Thus, one of the components of Fr V_O lies in a component U^+ of $F^{-1}(W^+)$, and the other component of Fr V_O lies in a component U^- of $F^{-1}(W^-)$. Any other component of $F^{-1}(W^+)$ or $F^{-1}(W^-)$ must lie entirely outside of V_O, since V_O does not contain any points of Γ. Thus, $V_O \cap F^{-1}(W^+) = V_O \cap U^+$ and $V_O \cap F^{-1}(W^-) = V_O \cap U^-$. We know that M_O is disjoint from U^+ and U^- and that M_O separates the components of Fr V_O. Hence, M_O separates U^+ and U^-. Thus, Fr $U^+ \cap$ Fr $U^- \subset M_O$. But $F^{-1}M =$ $\mathrm{Fr}(F^{-1}W^+) \cap \mathrm{Fr}(F^{-1}W^-)$, and we conclude that $F^{-1}M \cap V_O = M_O$. Thus, in the covering $F: V_O \to S^{n+1} - F(\Gamma)$, each point in M is covered just once. Thus, each $p \in S^{n+1} - F(\Gamma)$ is covered just once by F on V_O, i.e. L_p has only one minimum on M, and the embedding is O-taut.

Q.E.D.

<u>Lemma 9.24</u>: <u>Let M be a totally focal hypersurface in S^{n+1}. Then the principal curvatures are constant along the leaves of their principal foliations, i.e. M is a Dupin hypersurface.</u>

<u>Proof</u>: By Lemma 9.19, each of the principal curvatures has constant multiplicity. If a principal curvature λ has constant multiplicity $\nu > 1$, then its principal distribution T_λ is integrable, and λ is constant along the leaves of T_λ by Theorem 4.4. Thus, we only need to handle the case when $\nu = 1$. In doing this, we actually obtain more specific information about the multiplicities of all the principal curvatures.

Suppose that the g distinct principal curvatures of M are written as

$$\lambda_i = \cot \theta_i, \; 0 < \theta_i < \pi, \; 1 \leq i \leq g,$$

where the θ_i form an increasing sequence. Let ν_i be the multiplicity of λ_i. Let V_i be the component of NM$-\Gamma$ given by

$$V_i = \{(x,t) \mid \theta_i(x) < t < \theta_{i+1}(x)\},\ 1 \leq i \leq g-1,$$

and

$$V_0 = \{(x,t) \mid \theta_g(x) - \pi < t < \theta_1(x)\}.$$

Let Γ_i be the cross-section given by $t = \theta_i(x)$, and let Γ_0 be the cross-section $t = \theta_g(x) - \pi$. Then

$$\mathrm{Fr}\, V_i = \Gamma_i \cup \Gamma_{i+1},\ 0 \leq i \leq g-1.$$

For each V_i, one component of Fr V_i is mapped onto the component of the focal set lying in W^+, while the other is mapped onto the component of the focal set lying in W^-. If $\nu_i > 1$, then $F(\Gamma_i)$ is an immersed $(n-\nu_i)$-dimensional manifold by Theorem 4.6. If $\nu_i = 1$, then either $F(\Gamma_i)$ is an immersed $(n-1)$-dimensional manifold or $F(\Gamma_i)$ contains an open n-dimensional subset by Theorems 4.3 and 4.8. In any case, one can determine the multiplicity ν_i from the component of the focal set $F(\Gamma_i)$. Consequently, all the cross-sections which are mapped into W^+ have the same multiplicity, as do all those which are mapped into W^-, i.e.

$$\nu_{i+2} = \nu_i,\ \text{(subscripts mod } g).$$

(Note that if g is odd, then all the multiplicities must be equal.) Denote these two respective multiplicities by ν^+ and ν^-. Suppose now that ν^+ happens to be equal to 1. We consider the component V_0 on which the index is zero. We know that $F(\Gamma_1) \subset W^+$ and $F(\Gamma_0) \subset W^-$, so that λ_1 has multiplicity $\nu^+ = 1$ and λ_g has multiplicity ν^-. As in the Euclidean case, since the embedding is 0-taut, each non-degenerate L_p function has exactly one minimum, which must be an absolute minimum. Using this, one proves as in Lemma 1.24(a), that if p is the first focal point of (M,x) along a normal ray to M at x, then L_p has an absolute minimum at x. The spherical analogue of Lemma 7.1 then implies that this largest principal curvature λ_1 of multiplicity 1 must be constant along its lines of curvature. Thus, $F(\Gamma_1)$ is an immersed compact $(n-1)$-dimensional manifold by Theorem 4.10. Finally, all of the other principal curvatures λ_3, λ_5, ... of constant multiplicity $\nu^+ = 1$ must also be constant along their lines of curvature. For if such a λ_j is not constant along its lines of curvature, then the corresponding component $F(\Gamma_j) = F(\Gamma_1)$ of the focal set contains an open n-dimensional subset, contradicting the fact that $F(\Gamma_1)$ is (n-1)-dimensional. The same arguments apply to the other principal curvatures in case $\nu^- = 1$.

Q.E.D.

We now prove the main result of Carter-West [4].

<u>Theorem 9.25: Let M be a totally focal hypersurface in S^{n+1}. Then M has constant principal curvatures, i.e. M is isoparametric.</u>

<u>Proof</u>: We suppose that M has g distinct principal curvatures with constant multiplicities and use the notation of Lemma 9.24 by setting $\theta_{i+g} = \theta_i + \pi$. Label the cross-sections of Γ in $M \times [0,2\pi]$ as $\Gamma_1, \ldots, \Gamma_{2g}$ in increasing order of the t-variable. For $x \in M$, let $C_i(x)$ be the leaf through x of the principal foliation T_i of the principal curvature λ_i. We know that $C_i(x)$ is a ν^+-dimensional sphere for odd values of i and a ν^--dimensional sphere for even values of i. We first examine the location of the focal points along a given normal geodesic. The key is that the map F is a diffeomorphism on the component V_0 of NM-Γ, where the index is 0.

When a normal geodesic leaves a point x in the direction $\xi(x)$, it does not intersect M again until its has gone through the first focal point $p = F(x, \theta_1(x))$ of (M,x). It intersects M at the point $y = F(x, 2\theta_1(x))$, i.e. y is the point antipodal to x on the ν^+-sphere $C_1(x)$. A key observation is that this geodesic is also normal to M at y, since L_p has an absolute minimum at y. Thus, $p = F(y, \theta_1(y))$ also. Since f is totally focal, every focal point of (M,x) is also a focal point of (M,y) and vice-versa. As this normal geodesic continues past y, it enters the region W^-. The next focal point which it encounters is $p' = F(x, \theta_2(x)) = F(y, \theta_{2g}(y))$, i.e. it is the first focal point of (M,y) in the direction $-\xi(y)$. The function $L_{p'}$ has an absolute minimum at y and also at the next point of M on the geodesic, i.e. $F(x, 2(\theta_2 - \theta_1)(x))$. So the geodesic continues on, alternately meeting focal points and points of M, always orthogonal to M at points of intersection.

Note that the next focal point encountered is $F(x, \theta_3(x)) = F(y, \theta_{2g-1}(y))$. More generally,

$$F(x, \theta_i(x)) = F(y, \theta_{2g+2-i}(y)),\ 1 \leq i \leq 2g,\ \text{(subscripts mod } 2g).$$

We first show that the principal curvature λ_1 is constant. Let A be the subset of M on which λ_1 assumes its absolute maximum, i.e. where θ_1 has its absolute minimum value r. The set A is clearly closed, and we now show that it is open in M. First, we show that if $x \in A$, then every point on each leaf $C_i(x)$, $1 \leq i \leq g$, is also in A. For a fixed value of i, let $q = F(x, \theta_i(x))$. The leaf $C_i(x)$ is a sphere (of dimension ν^+ or ν^-) centered at the focal point q. For the point y constructed above, we have $q = F(y, \theta_j(y))$, where $j = 2g + 2 - i$. Let x_1 be any point on $C_i(x)$, and let x_t, $0 \leq t \leq 1$, be a path in $C_i(x)$ from x to x_1. Let $y_t = F(x_t, 2\theta_1(x_t))$. Then, as above, we have

$$q = F(x_t, \theta_i(x_t)) = F(y_t, \theta_j(y_t)).$$

Thus, the curve y_t lies on the spherical leaf $C_j(y)$ centered at q, and the function $\theta_j(y_t)$ is constant on $[0,1]$, as is the function $\theta_i(x_t)$. Thus,

$$\theta_i(x_1) - \theta_j(y_1) = \theta_i(x) - \theta_j(y) = 2r.$$

Since $\theta_i(x_1) - \theta_j(y_1)$ is the distance $2\theta_1(x_1)$ along the normal geodesic from x_1 from y_1, we have $\theta_1(x_1) = r$, as desired.

We now construct a neighborhood U of x which is contained in A. Let $\nu_k = \nu^+$ if k is odd, and $\nu_k = \nu^-$ is k is even. Let U_1 be an open ν_1-dimensional disk neighborhood of x in the leaf $C_1(x)$. Let U_2 be the union of open ν_2-disk (whose radii vary continuously over U_1) neighborhoods of leaves $C_2(z)$, for $z \in U_1$. Then U_2 is a trivial ν_2-disk bundle over U_1, and is thus diffeomorphic to R^m, $m = \nu_1 + \nu_2$. Proceed inductively, constructing U_k to be a trivial ν_k-disk bundle over U_{k-1}, whose fibres are disks in leaves of the foliation T_k. Eventually, one produces $U = U_g$, which is diffeomorphic to R^n, and is, therefore, an open neighborhood of x in M. Now let z be any point in U. We show that $z \in A$. By construction, there exists a point $z_{g-1} \varepsilon U_{g-1}$, such that z lies on the leaf of T_g through z_{g-1}. Then, there exists z_{g-2}, such that z_{g-1} lies on the leaf of T_{g-1} through z_{g-2}. Proceeding inductively, we obtain a sequence $z = z_g, z_{g-1}, \ldots, z_1, z_0 = x$, such that z_k lies on the leaf of T_k through z_{k-1}. By what we showed above, the fact that $x \varepsilon A$ implies that $z \varepsilon A$, as desired. Thus, θ_1 is constant on M, and $\Gamma_1 = M \times \{r\}$. If we let

$$\Gamma_{-1} = \{(x, \theta_g(x) - \pi) | x \varepsilon M\}$$

be the cross-section of first critical normals in the direction of $-\xi$, then the same argument shows that $\Gamma_1 = M \times \{-s\}$, for some $s > 0$. Now to show that θ_2 is constant, we simply use the fact shown earlier that for each $x \varepsilon M$, there exists $y \varepsilon M$ such that $y = F(x, 2\theta_1(x))$ and

$$F(x, \theta_2(x)) = F(y, \theta_{2g}(y)) = F(y, \theta_{-1}(y)).$$

Thus, $\theta_2(x) = 2r + s$, for all $x \varepsilon M$. In the next step, we consider the point w, such that

$$w = F(x, 2(\theta_2 - \theta_1)(x)) = F(x, 2(r+s)).$$

Then,

$$F(x, \theta_3(x)) = F(w, \theta_1(w))$$

and so

$$\theta_3(x) = 2(r+s) + r = 3r + 2s$$

for all x. By continuing this process, one shows that each principal curvature is constant.

Q.E.D.

10. TIGHT AND TAUT IMMERSIONS IN HYPERBOLIC SPACE.

In Euclidean space, there are two types of umbilic hypersurfaces, planes and spheres. The height functions and distance functions have level sets which are planes and spheres, respectively. There are three types of umbilic hypersurfaces in hyperbolic space H^m, spheres, horospheres and equidistant hypersurfaces (those at a fixed oriented distance from a hyperplane). These have constant sectional curvature which is positive, zero and negative, respectively. Hence, there are three natural types of distance functions

L_p, L_h, L_π which measure the distance from a given point p, horosphere h, or hyperplane π, respectively. These umbilic hypersurfaces themselves can be characterized in terms of the critical point behavior of these functions (Cecil-Ryan [3]), as was done for umbilic submanifolds of Euclidean space in Theorem 2.1. Specifically, [3, p. 73]:

Theorem 10.1: Let M be a connected, complete Riemannian n-manifold, $n \geq 2$, isometrically immersed in H^m. Every Morse function of the form L_p or L_π has index 0 or n at all of its critical points if and only if M is embedded as a sphere, horosphere, or equidistant hypersurface in a totally geodesic H^{n+1} in H^m.

An immersion $f:M \to H^m$ is called taut, horo-tight, tight, respectively, if every non-degenerate function L_p, L_h, L_π, respectively, has the minimum number of critical points. These definitions can also be phrased in terms of injectivity conditions on homology for appropriate "half-spaces" as was done for Euclidean tightness and tautness. In fact, one can consider the unit disk D^m in E^{m+1} to be endowed with the Euclidean metric $\langle \, , \, \rangle$ or the Poincaré metric g. Then, one can test an immersion $f:M \to D^m$ to see which of the following five properties it possesses: Euclidean tautness (E-taut), Euclidean tightness (E-tight), hyperbolic tautness (H-taut), hyperbolic tightness (H-tight), and horo- tightness. In Cecil-Ryan [4], the following implications among the properties were found:

(10.1)
E-taut ===> H-taut,
E-taut ===> H-tight ===> E-tight,
H-taut ===> horo-tight, H-tight ===> horo-tight.

Thus far, the published work on these notions in hyperbolic space is primarily contained in the papers, Cecil [2], Cecil-Ryan [3,4]. In this section, we will briefly survey the results of those papers. We also note that E. Teufel [4] has taken a different approach to the subject via his computation of total absolute curvature.

Treatment of the condition of tautness in hyperbolic space is complicated somewhat by the fact that if an eigenvalue λ of a shape operator A_ξ

satisfies $|\lambda| \leq 1$, then there is no focal point corresponding to λ along the normal geodesic to the submanifold M in the direction ξ (see Theorem 3.1). Thus, the standard arguments using the index theorem to treat tautness in Euclidean space cannot be applied without some modification. Nevertheless, Cecil [2] showed that a taut immersion $f:S^n \to H^m$ must be a metric sphere. Further, if $f:M \to H^3$ (Poincaré disk model) is a taut immersion of a compact, connected surface, then f embeds M as a cyclide of Dupin contained in the disk H^3 (Cecil-Ryan [4, p. 564]). In both cases, the immersion is sufficiently determined by its behavior at points where principal curvatures have absolute value greater than one to yield the desired result. Recently, Thorbergsson [2] has generalized the second result to show that a taut, compact, (k-1)-connected (not k-connected) hypersurface M^{2k} in H^{2k+1} is a cyclide of Dupin diffeomorphic to $S^k \times S^k$ (as in Theorem 8.4).

Classification of taut embeddings of non-compact manifolds in H^m seems more difficult at this point. Cecil [2] pointed out that if γ is any curve in H^2 whose curvature κ satisfies $|\kappa| \leq 1$ at all points and M is a cylinder in H^3 over γ, then M is taut since all critical points of distance functions are of index 0. Thus, any classification would have to permit this amount of imprecision.

As we noted earlier, for an immersion $f:M \to D^m$, tightness in the hyperbolic space $H^m = (D^m,g)$ implies the usual Euclidean (actually projective) tightness in $(D^m,\langle\ ,\ \rangle)$. In fact, it is shown in Cecil-Ryan [4, p. 566]) that $f:M \to (D^m,g)$ is (hyperbolic) tight if and only if $\Phi \circ f:M \to (D^m,\langle\ ,\ \rangle)$ is Euclidean tight for every isometry Φ of hyperbolic space (D^m,g). It is easy to construct examples of Euclidean convex hypersurfaces in D^m which do not remain Euclidean convex under every isometry of the Poincaré disk (D^m,g). Hence, they are Euclidean tight but not hyperbolically tight.

On the other hand, $f:S^n \to (D^{n+1},g)$ is hyperbolically tight if and only if all principal curvatures at all points satisfy $\lambda \geq 1$. Thus, a suitably chosen small deformation of a metric sphere in (D^{n+1},g) produces a hyperbolically tight hypersurface which is diffeomorphic to S^n, but is not a metric sphere. Hence, the class of hyperbolically tight embeddings of S^n into (D^{n+1},g) lies strictly between the class of Euclidean convex hypersurfaces and the class of round spheres.

A cyclide of Dupin M in (D^3,g) is Euclidean taut, and hence, hyperbolically tight. One can obtain a hyperbolically tight embedding of a torus into (D^3,g) which is not a cyclide by modifying a cyclide slightly on the part of the cyclide where both principal curvatures (in (D^3,g)) are greater than one, while keeping fixed the part where either of the principal curvatures has absolute value less than or equal to one. This type of variation, as done with the sphere above, is always possible. However, at this time, we do not know whether or not there exist hyperbolically tight immersions of the other surfaces into H^3, except, of course, that there do not exist tight immersions of P^2 and K^2 into H^3 by Kuiper's results in Euclidean space. Thus, we ask:

Problem 10.2: Which compact surfaces admit tight immersions into H^3?

Every hyperbolically taut sphere and every taut surface in H^3 is also tight in H^3. Thus, we state:

Problem 10.3: Investigate the relationship between tautness and tightness in H^m. In particular, does tautness imply tightness for compact manifolds?

Finally, the class of horo-tight immersions has actually not been studied at all. As we noted, both hyperbolic tautness and hyperbolic tightness imply horo-tightness, which may turn out to be about as strong as Euclidean tightness. Thus, we state:

Problem 10.4: Study the class of horo-tight immersions. In particular, what are the horo-tight immersions of S^n? Which compact surfaces admit horo-tight immersions into H^3?

Isoparametric hypersurfaces

0. INTRODUCTION.

According to the original definition, a family of hypersurfaces M_t in a real space form $\tilde{M}^{n+1}(c)$ of constant sectional curvature c is called <u>isoparametric</u> if each M_t is equal to $V^{-1}(t)$, where V is a non-constant real-valued function on a domain in $\tilde{M}^{n+1}(c)$ which satisfies a system of differential equations of the form

$$|\text{grad } V|^2 = T \circ V, \quad \Delta V = S \circ V \text{ (Laplacian)},$$

for some smooth real-valued functions S, T. Thus, the two classical Beltrami differential parameters are functions of V itself, whence the name isoparametric. Equivalently, an isoparametric family of hypersurfaces can be characterized as a family of parallel hypersurfaces, each of which has constant principal curvatures (see Section 5).

We will say that a hypersurface is <u>isoparametric</u> if it has constant principal curvatures. It follows from the formulas for shape operators of tubes (Theorem 3.2 of Chapter 2), that if M is isoparametric, then so are all of its parallel hypersurfaces. An isoparametric hypersurface in E^{n+1} is either umbilic or an open subset of a standard spherical cylinder $S^k \times E^{n-k}$. This was proven by Levi-Civita [1] for n = 2 and by B. Segre [1] for arbitrary n. Shortly after the papers of Levi-Civita and Segre, E. Cartan [1]-[4] took up the study of isoparametric hypersurfaces in $\tilde{M}^{n+1}(c)$ for $c \neq 0$. Suppose the isoparametric hypersurface M has g distinct principal curvatures $\lambda_1, \ldots, \lambda_g$ with respective multiplicities $m_1, \ldots, m_g$. Then Cartan established the following key identity which is the cornerstone of his work, namely that for each i, $1 \leq i \leq g$,

$$\sum_{j \neq i} m_j \frac{c + \lambda_i \lambda_j}{\lambda_i - \lambda_j} = 0.$$

From this, he was rather quickly able to determine all isoparametric hypersurfaces in the cases $c \leq 0$. For example, if $c = 0$, and λ_i is the smallest positive principal curvature, each term in the summand is non-positive, and hence must equal zero. Thus there are at most two distinct principal curvatures, and if there are two, then one must be zero. The classification then follows from standard methods. In the case $c = -1$, take λ_i to be the largest principal curvature between 0 and 1 or the smallest curvature greater than or equal to 1 such that no principal curvature lies between λ_i and $1/\lambda_i$. Then each term

$$\frac{-1 + \lambda_i \lambda_j}{\lambda_i - \lambda_j}$$

is negative unless $\lambda_j = 1/\lambda_i$. Hence, there are at most two distinct principal curvatures, and if there are two, then they are reciprocals of each other. From this, one can show that if M is not umbilic, then M is an open subset of a standard product $S^k \times H^{n-k}$ in H^{n+1} (see Ryan [2, pp. 252-253]).

For hypersurfaces of the sphere S^{n+1}, however, the basic identity does not lead to such severe restrictions, and Cartan produced some interesting examples with more than two distinct principal curvatures. He showed that all isoparametric hypersurfaces in S^{n+1} with two distinct principal curvatures are standard products of two spheres, and he found that those with 3 principal curvatures are precisely the tubes of constant radius over the standard Veronese embeddings of FP^2 for $F = R, C, Q$ (quaternions), O (Cayley numbers) in S^4, S^7, S^{13}, S^{25}, respectively, which were discussed in Section 9 of Chapter 1. For each of these hypersurfaces, the focal set consists of a pair of antipodal Veronese embeddings of FP^2 (see Example 7.3). In the process, he proved the general result that any isoparametric family with g distinct principal curvatures of the same multiplicity can be defined by the equation

$$F = \cos gt \text{ (restricted to } S^{n+1}),$$

where F is a harmonic homogeneous polynomial of degree g on E^{n+2} satisfying

$$|\operatorname{grad} F|^2 = g^2 r^{2g-2} \quad \text{(gradient in } E^{n+2}).$$

It follows that a piece of isoparametric hypersurface with g distinct principal curvatures of the same multiplicity can be extended to a compact hypersurface with the same property.

Cartan also produced isoparametric hypersurfaces in S^5 and S^9 with four distinct principal curvatures of the same multiplicity. He noted that each of his examples is homogeneous. In fact, each is the orbit of a point under an appropriate closed subgroup of SO(n + 2). Of course, such orbit hypersurfaces have constant principal curvatures. Based on the properties of his example, Cartan asked the following questions:

Question 1: For each positive integer g, does there exist an isoparametric family with g distinct principal curvatures of the same multiplicity?

Question 2: Does there exist an isoparametric family with more than 3 distinct principal curvatures not of the same multiplicity?

Question 3: Does every isoparametric family of hypersurfaces admit a transitive group of isometries?

The subject was then virtually ignored for thirty years until Nomizu [1, 2] revived it with a survey article of Cartan's results. Nomizu generalized one of Cartan's examples with four principal curvatures (Example 7.4) to produce an example whose principal curvatures do not all have the same multiplicity, thus answering Question 2 affirmatively. This answer was also obtained at approximately the same time by Takagi and Takahashi [1], who used the work of Hsiang and Lawson [1] to determine all homogeneous isoparametric hypersurfaces of the sphere. Specifically, they showed that each homogeneous isoparametric hypersurface is an orbit of the isotropy representation of a Riemannian symmetric space of rank 2, and they gave a list of examples, Takagi-Takahashi [1, p. 480]. The list includes some examples with 6 distinct principal curvatures, as well as those with g = 1, 2, 3, 4 principal curvatures.

Then in the early 1970's (although published later), H.F. Münzner [1] produced a far-reaching generalization of Cartan's work. Through a geometric study of the focal submanifolds and their second fundamental forms, he showed that if M is an isoparametric hypersurface with principal curvatures cot θ_i, $0 < \theta_1 < \ldots < \theta_g < \pi$, with multiplicities m_i, then

$$\theta_k = \theta_1 + \frac{(k-1)}{g}\pi, \qquad 1 \leq k \leq g,$$

and the multiplicities satisfy $m_i = m_{i+2}$ (subscripts mod g). As a consequence, one sees that if g is odd, then all of the multiplicities must be equal. Münzner's computation also shows that the focal submanifolds must be minimal (which Nomizu [1, 2] had shown by a different calculation), and further, that Cartan's basic identity is equivalent to the minimality of the focal submanifolds (see Remark 2.6).

Beginning with these facts, he then showed that the hypersurfaces of any isoparametric family with g distinct principal curvatures in S^{n+1} can be represented as open subsets of level hypersurfaces in S^{n+1} of a homogeneous polynomial F of degree g on E^{n+2} which satisfies the differential equations

$$|\text{grad } F|^2 = g^2 r^{2g-2}$$

$$\Delta F = c r^{g-2}$$

where $c = g^2(m_2 - m_1)/2$, and m_1, m_2 are the two (possibly equal) multiplicities of the principal curvatures. As with Cartan's result for the case when all multiplicities are equal, Münzner's result implies that any piece of isoparametric hypersurface can be extended to a compact isoparametric hypersurface. Münzner's construction also shows that there are only two focal submanifolds, even though there may be more than two principal curvatures. The focal submanifold corresponding to the principal curvature λ_i is the same as that for λ_{i+2} (subscripts mod g). Further, each hypersurface of the family divides the ambient sphere into two ball bundles over the respective focal submanifolds. From this topological information, Münzner is able to determine the cohomology rings of the hypersurfaces and the focal submanifolds, and then by using delicate cohomological and algebraic arguments, he obtains the splendid result that: <u>the number g of distinct principal curvatures must satisfy g = 1, 2, 3, 4 or 6</u>.

Of course, Münzner's work answered Cartan's Question 1, and in combination with the list of Takagi and Takahashi of homogeneous examples, it seemed to indicate that the answer to Question 3 was affirmative. However, beginning with the differential equations of Münzner for the homogeneous polynomial F, Ozeki and Takeuchi [1] produced two infinite series of isoparametric families which are not on the list of Takagi and Takahashi and are, therefore, inhomogeneous. These examples all have 4 distinct principal curvatures.

Then, Ferus, Karcher and Münzner [1] gave a construction of isoparametric hypersurfaces with $g = 4$ using representations of Clifford algebras, which includes all known examples, except two. They were able to show geometrically that many of their examples are not homogeneous by an examination of the second fundamental forms of the focal submanifolds, which are quite accessible in their presentation. This paper is a thorough study of the known examples and their various properties, and is the starting point of any further research in the case $g = 4$. See also the notes of Ferus [4]. J. Dorfmeister and E. Neher [1]-[5] have extended the work of Ferus, Karcher and Münzner to a more algebraic setting involving triple systems.

As for the case $g = 6$, not much has been known until quite recently. The list of homogeneous isoparametric hypersurfaces of Takagi and Takahashi contains an example with $g = 6$ and $m_1 = m_2 = 1$, and one with $g = 6$ and $m_1 = m_2 = 2$. In part 2 of Münzner [1], it was shown that $m_1 = m_2$ always holds in the case $g = 6$. In his thesis, U. Abresch [1,2] has shown that the common multiplicity of m_1 and m_2 must be either 1 or 2. This leads to the natural conjecture that every compact isoparametric hypersurface with $g = 6$ is homogeneous. In a recent preprint, Dorfmeister and Neher [5] have shown that the conjecture holds in the case $m_1 = m_2 = 1$, and remark that a large portion of their work can be carried over to the case $m_1 = m_2 = 2$.

In 1979, Cecil and Ryan [6] showed, using Münzner's result, that all isoparametric hypersurfaces and their focal submanifolds in the sphere are taut. Thorbergsson [1] provided a different proof of the tautness of isoparametric hypersurfaces as a corollary of his result on the tautness of Dupin hypersurfaces (see Section 6 of Chapter 2). From Münzner's work, it is also easy to show that every isoparametric hypersurface in the sphere is totally focal as defined by Carter and West [2], i.e. every distance function is either non-degenerate or has only degenerate critical points. Carter and West [4] proved that every totally focal hypersurface in the sphere is isoparametric (see Section 9 of Chapter 2).

In another direction, Nomizu [4] has recently begun a study of isoparametric hypersurfaces in Lorentzian space forms by proving a generalization of Cartan's basic identity to that situation. Hahn [1]-[2] and Magid [1] have done further work in this area.

There is also a theory of isoparametric (constant principal curvatures) real hypersurfaces in complex projective space CP^m endowed with the Fubini-Study metric. R. Takagi [2] classified those homogeneous real hypersurfaces in CP^m which are orbits under an analytic subgroup of the the projective unitary group $PU(m + 1)$, and they are listed in Takagi [3, part I, p. 47]. As in the spherical case, there is a one-to-one correspondence between these homogeneous isoparametric families of real hypersurfaces in CP^m and the isotropy representations of various Hermitian symmetric spaces of rank two (see Takagi [3, part I, pp. 43-48]). Takagi showed that the geodesic hyperspheres are the only connected, complete real hypersurfaces in CP^m with two constant principal curvatures. In [3, part II] he found that the only connected, complete real isoparametric hypersurfaces in CP^m with 3 distinct principal curvatures are the tubes over a totally geodesic CP^k, $0 < k < m-1$, and the tubes over the complex quadric Q^{m-1} in CP^m. Using results concerning focal sets and shape operators of tubes, Cecil and Ryan [7] showed that all of the homogeneous examples of Takagi are tubes over complex submanifolds. They were also able to generalize Takagi's characterization of geodesic hypersurfaces to show that if M is a connected real hypersurface in CP^m $(m \geq 3)$ with at most two distinct principal curvatures at each point, then M is an open subset of a geodesic hypersphere. There are no Einstein real hypersurfaces in CP^m, but the geodesic hyperspheres and certain of the tubes over a CP^k, and over Q^{m-1} can be distinguished by means of a slightly less restrictive requirement on their Ricci tensor (see Maeda [1], Kon [1], Cecil-Ryan [7]). Finally, returning to critical point theory, a totally geodesic CP^n and the complex quadric $Q^n \subset CP^{n+1} \subset CP^m$ were shown to be the only complex n-dimensional submanifolds in CP^m satisfying the condition that every distance function in CP^m has index 0 or n at each of its non-degenerate critical points (Cecil [1]). This uses the characterization of hyperplanes and quadrics as the only complex Einstein hypersurfaces, due to Smyth [1] (see also Nomizu-Smyth [1]).

Finally, we note that an isoparametric family of hypersurfaces in S^{m+1} and its two focal submanifolds form what J. Bolton [1] defines to be a transnormal system in that any geodesic in S^{n+1} meets the submanifolds orthogonally at either none or all of its points. With just this geometric hypothesis, Bolton can recover many facts about an isoparametric family, in particular, that there are only two submanifolds of codimension greater than one in a transnormal system which contains at least one hypersurface. Bolton's idea generalizes the notion of a transnormal manifold which was introduced by S.A. Robertson [1], who has recently written a survey article on the topic [2]. In a recent preprint, Carter and West [5] have generalized the idea of an isoparametric function to an isoparametric map between any Riemannian manifolds and have linked this with the idea of a system of transnormal submanifolds (primarily of codimension greater than one). They find that systems of isoparametric manifolds give rise to kaleidoscopic patterns and thus lead to some interesting examples. They also find that each isoparametric manifold is transnormal and totally focal. See also Terng [1].

This chapter is primarily based on Part 1 of Münzner's paper, which is developed in Section 1-6, along with some insights from the work of Cartan and Nomizu. We will not give the algebraic topological part of Münzner's proof that g = 1, 2, 3, 4 or 6, and the reader is referred to Part 2 of a Münzner [1]. In Section 7, we study Cartan's examples in the cases g = 1, 2, 3 and Nomizu's generalization of one of Cartan's examples with g = 4, which is also noteworthy because it projects to an isoparametric family in complex projective space.

1. PARALLEL HYPERSURFACES IN THE SPHERE.

In order to make the chapter more self-contained, we begin by briefly deriving the formulas for the shape operators of a family of parallel hypersurfaces in the sphere. This is just a special case of the formulas for shape operators of tubes given in Chapter 2.

Let M be a connected oriented n-dimensional manifold embedded in the sphere S^{n+1}. As in the introduction, we write the principal curvatures of M in the form

$$\lambda_i = \cot \theta_i, \quad 0 < \theta_i < \pi, \quad 1 \leq i \leq g,$$

where the θ_i form an increasing sequence. Although we will be chiefly interested in the case where the λ_i are constant on M, it is sufficient initially to assume that they have constant multiplicity. Then, by Theorem 4.4 of Chapter 2, the principal distributions

$$T_i(x) = \{X \in T_xM \mid AX = \lambda_i X\}$$

are integrable.

Let ξ be a smooth field of unit normals to M. For each $x \in M$ and $t \in R$, let $\phi_t : M \to S^{n+1}$ be defined by

$$\phi_t(x) = \cos t\, x + \sin t\, \xi_x,$$

i.e. $\phi_t(x)$ is the point at an oriented distance t along the normal geodesic to M through x. The map ϕ_t is certainly differentiable on M. We wish to find when it is an immersion.

Given a vector X tangent to M at a point p, let x_u be a curve in M with initial position p and initial tangent vector $\vec{x}_0 = X$. Then $(\phi_t)_* X$ is the initial tangent vector to the curve $\phi_t(x_u)$, i.e.

$$(\phi_t)_* X = \cos t\, \vec{x}_0 + \sin t\, \vec{\xi}_p = \cos t\, \vec{x}_0 - \sin t\, A_p \vec{x}_0$$

$$= (\cos t\, I - \sin t\, A_p) X,$$

where A_p denotes the shape operator of M at p, and we are identifying the vector $(\cos t\, I - \sin t\, A_p)X$ in T_pM with its Euclidean parallel translate at

$\phi_t(p)$. If $X \in T_i(p)$, then

$$(1.1) \qquad (\phi_t)_* X = (\cos t - \sin t \cot \theta_i) X = \frac{\sin(\theta_i - t)}{\sin \theta_i} X,$$

for $1 \leq i \leq g$. Thus, rank $(\phi_t)_* = n$ at p, if $t \neq \theta_i (\text{mod } \pi)$ for any i.

The above discussion may be interpreted locally. For each $x \in M$, there is a neighborhood U_x on which ϕ_t is an embedding for certain ranges of values of t. For example, if $\overline{U}_x$ is compact and the minimum value of $\theta_1(y)$ and $\pi - \theta_g(y)$ is $\varepsilon > 0$ on $\overline{U}_x$, then ϕ_t is an embedding on U_x for $|t| < \varepsilon$. Other values of t lying between adjacent θ_i also give a local embedding. The immersed hypersurface $\phi_t : M \to S^{n+1}$ so defined is called the <u>parallel hypersurface at (oriented) distance t</u>.

In case the θ_i are constant on M, each ϕ_t is an immersion on all of M, except for values of t congruent to some $\theta_i (\text{mod } \pi)$. Thus, each isoparametric hypersurface gives rise to a family of parallel hypersurfaces, each of which, as we shall soon see, is also isoparametric. We now find the shape operator of a parallel hypersurface, and therefore its principal curvatures. This is just a special case of Theorem 3.2 of Chapter 2.

<u>Theorem 1.1</u>: <u>If a hypersurface M in S^{n+1} has a principal curvature $\lambda = \cot \theta$ at a point $x \in M$, then the parallel hypersurface $\phi_t(M)$ has a principal curvature $\tilde{\lambda} = \cot(\theta - t)$ at $\phi_t(x)$ having the same multiplicity and (up to parallel translation in E^{n+2}) the same principal subspace.</u>

<u>Proof</u>: For $y \in M$, one can easily check that

$$\tilde{\xi}_y = -\sin t \, y + \cos t \, \xi_y,$$

when translated to $\tilde{y} = \phi_t(y)$, is a unit normal vector to $\phi_t(M)$ at $\tilde{y}$. For $X \in T_i(x)$, we take a curve x_u with $x_0 = x$, $\vec{x}_0 = X$, and compute

$$D_{(\phi_t)_*X}\tilde{\xi} = \frac{d}{du}(\tilde{\xi}_{x_u})|_{u=0} = -\sin t\, X - \cos t \cot \theta_i X$$

$$= -\frac{\cos(\theta_i - t)}{\sin \theta_i} X,$$

as a vector in E^{n+2}, where D is Euclidean covariant differentiation. But by definition of the shape operator A_t of the immersion $\phi_t: M \to S^{n+1}$, we have

$$D_{(\phi_t)_*X}\tilde{\xi} = -(\phi_t)_* A_t X,$$

so that using (1.1)

$$(\phi_t)_* A_t X = \frac{\cos(\theta_i - t)}{\sin(\theta_i - t)} X = \cot(\theta_i - t) X.$$

In other words,

$$A_t X = \cot(\theta_i - t) X \tag{1.2}$$

for $X \in T_i$, and thus the proof is complete.

Q.E.D.

<u>Corollary 1.2</u>: <u>Let M be an isoparametric hypersurface having g distinct principal curvatures cot θ_i, $1 \leq i \leq g$ with respective multiplicities m_i. If t is any real number not congruent to any θ_i (mod π), then ϕ_t immerses M as an isoparametric hypersurface with principal curvatures cot $(\theta_i - t)$, $1 \leq i \leq g$, with the same multiplicities. Furthermore, the principal distributions induced by the ϕ_t all coincide with the corresponding original principal distributions</u>.

We remark that when M is an embedded isoparametric hypersurface, each hypersurface ϕ_t will be an embedding, not merely an immersion. However, to prove this, we will need the alternate presentation of isoparametric hypersurfaces as level sets of a function from S^{n+1} to R given in Sections 4 and 5. The collection of hypersurfaces $\phi_t(M)$, $t \neq \theta_i \pmod{\pi}$, is called an <u>isoparametric family of hypersurfaces</u>.

2. <u>FOCAL SUBMANIFOLDS</u>.

We recall some facts from Chapter 2 concerning conditions under which the focal set is a submanifold. We now assume that M is an embedded isoparametric hypersurface with principal curvatures $\lambda_i = \cot \theta_i$ as in Section 1. Suppose $t = \theta_i$. Then the Jacobian $(\phi_t)_*$ collapses T_i to 0, and $\phi_t(M)$ consists entirely of focal points. The leaves of T_i on M are open subsets of m_i-spheres and the leaf-space M/T_i with the quotient topology is an $(n-m_i)$-dimensional manifold (see Theorems 4.6 and 4.10 of Chapter 2). Furthermore, there is an immersion

$$\psi: M/T_i \to S^{n+1}$$

with $\psi \circ \pi = \phi_t$, where π is the projection from M onto M/T_i.

For each $x \in M$, we have the orthogonal decomposition

$$T_xM = T_i(x) \oplus T_i^{\perp}(x).$$

Then $(\phi_t)_*$ is injective on $T_i^{\perp}(x)$ and $(\phi_t)_* = 0$ on $T_i(x)$ (as shown in Chapter 2). See also Nomizu [2] for these facts.

Define $h: M \to S^{n+1}$ by

$$h(x) = -\sin t\, x + \cos t\, \xi_x,$$

and let $p = \phi_t(x)$. Since $\langle \phi_t(x), h(x) \rangle = 0$, we see that h(x) is tangent to S^{n+1} at p. Further $\langle h(x), X \rangle = 0$ for $X \in T_i(x)$, so h(x) is normal to the focal submanifold V_i, given by the immersion $\psi: M/T_i \to S^{n+1}$, at the point p.

We make use of the fact that each leaf of T_i in M is an open subset of an m_i-sphere, which is mapped to a single point by ϕ_t, to find a convenient representation of the normal space V_i and the corresponding shape operators.

Specifically, let p be an arbitrary point of $\phi_t(M)$. We have seen that we can consider

$$h:\phi_t^{-1}(p) \to S_p^{\perp}V_i = \{\eta \in T_p^{\perp}V_i \mid |\eta| = 1\},$$

where the domain is an open subset C of an m_i-sphere, and the range is contained in the m_i-sphere $S_p^{\perp}V_i$. For $x \in \phi_t^{-1}(p)$ and $X \in T_i(x)$, we have (recalling $t = \theta_i$),

$$h_*X = -\sin t\, X + \cos t\, (-AX) =$$

$$-\sin \theta_i\, X + \cos \theta_i(-\cot \theta_i)X = \frac{-1}{\sin \theta_i} X \neq 0.$$

Thus, h_* has full rank m_i on C and is a local diffeomorphism of open subsets of m_i-spheres.

<u>Theorem 2.1</u>: <u>(Münzner [1]). Let p ε</u> V_i <u>be a focal point of an isoparametric hypersurface M. Let η be a unit normal to</u> V_i <u>at p and suppose</u> $\eta = h(x)$ <u>for some</u> $x \in \phi_t^{-1}(p)$, $t = \theta_i$. <u>Then the shape operator</u> A_η <u>of</u> V_i <u>is given in terms of its eigenvectors by</u>

$$A_\eta X = \cot(\theta_j - \theta_i)X, \quad \text{for } X \in T_j(x),\ j \neq i.$$

<u>(As before, we are identifying</u> $T_j(x)$ <u>with its parallel translate at p).</u>

<u>Proof</u>: Let $\eta = h(x)$ for $x \in \phi_t^{-1}(p)$. The same calculation used in proving Theorem 1.1. is valid here, and it leads to (1.2), which we write as

$$A_\eta X = \cot(\theta_j - \theta_j)X, \quad \text{for } X \in T_j(x),\ j \neq i.$$

Q.E.D.

<u>Corollary 2.2</u>: <u>Let p ε</u> V_i <u>be a focal point of an isoparametric hypersurface M. Every shape operator</u> A_η, $\eta \in S_p^{\perp}V_i$, <u>has eigenvalues</u> $\cot(\theta_j - \theta_i)$ <u>with</u>

multiplicities m_j, $j \neq i,\ 1 \leq j \leq g$.

Proof: By Theorem 2.1, the result holds on the open subset $h(C)$ of $S_p^{\perp}V_i$. Consider the characteristic polynomial $\chi_\tau(\eta) = \det(A_\eta - \tau I)$, as a function of η on $T_p^{\perp}V_i$. Since A_η is linear in η, we have for each fixed τ, that $\chi_\tau(\eta)$ is a polynomial of degree $n - m_i$ as a function on the vector space $T_p^{\perp}V_i$. Consequently, the restriction of $\chi_\tau(\eta)$ to the sphere $S_p^{\perp}V_i$ is an analytic function of η. Since $\chi_\tau(\eta)$ is constant on the open set $h(C)$, it is constant on $S_p^{\perp}V_i$.

Q.E.D.

The next result shows that the curvatures and their multiplicities must satisfy some very restrictive conditions.

Theorem 2.3: _Let M be an isoparametric hypersurface with principal curvatures cot_ θ_k, $0 < \theta_1 < \ldots < \theta_g < \pi$, _with multiplicities_ m_k. _Then_

$$\theta_k = \theta_1 + \frac{(k-1)}{g}\pi, \quad 1 \leq k \leq g,$$

and the multiplicities satisfy $m_k = m_{k+2}$, _(subscripts mod g)_.

Proof: Corollary 2.2 shows that for each fixed i, the set $\{\cot(\theta_j - \theta_i)\}$ of eigenvalues of the shape operator A_η of the focal submanifold V_i is independent of the choice of unit normal η. In particular, since $A_{-\eta} = -A_\eta$, the sets

$$\{\cot(\theta_j - \theta_i)\}_{j \neq i} \quad \text{and} \quad \{-\cot(\theta_j - \theta_i)\}_{j \neq i}$$

coincide for each fixed i.

Let Γ be the set of unit complex numbers, and let Ω be the subset consisting of those of the form $\exp(2i\theta)$, where $\cot\theta$ is a principal curvature. Then Ω is invariant by reflections $z \to w^2\bar{z}$, for all $w \in \Omega$. This implies that

$$\theta_{j+1} - \theta_j = \theta_j - \theta_{j-1}, \quad m_{j+1} = m_{j-1}, \text{ for } 2 \leq j \leq g-1, \tag{2.1}$$

and

(2.2) $$\theta_2 - \theta_1 = \theta_1 - (\theta_g - \pi), \quad m_2 = m_g,$$

provided $g > 1$. (If $g = 1$, there is nothing to prove.) Writing $\theta_2 - \theta_1 = \delta$, we get from (2.1) that $\theta_g - \theta_1 = (g-1)\delta$ while from (2.2) $\theta_g - \theta_1 = \pi - \delta$, from which the result follows.

Q.E.D.

<u>Corollary 2.4</u>: <u>Let M be an isoparametric hypersurface with g distinct principal curvatures. If g is odd, then all of the principal curvatures have the same multiplicity. If g is even, there are at most two distinct multiplicities.</u>

We are following the development of Münzner [1]. The following result was also obtained independently by Nomizu [1].

<u>Corollary 2.5</u>: <u>Each focal submanifold</u> V_i <u>of an isoparametric hypersurface is a minimal submanifold of</u> S^{n+1}.

<u>Proof</u>: For a given unit normal η to V_i,

$$\text{trace } A_\eta = \sum_{j \neq i} m_j \cot(\theta_j - \theta_i) = \sum_{k=1}^{g-1} m_k \cot\left(\frac{k\pi}{g}\right)$$

$$= -\sum_{k=1}^{g-1} m_k \cot\left(\pi - \frac{k\pi}{g}\right) = -\sum_{k=1}^{g} m_k \cot \frac{(g-k)}{g}\pi$$

$$= -\sum_{j=1}^{g-1} m_{g-j} \cot \frac{j\pi}{g}.$$

If g is even, then $m_{g-j} = m_j$ for all j, while if g is odd, all the multiplicities are equal by Theorem 3.3. In either case, the last expression is equal to $-\text{trace } A_\eta$, so that $\text{trace } A_\eta = 0$, and V_i is minimal.

Q.E.D.

Remark 2.6: Cartan's identity.

We remark that Corollary 2.5 allows one to prove and gives a geometric meaning to Cartan's identity,

$$\sum_{j\neq i} m_j \frac{c + \lambda_i \lambda_j}{\lambda_i - \lambda_j} = 0.$$

in the case $c = 1$. With $\lambda_i = \cot \theta_i$, we have

$$\sum_{j\neq i} m_j \frac{1 + \lambda_i \lambda_j}{\lambda_i - \lambda_j} = \sum_{j\neq i} m_j \cot(\theta_j - \theta_i) = \text{trace } A_\eta,$$

for η normal to the focal manifold V_i. Hence, Corollary 2.5 not only proves the Cartan identity, but shows that it is equivalent to the minimality of the focal submanifolds.

3. CALCULUS OF HOMOGENEOUS FUNCTIONS.

Cartan originally defined a family of isoparametric hypersurfaces as level sets of a function $F: S^{n+1} \to R$ such that both of the differential operators $|\text{grad } F|^2$ and Laplacian of F are functions of F itself. We will see that this is equivalent to the definition in terms of parallel hypersurfaces (Section 5).

Let $F: E^{n+2} \to R$ be a homogeneous function of degree g. For $z \in E^{n+2}$, we have by Euler's theorem

$$\langle z, \text{grad}^E F \rangle = g\, F(z). \tag{3.1}$$

The superscript E is used to denote the Euclidean gradient $\text{grad}^E F$, while the gradient of the restriction of F to the unit sphere S^{n+1} will be denoted $\text{grad}^S F$. Similarly, the two Laplacians will be denoted by $\Delta^E F$ and $\Delta^S F$, respectively.

Theorem 3.1: Let $F: E^{n+2} \to R$ be a homogeneous function of degree g. Then

(a) $|\text{grad}^S F|^2 = |\text{grad}^E F|^2 - g^2 F^2$

(b) $\Delta^S F = \Delta^E F - g(g-1)F - g(n+1)F.$

Proof: (a) For $z \in S^{n+1}$,

$$\operatorname{grad}^S F = \operatorname{grad}^E F - \langle \operatorname{grad}^E F, z\rangle z. \tag{3.2}$$

Thus

$$\begin{aligned}|\operatorname{grad}^S F|^2 &= |\operatorname{grad}^E F|^2 - 2gF(z)\langle \operatorname{grad}^E F, z\rangle + g^2F^2(z)|z|^2 \\ &= |\operatorname{grad}^E F|^2 - 2g^2F^2(z) + g^2F^2(z) = |\operatorname{grad}^E F|^2 - g^2F^2(z).\end{aligned}$$

(b) Let ∇ and D denote the respective Levi-Civita connections on S^{n+1} and E^{n+2}. The Laplacian $\Delta^S F$ is the trace of the operator on T_zS^{n+1} given by

$$V \to \nabla_V \operatorname{grad}^S F.$$

For $V \in T_zS^{n+1}$,

$$\nabla_V \operatorname{grad}^S F = D_V \operatorname{grad}^S F - \langle D_V \operatorname{grad}^S F, z\rangle z.$$

Differentiating (3.2) and using (3.1), we get

$$D_V \operatorname{grad}^S F = D_V \operatorname{grad}^E F - gFV - g(VF)z. \tag{3.3}$$

Taking the component of (3.3) tangent to S^{n+1}, we have

$$\nabla_V \operatorname{grad}^S F = D_V \operatorname{grad}^E F - \langle D_V \operatorname{grad}^E F, z\rangle z - gFV. \tag{3.4}$$

From this, we can compute the Laplacian. Let $V_1, \ldots, V_{n+1}$ be an orthonormal basis for T_zS^{n+1}. Then $V_1, \ldots, V_{n+1}, z$ is an orthonormal basis for T_zE^{n+2}. First,

$$(3.5) \qquad \sum_{i=1}^{n+1} \langle D_{V_i} \operatorname{grad}^E F, V_i \rangle = \Delta^E F - \langle D_z \operatorname{grad}^E F, z \rangle.$$

Recall that

$$\langle D_z \operatorname{grad}^E F, z \rangle = D_z \langle \operatorname{grad}^E F, z \rangle - \langle \operatorname{grad}^E F, z \rangle.$$

Using (3.1), we have

$$D_z \langle \operatorname{grad}^E F, z \rangle = D_z(gF) = g \langle \operatorname{grad}^E F, z \rangle.$$

Thus

$$(3.6) \qquad \langle D_z \operatorname{grad}^E F, z \rangle = (g-1) \langle \operatorname{grad}^E F, z \rangle = (g-1)gF(z).$$

Now

$$\sum_{i=1}^{n+1} \langle D_{V_i} \operatorname{grad}^E F, z \rangle \langle z, V_i \rangle = 0,$$

while the trace of the map $V \to -gFV$ on $T_z S^{n+1}$ is clearly $-(n+1)gF$. From this and (3.4) - (3.6), we have

$$\Delta^S F = \Delta^E F - g(g-1)F - (n+1)gF.$$

Q.E.D.

We now consider some examples of isoparametric homogeneous polynomials, i.e. polynomials such that $|\operatorname{grad}^S F|^2$ and $\Delta^S F$ are both functions of F itself.

<u>Examples 3.2</u>:

(a) $F: E^{n+2} \to R$ is a linear function $F(z) = \langle z, p \rangle$, $p \in E^{n+2}$, $|p| = 1$. F is homogeneous of degree $g = 1$ and

$$|\mathrm{grad}^E F|^2 = 1, \qquad \Delta^E F = 0.$$

Thus by Theorem 3.1,

$$|\mathrm{grad}^S F|^2 = 1 - F^2, \qquad \Delta^S F = -(n+1)F.$$

The level sets

$$M_t = \{z \in S^{n+1} | F(z) = t\}, \qquad -1 < t < 1,$$

are all small n-spheres, except for t = 0 which gives a great n-sphere. The focal submanifolds are the two antipodal points $\pm$ p.

(b) Let $E^{n+2} = E^{j+1} \times E^{k+1}$, where $j+k = n$.

For any $z = (x,y)$ in $E^{j+1} \times E^{k+1}$, set

$$F(z) = |x|^2 - |y|^2.$$

Then F is homogeneous of degree g=2 and for $r^2 = |x|^2 + |y|^2$, we have

$$\mathrm{grad}^E F = 2(x,-y), \qquad |\mathrm{grad}^E F|^2 = 4r^2,$$

$$\Delta^E F = 2(j-k).$$

Thus by Theorem 3.1,

$$|\mathrm{grad}^S F|^2 = 4(1-F^2)$$

$$\Delta^S F = 2(j-k)-2(n+2)F.$$

Consider the level sets,

$$M_t = \{z \in S^{n+1} | F(z) = \cos(2t)\}, \qquad 0 < t < \frac{\pi}{2}.$$

Except for the two focal submanifolds

$$M_0 = \{(x,y) \mid y=0\} = S^j \times \{0\},$$

and

$$M_{\frac{\pi}{2}} = \{(x,y) \mid x=0\} = \{0\} \times S^k,$$

each level set M_t is the Cartesian product of a j-sphere of radius cos t and a k-sphere of radius sin t.

We remark that one can obtain the same family of level hypersurfaces from the polynomial $G(x,y) = |x|^2$, and note that $F = 2G-1$ on S^{n+1}.

4. ISOPARAMETRIC HYPERSURFACES AS ALGEBRAIC SUBMANIFOLDS.

In this section, we prove Münzner's [1] key result that isoparametric hypersurfaces are algebraic in the following sense. Let M be a connected oriented isoparametric hypersurface embedded in S^{n+1} with g distinct principal curvatures. Then there exists a homogeneous polynomial F of degree g on E^{n+2} such that M is an open subset of a level set of the restriction of F to S^{n+1}, and F satisfies

$$|\mathrm{grad}^E F|^2 = g^2 r^{2g-2} \tag{4.1}$$

$$\Delta^E F = c r^{g-2} \tag{4.2}$$

where $c = g^2(m_2-m_1)/2$ and m_1, m_2 are the two possible distinct multiplicities of the principal curvatures (see Theorem 2.3). If all the multiplicities are equal, then c = 0 and F is harmonic on E^{n+2}. This generalizes the work of Cartan, who proved the result in the case where all the multiplicities are equal. Münzner calls F a <u>Cartan polynomial</u>.

The normal bundle NM is trivial and we consider the normal exponential map $\phi: M \times R \to S^{n+1}$ defined by

$$\phi(x,t) = \phi_t(x) = \cos t \, x + \sin t \, \xi_x.$$

We know that ϕ has rank n+1, except where cot t is a principal curvature of M. In particular, for any regular point (x,t) of ϕ, there exists an open neighborhood U of (x,t) such that ϕ is a diffeomorphism of U onto $\hat{U} = \phi(U)$. We define $\tau:\hat{U} \to R$ by

$$\tau(p) = \theta_1 - \pi_2(\phi^{-1}(p)),$$

where π_2 is projection onto the second coordinate. In other words, if $p = \phi(x,t)$ then $\tau(p) = \theta_1 - t$. Further, define $V:\hat{U} \to R$ by

$$V(p) = \cos g\ \tau(p).$$

Note that $\tau(p)$ is just the oriented distance to the first focal point along the normal geodesic to the parallel hypersurface M_t through p. Hence, if we begin the construction with a parallel hypersurface near M rather than M itself, we are led to the same functions τ and V on $\hat{U}$. Further, choosing the opposite orientation $-\xi$ on M has the effect of replacing V by $-V$. We next extend V to a homogeneous function F of degree g on the cone in E^{n+2} over $\hat{U}$ by

$$F(rp) = r^g \cos g\ \tau(p), \qquad p \in \hat{U},\ r > 0.$$

Let m_1, m_2 be the respective multiplicities of the principal curvatures λ_1, λ_2.

<u>Theorem 4.1</u>: <u>The function F defined on the cone over $\hat{U}$ satisfies</u>

$$|\mathrm{grad}^E F|^2 = g^2 r^{2g-2}$$

$$\Delta^E F = c r^{g-2},$$

<u>where $r(z) = |z|$ and $c = g^2(m_2 - m_1)/2$.</u>

<u>Proof</u>: Suppose $p = \cos t\ x + \sin t\ \xi_x$ is a point in $\hat{U}$. The vector

$$\tilde{\xi}_p = -\sin t\ x + \cos t\ \xi_x$$

is a unit normal to the parallel hypersurface $\phi_t(M)$ at p. Moreover,

$$\tilde{\xi}_p = (\phi_*)_{(x,t)}(\frac{\partial}{\partial t}),$$

and it is easy to check that

$$\tilde{\xi}_p = -\text{grad}^S \tau. \tag{4.3}$$

The function F on the cone in E^{n+2} over $\hat{U}$ is given by

$$F(z) = |z|^g \cos g\ \tau(\frac{z}{|z|}).$$

If $\sigma: E^{n+2} - \{0\} \to S^{n+1}$ is defined by

$$\sigma(z) = \frac{z}{|z|} = \frac{z}{r},$$

we may write

$$F = r^g \cos g(\tau \circ \sigma).$$

Now $\text{grad}^E r = \sigma$. On the other hand, let $Z \in T_z E^{n+2}$. Then

$$\langle Z, \text{grad}^E(\tau \circ \sigma) \rangle = \tau_* \sigma_* Z.$$

But

$$\sigma_* Z = \frac{1}{r}(Z - \langle Z, \sigma \rangle \sigma), \tag{4.4}$$

and $\tau_*(\sigma_* Z) = \langle \text{grad}^S \tau, \sigma_* Z \rangle$.

Using (4.3) and (4.4), and the fact that $\langle \operatorname{grad}^S \tau, \sigma \rangle = 0$, we get

$$\tau_*(\sigma_* Z) = -\frac{1}{r} \langle \tilde{\xi} \circ \sigma, Z \rangle.$$

Thus

$$\operatorname{grad}^E(\tau \circ \sigma) = -\frac{1}{r} \tilde{\xi} \circ \sigma. \tag{4.5}$$

Hence

$$\operatorname{grad}^E F = g r^{g-1}(\cos g(\tau \circ \sigma)\, \sigma + \sin g(\tau \circ \sigma)\, \tilde{\xi} \circ \sigma) \tag{4.6}$$

and

$$|\operatorname{grad}^E F|^2 = g^2 r^{2g-2},$$

since $\sigma(z)$ and $\tilde{\xi}(\sigma(z))$ are orthonormal at $\sigma(z)$.

We now compute $\Delta^E F$. Write $\operatorname{grad}^E F = g r^{g-1} W$, where

$$W = \cos g(\tau \circ \sigma)\sigma + \sin g(\tau \circ \sigma)\, \tilde{\xi} \circ \sigma .$$

In the rest of the proof, all gradients are with respect to E^{n+2}. We have

$$\Delta^E F = \operatorname{div} \operatorname{grad}^E F = \langle \operatorname{grad}(g r^{g-1}), W \rangle + g r^{g-1} \operatorname{div} W. \tag{4.7}$$

The first term in the preceding line is

$$\begin{aligned} \langle \operatorname{grad}(g r^{g-1}), W \rangle &= g(g-1) r^{g-2} \langle \operatorname{grad} r, W \rangle = g(g-1) r^{g-2} \langle \sigma, W \rangle \\ &= g(g-1) r^{g-2} \cos g(\tau \circ \sigma). \end{aligned} \tag{4.8}$$

On the other hand,

$$(4.9)\qquad \operatorname{div} W = \langle \operatorname{grad} \cos g(\tau \circ \sigma), \sigma\rangle + \langle \operatorname{grad} \sin g(\tau \circ \sigma), \tilde{\xi} \circ \sigma\rangle$$

$$+ \cos g(\tau \circ \sigma) \operatorname{div} \sigma + \sin g(\tau \circ \sigma) \operatorname{div} (\tilde{\xi} \circ \sigma).$$

We handle the terms on the right one at a time. First, using (4.5),

$$(4.10)\qquad \langle \operatorname{grad} \cos g(\tau \circ \sigma), \sigma\rangle = -g \sin g(\tau \circ \sigma) \langle \operatorname{grad} \tau \circ \sigma, \sigma\rangle$$

$$= -g \sin g(\tau \circ \sigma) \langle -\frac{1}{r} \tilde{\xi} \circ \sigma, \sigma \rangle = 0,$$

and

$$(4.11)\quad \langle \operatorname{grad} \sin g(\tau \circ \sigma), \tilde{\xi} \circ \sigma\rangle = g \cos g(\tau \circ \sigma) \langle -\frac{1}{r} \tilde{\xi} \circ \sigma, \tilde{\xi} \circ \sigma\rangle$$

$$= \frac{-g}{r} \cos g(\tau \circ \sigma).$$

Next,

$$(4.12)\qquad \operatorname{div} \sigma = -\frac{1}{r^2} \langle \operatorname{grad} r, z\rangle + \frac{1}{r} \operatorname{div} z = -\frac{1}{r} + \frac{n+2}{r} = \frac{n+1}{r}.$$

For the final term in (4.9), we need the following

<u>Lemma 4.2</u>: <u>(a) $\operatorname{div}(\tilde{\xi} \circ \sigma) = \frac{1}{r} (\operatorname{div} \tilde{\xi}) \circ \sigma$</u>

(b) <u>$\operatorname{div} \tilde{\xi} = - \sum_{i=1}^{g} m_i \cot (\tau + (i-1)\pi/g)$</u>

(c) <u>$\operatorname{div} \tilde{\xi} = -(n \cot g\tau + (m_1-m_2)g/(2 \sin g\tau))$</u>

<u>Remark 4.3</u>: Since $\operatorname{grad}^S \tau = - \tilde{\xi}$, we have $\Delta^S \tau = n \cot g\tau + \frac{(m_1-m_2)}{2 \sin g\tau} g$.

<u>Proof of Lemma 4.2</u>: (a) Let Z be any tangent vector to E^{n+2} at any arbitrary point and let z_u be a curve in E^{n+2} with initial tangent vector Z. Then

$$D_Z(\hat{\xi} \circ \sigma) = \alpha'(0), \quad \text{where } \alpha(u) = \hat{\xi} \circ \sigma (z_u).$$

But $\alpha'(0) = D_Y\hat{\xi}$, where $Y = \sigma_* Z$. We recall (4.4)

$$\sigma_* Z = \frac{1}{r} (Z - \langle Z, \sigma\rangle \sigma).$$

Thus, for $Z = \frac{z}{r}$, $\sigma_* Z = 0$ while for $Z \in TS^{n+1}$ (modulo parallel translation), we have $\sigma_* Z = \frac{1}{r} Z$. Hence

$$D_Z \hat{\xi} \circ \sigma = 0, \qquad \text{for } Z = \frac{z}{r},$$

and

$$D_Z \hat{\xi} \circ \sigma = \frac{1}{r} (D_Z\hat{\xi}) \circ \sigma \text{ for } Z \in TS^{n+1}.$$

We conclude that

$$\text{div}(\hat{\xi} \circ \sigma) = \frac{1}{r} \sum_{i=1}^{n+1} \langle (D_{e_i} \hat{\xi}) \circ \sigma, e_i\rangle e_i,$$

where $e_1, \ldots, e_{n+1}$ are orthonormal and tangent to S^{n+1}. Thus

$$\text{div } (\hat{\xi} \circ \sigma) = \frac{1}{r} (\text{div } \hat{\xi}) \circ \sigma,$$

proving (a).

(b) Let $\hat{\nabla}$ be the Levi-Civita connection on S^{n+1} and let $p = \phi(x,t)$ be a point in $\hat{U}$. Let $e_1, \ldots, e_n$ be an orthonormal basis for T_pM_t, where $M_t = \phi_t(M)$. Then

$$\operatorname{div} \tilde{\xi} = \sum_{i=1}^{n} \langle \hat{\nabla}_{e_i} \tilde{\xi}, e_i \rangle + \langle \hat{\nabla}_{\tilde{\xi}} \tilde{\xi}, \tilde{\xi} \rangle.$$

The last term is clearly zero, and if we choose the e_i to be principal vectors for the shape operator A_t of M_t, we have

$$\operatorname{div} \tilde{\xi} = - \sum_{i=1}^{n} \langle A_t e_i, e_i \rangle = - \operatorname{trace} A_t.$$

Thus by Corollary 1.2 and Theorem 2.3,

$$\operatorname{div} \tilde{\xi} = - \sum_{i=1}^{g} m_i \cot (\theta_i - t) =$$

$$- \sum_{i=1}^{g} m_i \cot((\theta_i - \theta_1) - (t - \theta_1)) =$$

$$- \sum_{i=1}^{g} m_i \cot (\tau + \frac{(i-1)}{g} \pi).$$

(c) If all of the multiplicities are equal, then

$$- \sum_{i=1}^{g} m_i \cot (\tau + \frac{(i-1)}{g} \pi) =$$

$$- \frac{n}{g} \sum_{i=1}^{g} \cot (\tau + \frac{(i-1)}{g} \pi) = (\frac{-n}{g}) g \cot g\tau = -n \cot g\tau,$$

thus proving (c) in this case. If all of the multiplicities are not equal, then g is even and $m_1 = m_3 = \ldots = m_{g-1}$, $m_2 = m_4 = \ldots = m_g$. Setting $g = 2\ell$,

$$\sum_{i=1}^{g} m_i \cot (\tau + \frac{(i-1)}{g} \pi) = \tag{4.13}$$

$$m_1 \sum_{j=1}^{\ell} \cot (\tau + \frac{(j-1)}{\ell} \pi) + m_2 \sum_{j=1}^{\ell} \cot (\tau + \frac{\pi}{2\ell} + \frac{(j-1)\pi}{\ell}) =$$

$$m_1\ell \cot(\ell\tau) + m_2\ell \cot(\ell\tau + \tfrac{\pi}{2}) =$$

$$m_1\ell \cot(\ell\tau) - m_2\ell \tan(\ell\tau).$$

Now set $m_1/(m_1 + m_2) = \cos^2\omega$ and $m_2/(m_1 + m_2) = \sin^2\omega$, and note that $(m_1 + m_2)g = 2n$. The expression (4.13) becomes

$$\frac{2n\ell}{g}(\cos^2\omega \cot \ell\tau - \sin^2\omega \tan \ell\tau)$$

$$= n\,\frac{\cos^2\omega\,(1 + \cos g\tau) - \sin^2\omega\,(1 - \cos g\tau)}{\sin g\tau}$$

$$= n\,\frac{(\cos 2\omega + \cos g\tau)}{\sin g\tau} = n \cot g\tau + \frac{g(m_1 - m_2)}{2 \sin g\tau}.$$

Q.E.D.

We complete the computation of $\Delta^E F$ in the proof of Theorem 4.1. From (4.7)-(4.12) and Lemma 4.2 we have

$$\Delta^E F = g(g-1)r^{g-2}\cos g(\tau \circ \sigma) + gr^{g-1}\left(-\frac{g}{r}\cos g(\tau \circ \sigma)\right)$$

$$+ g\,r^{g-1}\cos g(\tau \circ \sigma)\left(\frac{n+1}{r}\right)$$

$$+ gr^{g-1}\sin g(\tau \circ \sigma)\left(\frac{1}{r}\right)\left(-n \cot g(\tau \circ \sigma) - \frac{(m_1 - m_2)g}{2 \sin g(\tau \circ \sigma)}\right)$$

$$= gr^{g-2}\cos g(\tau \circ \sigma)(g-1-g + n+1 - n) + gr^{g-2}\,\frac{(-(m_1 - m_2))g}{2}$$

$$= g^2 r^{g-2}\,\frac{(m_2 - m_1)}{2}.$$

Q.E.D.

The fact that F satisfies the differential equations of Theorem 4.1 allows one to prove:

Theorem 4.4: The homogeneous function F of Theorem 4.1 is the restriction of a polynomial in E^{n+2} to the cone over $\hat{U}$.

We first need the following elementary calculation.

Lemma 4.5: $\Delta r^k = k(k+n)r^{k-2}$ on E^{n+2} for any positive integer k.

Proof: One computes grad $r^k = kr^{k-1}$ grad $r = kr^{k-1}\sigma$, where $\sigma(z) = z/r$ as in Theorem 4.1. Then

$$\Delta r^k = \text{div grad } r^k = k \langle \text{grad } r^{k-1}, \sigma\rangle + kr^{k-1} \text{div } \sigma$$

$$= k(k-1)r^{k-2} \langle\sigma, \sigma\rangle + kr^{k-1}(n+1)/r$$

$$= k(k+n)r^{k-2}.$$

Q.E.D.

We now begin the proof of Theorem 4.4. From Theorem 4.1 and Lemma 4.5, we see that for

$$a = \frac{g}{g+n} \frac{(m_2 - m_1)}{2},$$

the function $G = F - ar^g$ is harmonic on its domain of definition, the cone over $\hat{U}$. Thus all the partial derivatives of G of all orders are also harmonic.

Lemma 4.6: $\Delta^g|\text{grad } G|^2 = 0$, where $\Delta^1 = \Delta$, $\Delta^k = \Delta \circ \Delta^{k-1}$.

Proof: As in Lemma 4.5,

$$\text{grad } G = \text{grad } F - a\, g\, r^{g-1}\sigma.$$

Therefore,

$$|\text{grad } G|^2 = |\text{grad } F|^2 + a^2 g^2 r^{2g-2} - 2a\, gr^{g-1} \langle\text{grad } F, \sigma\rangle.$$

Using (4.6) for grad F, the above becomes

$$|\text{grad } G|^2 = g^2 r^{2g-2}(1+a^2) - (2a\, g r^{g-1})(g r^{g-1} \cos g(\tau \circ \sigma))$$

$$= g^2 r^{2g-2}(1+a^2) - 2a\, g^2 r^{g-2} F.$$

Thus

$$|\text{grad } G|^2 + 2a\, g^2 r^{g-2} G = g^2 r^{2g-2}(1 + a^2) - 2a\, g^2 r^{g-2} a r^g \tag{4.14}$$

$$= g^2 r^{2g-2}(1-a^2).$$

Now using Lemma 4.5,

$$\Delta(r^k G) = \text{div}(r^k \text{grad } G) + \text{div}(G \text{ grad } r^k)$$

$$= 2 \langle \text{grad } r^k, \text{grad } G \rangle + r^k \Delta G + G(\Delta r^k)$$

$$= 2k\, r^{k-1} \langle \sigma, \text{grad } F - a\, g\, r^{g-1} \sigma \rangle + k(k+n) r^{k-2} G$$

$$= 2kg r^{g+k-2} \cos g(\tau \circ \sigma) - 2kag r^{k+g-2} + k(k+n) r^{k-2} G$$

$$= 2kg r^{k-2}(F - a r^g) + k(k+n) r^{k-2} G$$

$$= 2kg r^{k-2} G + k(k+n) r^{k-2} G$$

$$= k r^{k-2} G\, (2g + k(k+n)).$$

Referring to (4.14), suppose the multiplicities m_1 and m_2 are not equal. Then g is even, and g/2 applications of Δ will reduce the term $2ag^2 r^{g-2} G$ on the left to zero. The term on the right hand side, $g^2(1-a^2) r^{2g-2}$ will be reduced to zero by g applications of Δ, and the result follows. On the other hand, if all of the multiplicities are equal, then a = 0, and the term on the left-hand side involving r^{g-2} does not appear. As before, g applications of Δ reduce the right-hand side to zero, and the proof is complete.

Q.E.D.

The following lemma will allow us to complete the proof of Theorem 4.4.

Lemma 4.7: *For any harmonic function G,*

$$\Delta^g |\text{grad } G|^2 = \Sigma \left(\frac{\partial^{g+1} G}{\partial x_{i_1} \partial x_{i_2} \cdots \partial x_{i_{g+1}}}\right)^2,$$

where the sum takes place over all g+1-tuples $(i_1, \ldots, i_{g+1})$ *with each* $i_j \in \{1, \ldots, n+2\}$. *(The* i_j*'s need not be distinct from one another).*

Proof: We begin by noting that $|\text{grad } G|^2 = \sum_{i=1}^{n+2} \left(\frac{\partial G}{\partial x_i}\right)^2$.
For arbitrary $f: E^{n+2} \to R$, one has the identity $\Delta f^2 = 2|\text{grad } f|^2 + 2f\Delta f$.
Using this and the fact that $\Delta G = 0$, we get

$$\Delta |\text{grad } G|^2 = \sum_{i=1}^{n+2} 2\left|\text{grad } \frac{\partial G}{\partial x_i}\right|^2 = \sum_{i=1}^{n+2} \sum_{j=1}^{n+2} \left(\frac{\partial^2 G}{\partial x_i \partial x_j}\right)^2.$$

The sums in the rest of the proof are all from 1 to n+2. We repeat the above step, using the fact that all partials of G are harmonic.

$$\Delta^2 |\text{grad } G|^2 = 2^2 \Sigma\, \Sigma \left|\text{grad } \frac{\partial^2 G}{\partial x_i \partial x_j}\right|^2$$

$$= 2^2 \Sigma\, \Sigma\, \Sigma \left(\frac{\partial^3 G}{\partial x_i \partial x_j \partial x_k}\right)^2.$$

Continuing the process, we obtain

$$\Delta^g |\text{grad } G|^2 = 2^g \Sigma \left(\frac{\partial^{g+1} G}{\partial x_{i_1} \cdots \partial x_{i_{g+1}}}\right)^2$$

over all g+1-tuples $(i_1, \ldots, i_{g+1})$ with each i_j in $\{1, \ldots, n+2\}$.

Q.E.D.

We now complete the proof of Theorem 4.4.

By Lemmas 4.6 and 4.7, all (g+1)-th partial derivatives of G are zero, while we show that $\Delta^{g-1}|\text{grad } G|^2$ is a non-zero constant, as follows.

If $a = 0$, then from (4.14)

$$|\text{grad } G|^2 = g^2 r^{2g-2}$$

and

$$\Delta^{g-1}|\text{grad } G|^2 = (2g)^{g-1}(g-1)!\ (n+2(g-1))(n+2(g-2))\ldots(n+2). \tag{4.15}$$

On the other hand, if $a \neq 0$, then $g = 2\ell$, and the second term on the left-hand side of (4.14) is handled by noting that

$$\Delta^{\ell}(r^{g-2}G) = 0,$$

so that $\Delta^{g-1}|\text{grad } G|^2$ is just $1-a^2$ times the term on the right side of (4.15), and is thus a non-zero constant. (Note $|a| < 1$).

These facts imply that G is a polynomial of degree exactly g. Since G is a homogeneous function on the cone $\hat{U}$, it must be a homogeneous polynomial. This completes the proof of Theorem 4.4.

Summary of Section 4:

Given a connected oriented isoparametric hypersurface M in S^{n+1} with principal curvatures $\cot \theta_i$,

$$0 < \theta_1 < \ldots < \theta_g < \pi,$$

we have constructed a homogeneous polynomial F of degree g with the following properties:

$$|\text{grad}^E F|^2 = g^2 r^{2g-2} \tag{4.16}$$

$$(4.17) \qquad \Delta^E F = \frac{m_2 - m_1}{2} g^2 r^{g-2}.$$

This polynomial is called the Cartan polynomial of M. By Theorem 3.1, the restriction V of F to S^{n+1} satisfies

$$(4.18) \qquad |\mathrm{grad}^S V|^2 = g^2(1-V^2)$$

$$(4.19) \qquad \Delta^S V = g(\frac{(m_2 - m_1)}{2} g - (n+g)V),$$

and thus V is isoparametric in the sense that $|\mathrm{grad}^S V|^2$ and $\Delta^S V$ are both functions of V itself.

On M, the function V is identically equal to $\cos g\theta_1$, while if z is on the parallel hypersurface M_t, $V(z) = \cos g(\theta_1 - t)$. Thus each M_t lies on a level hypersurface of V.

Of course, one would like to show that the whole level hypersurface of such a function V is isoparametric. Note that any non-constant function satisfying (4.18) must have $[-1, 1]$ as its range, since it must have a maximum and a minimum on S^{n+1}. Moreover, its gradient can vanish only when $V = \pm 1$. Thus, for any $s \in (-1, 1)$, the level set

$$M_s = \{z \in S^{n+1} | V(z) = s\}$$

is a compact hypersurface. Further, we will show in Section 6 that each M_s is connected. Thus each of the original hypersurfaces can be extended to a compact connected hypersurface of S^{n+1} in this way.

When $s = \pm 1$, M_s is not a hypersurface since grad $V = 0$ there. We shall see that M_1 and M_{-1} are the focal submanifolds discussed in Section 2. This gives the surprising result that there are only two focal submanifolds, even if there are $g > 2$ distinct principal curvatures. Along each normal geodesic to an isoparametric hypersurface, there are 2g focal points, which lie alternately on one focal submanifold and then the other (see Section 6).

In view of the construction of this section, it is not surprising that the M_s form a parallel family of isoparametric hypersurfaces along with their focal submanifolds. This we will prove in the next two sections.

5. GEOMETRY OF LEVEL HYPERSURFACES.

In this section, we investigate the relationship between a real valued function F and the geometry of its level hypersurfaces. Although we are primarily interested in the sphere S^{n+1} as our ambient space, our initial theorems are valid in any real space form. We always assume that $\hat{M}$ has dimension n+1. The treatment here follows Nomizu [2] and Münzner [1].

Theorem 5.1: Let $F:\hat{M} \to R$ be a real valued smooth function defined on a space $\hat{M}$ of constant sectional curvature c. Suppose that grad F does not vanish on the set $M = F^{-1}(0)$. Then the shape operator A of the hypersurface M satisfies

$$\langle AX, Y\rangle = -\frac{H_F(X, Y)}{|\text{grad } F|},$$

where X and Y are tangent vectors to M and H_F denotes the Hessian of the function F.

Proof: First note that $\xi = \text{grad } F / |\text{grad } F|$ is a field of unit normals on M, and that the shape operator of M is determined by

$$\langle AX, Y\rangle = -\langle \hat{\nabla}_X \xi, Y\rangle,$$

where $\hat{\nabla}$ is the Levi-Civita connection on $\hat{M}$. Then writing $|\text{grad } F| = \rho$, we have

$$(5.0) \qquad \langle \hat{\nabla}_X(\rho\xi), Y\rangle = (X\rho)\langle \xi, Y\rangle + \rho\langle \hat{\nabla}_X \xi, Y\rangle = -\rho\langle AX, Y\rangle.$$

On the other hand,

$$\langle \hat{\nabla}_X(\text{grad } F), Y\rangle = \hat{\nabla}_X \langle \text{grad } F, Y\rangle - \langle \text{grad } F, \hat{\nabla}_X Y\rangle$$

$$= X(YF) - (\hat{\nabla}_X Y)(F) = H_F(X,Y).$$

Q.E.D.

Remark 5.2: The last line of the preceding proof may be taken as the definition of the Hessian. One can check that for any smooth function F, H_F is a symmetric (0,2) tensor field. At a critical point of F, this definition reduces to the one made in Section 1 of Chapter 1. In the calculation above, F is constant on M, so that $X(YF) \equiv 0$. However,

$$(\tilde{\nabla}_X Y)F = (\nabla_X Y)F + \langle AX, Y\rangle \xi F$$

$$= \langle AX, Y\rangle \langle \text{grad } F, \xi\rangle = \langle AX, Y\rangle \rho,$$

need not be zero.

Theorem 5.3: With the notation of Theorem 5.1, the mean curvature $h = (\text{trace } A)/n$ satisfies

$$h = \frac{1}{n\rho^2}\left(\langle \text{grad } F, \text{grad } \rho\rangle - \rho\, \Delta F\right).$$

Proof: Let $e_1, \ldots, e_n$ be an orthonormal basis for the tangent space T_xM at a given point x. Using the formula (5.0) we compute

$$nh = -\frac{1}{\rho}\sum_{i=1}^{n} \langle \tilde{\nabla}_{e_i} \text{grad } F, e_i\rangle$$

$$= \frac{-1}{\rho}\left(\Delta F - \langle \tilde{\nabla}_{\xi} \text{grad } F, \xi\rangle\right)$$

$$= \frac{-1}{\rho}\left(\Delta F - \frac{1}{\rho^2}\langle \tilde{\nabla}_{\text{grad } F} \text{grad } F, \text{grad } F\rangle\right)$$

$$= \frac{-1}{\rho}\left(\Delta F - \frac{1}{2\rho^2}\tilde{\nabla}_{\text{grad } F}|\text{grad } F|^2\right)$$

$$= \frac{-1}{\rho}\left(\Delta F - \frac{2\rho}{2\rho^2}\tilde{\nabla}_{\text{grad } F}\,\rho\right)$$

$$= \frac{-1}{\rho}\left(\Delta F - \frac{1}{\rho}\langle \text{grad } F, \text{grad } \rho\rangle\right)$$

$$= \frac{1}{\rho^2} (\langle \text{grad } F, \text{grad } \rho \rangle - \rho\, \Delta F).$$

Q.E.D.

So far, the results in this section are valid for any level hypersurface. We now specialize to the case where the function F has a special form, as in Cartan's original definition of isoparametric.

<u>Definition 5.4</u>: Let $\hat{M}$ be a space of constant curvature. A smooth real valued function F on $\hat{M}$ is called an <u>isoparametric function</u> if there exist smooth real-valued functions T and S such that

$$|\text{grad } F|^2 = T \circ F \quad \text{and} \quad \Delta F = S \circ F.$$

<u>Remark 5.5</u>: As shown in (4.18) and (4.19), the Cartan polynomial V on S^{n+1} constructed in the last section is an isoparametric function with

$$T(u) = g^2(1-u^2)$$

$$S(u) = g\, (\frac{(m_2 - m_1)}{2} g - (n + g)u)).$$

The functions giving rise to the isoparametric families in Example 3.2 are likewise isoparametric when restricted to the sphere.

We are working toward proving that every isoparametric function is related in a similar way to an isoparametric family of hypersurfaces.

<u>Theorem 5.6</u>: <u>If $F:\hat{M} \to R$ is an isoparametric function, then each level hypersurface of F has constant mean curvature.</u>

<u>Proof</u>: Using Theorem 5.3 to compute the mean curvature, we get

$$nh = \frac{1}{\rho^2} (\langle \text{grad } F, \text{grad } \rho \rangle - \rho\, \Delta F)$$

$$= \frac{1}{\rho^2} (\langle \text{grad } F, \frac{T' \circ F}{2\rho} \text{grad } F \rangle - \rho\, S \circ F)$$

$$= \frac{T' \circ F}{2\rho} - \frac{S \circ F}{\rho} = \left(\frac{1}{2\sqrt{T}}(T' - 2S)\right) \circ F.$$

Thus, on each level hypersurface M_s, the mean curvature h is constant.

Q.E.D.

The vector field $\xi = \text{grad } F / |\text{grad } F|$ is defined on the open subset of $\hat{M}$ on which $|\text{grad } F| \neq 0$. We now show

Theorem 5.7: Let $F:\hat{M} \to R$ be a function for which $|\text{grad } F|$ is a function of F. Then the integral curves to the vector field $\xi = \text{grad } F/|\text{grad } F|$ are geodesics in $\hat{M}$.

Proof: For this local calculation, let X be a vector field in a neighborhood of a point where $|\text{grad } F| \neq 0$ such that X is orthogonal to ξ at each point. Thus X is everywhere tangent to a level surface of F. We need to show that $\langle \hat{\nabla}_\xi \xi, X \rangle = 0$, since we know $\langle \hat{\nabla}_\xi \xi, \xi \rangle = 0$ already. We have

$$\langle \hat{\nabla}_\xi \xi, X \rangle = - \langle \xi, \hat{\nabla}_\xi X \rangle = - \langle \xi, \hat{\nabla}_X \xi + [X, \xi] \rangle,$$

where [,] is the Lie bracket. However, $\langle \xi, \hat{\nabla}_X \xi \rangle = 0$, so

$$\langle \hat{\nabla}_\xi \xi, X \rangle = - \langle \xi, [X, \xi] \rangle.$$

We complete the proof by showing that $[X, \xi]F = 0$ at all points, and hence $\langle \xi, [X, \xi] \rangle = 0$. We have

$$[X, \xi]F = X(\xi F) - \xi(XF).$$

Since X is orthogonal to grad F at all points, XF is identically zero, and so $\xi(XF) = 0$. On the other hand, since $\xi F = |\text{grad } F|$, and $|\text{grad } F|$ is a function of F, we have $X(\xi F) = 0$.

Q.E.D.

We remark that Theorem 5.7 shows that a family of level hypersurfaces of an isoparametric function is a family of parallel hypersurfaces, modulo reparametrization to take into account the possibility that $|\text{grad } F|$ is not identically equal to one.

Since we know that the level surfaces of an isoparametric function all have constant mean curvature by Theorem 5.6, the next result will imply that the level surfaces of an isoparametric function all have constant principal curvatures, i.e. they are isoparametric as defined earlier.

<u>Theorem 5.8</u>: <u>Let</u> $\phi_t : M \to \hat{M}$, $-\varepsilon < t < \varepsilon$, <u>be a family of parallel hypersurfaces in a real space form. Then</u> $\phi_0 M$ <u>has constant principal curvatures if and only if each</u> $\phi_t M$ <u>has constant mean curvature</u>.

<u>Proof</u>: See Nomizu [2, p. 192] for the proof when $\hat{M} = E^{n+1}$. We provide a proof here for the case $\hat{M} = S^{n+1}$, and the proof when $\hat{M}$ has constant negative curvature is likewise very similar.

Let $\lambda_i = \cot \theta_i$, $1 \leq i \leq n$, be the principal curvatures of ϕ_0, where we are not assuming that the λ_i are necessarily distinct. By Theorem 1.1, the principal curvatures of $\phi_t(M)$ are $\cot(\theta_i - t)$, $1 \leq i \leq n$. Obviously, if the λ_i are all constant, then ϕ_t has constant principal curvatures, and therefore constant mean curvature.

Conversely, suppose each ϕ_t has constant mean curvature, so that the function,

$$a(t) = \sum_{i=1}^{n} \cot(\theta_i(x) - t),$$

is a function of t alone, even though the $\theta_i(x)$ are assumed to depend on the point $x \in M$. For $t = 0$, we have

$$a(0) = \sum_{i=1}^{n} \cot \theta_i(x) = C_1, \text{ a constant.} \tag{5.1}$$

If we evaluate $a'(0)$, we get

$$a'(0) = \sum_{i=1}^{n} \csc^2\theta_i(x) = \sum_{i=1}^{n} 1 + \cot^2\theta_i(x),$$

and hence

$$\sum_{i=1}^{n} \cot^2\theta_i(x) = C_2, \qquad \text{a constant.}$$

Next,

$$a''(0) = \sum_{i=1}^{n} 2\csc^2\theta_i(x) \cot\theta_i(x) = 2\sum_{i=1}^{n} \cot\theta_i(x) + \cot^3\theta_i(x).$$

Using (5.1), we find

$$\sum_{i=1}^{n} \cot^3\theta_i(x) = C_3, \qquad \text{a constant.}$$

Continuing this process, by evaluating $a^{(3)}(0),\ldots, a^{(n-1)}(0)$, one finds that the functions

$$s_k(x) = \sum_{i=1}^{n} \cot^k\theta_i(x) = \sum_{i=1}^{n} \lambda_i^{\,k}(x), \qquad 1 \leq k \leq n,$$

are constant. By Newton's identities (see Van der Waerden [1, p. 81]) the coefficients of the characteristic polynomial of A at x are polynomials in the $s_k(x)$ which we have just shown to be constant. Thus the principal curvatures themselves are constant.

Q.E.D.

For the sake of completeness, we state the following consequence of Theorems 5.6 – 5.8.

Corollary 5.9: If $F:\tilde{M} \to R$ is an isoparametric function on a real space form, then each level hypersurface of F has constant principal curvatures.

On the other hand, we saw in the previous section how to begin with a connected hypersurface M with constant principal curvatures in S^{n+1}, and produce an isoparametric polynomial function V, such that M is a piece of a level set M_s of V. By Corollary 5.9, this level hypersurface M_s has constant principal curvatures and is thus an extension of M to a compact isoparametric hypersurface. In the next section, we will see that M_s is connected.

We now see that any solution of the Münzner differential equations (4.1) and (4.2) is related to a Cartan polynomial.

Theorem 5.10 (Münzner [1]): $\hat{F}:E^{n+2} \to R$ be a homogeneous polynomial of degree $\tilde{g}$ which satisfies the Münzner differential equations (4.1) and (4.2) in Theorem 4.1 with parameters $\tilde{g}$ and $\tilde{c}$ such that the restriction $\hat{V}$ of $\hat{F}$ to S^{n+1} is not constant. Then 0 is a regular value of $\hat{V}$ and $\hat{V}^{-1}(0)$ is an oriented isoparametric hypersurface with normal $\text{grad}^S\hat{V}$. Let F be the Cartan polynomial of a connected component of $\hat{V}^{-1}(0)$. Then, either $\hat{F} = F$ or $\hat{F} = \pm(2F^2 - r^{2g})$, in which case $c = 0$, $\tilde{g} = 2g$, $\tilde{c} = \mp\tilde{g}n$.

Remark 5.11: It is possible to find other polynomials which when restricted to S^{n+1} have a family of isoparametric hypersurfaces as level sets, for example, the polynomial G in Example 3.2.b defined on $E^{n+2} = E^{j+1} \times E^{k+1}$ by

$$G(x,y) = |x|^2.$$

Recall that G gives the same family of products of spheres as the Cartan polynomial

$$F(x,y) = |x|^2 - |y|^2,$$

since $F = 2G - 1$ on S^{n+1}. Note, however, that G does not satisfy the differential equations of Münzner.

Corollary 5.12: Let $F:E^{n+2} \to R$ be a homogeneous polynomial of degree g which satisfies the Münzner differential equations with parameters c and g. Suppose that the restriction V of F to S^{n+1} is not constant. Then the following statements are equivalent:

(a) <u>F is the Cartan polynomial of a connected, oriented isoparametric hypersurface.</u>

(b) <u>$c \neq \pm gn$.</u>

(c) <u>F is the Cartan polynomial of each component of $V^{-1}(0)$.</u>

<u>Proof of Corollary 5.12</u>: Note that $n = (m_1 + m_2)g/2$, and hence

$$\pm gn = \pm(m_1 + m_2)g^2/2.$$

But if F is a Cartan polynomial, then by Theorem 4.1,

$$c = (m_2 - m_1)g^2/2.$$

Thus, $c = \pm gn$ is not possible, since m_1 and m_2 are non-zero. So (a) implies (b). Next, if we assume (b), then by Theorem 5.10, F must be the Cartan polynomial of each connected component of $V^{-1}(0)$. Thus (b) implies (c) and, obviously, (c) implies (a).

Q.E.D.

<u>Proof of Theorem 5.10</u>: We are assuming that $\hat{F}$ satisfies the differential equations of Münzner

$$|\mathrm{grad}^E \hat{F}|^2 = \tilde{g}^2 r^{2\tilde{g}-2}, \tag{5.2}$$

$$\Delta^E \hat{F} = \tilde{c}\, r^{\tilde{g}-2},$$

for some constants $\tilde{c}$, $\tilde{g}$. By Theorem 3.1, the restriction $\hat{V}$ of $\hat{F}$ to S^{n+1} satisfies

$$|\mathrm{grad}^S \hat{V}|^2 = \tilde{g}^2(1 - \hat{V}^2), \tag{5.3}$$

$$\Delta^S \hat{V} = \tilde{c} - \tilde{g}(\tilde{g} + n)\hat{V}.$$

As we remarked earlier, assuming $\hat{V}$ is not constant, it must have a maximum value of 1 and a minimum value of -1 on S^{n+1} by (5.3).

Let M_0 be a connected component of the level hypersurface $\hat{V}^{-1}(0)$. Pick $x_0 \in M_0$, and a neighborhood U_0 of x_0 in S^{n+1} on which $|\hat{V}| \neq 1$. Since M_0 has constant principal curvatures by Corollary 5.9, we may construct its Cartan polynomial F, as in the preceding section. In particular, we can define the function τ on the cone over U_0. We have $\tau(x_0) = \theta_1$, and $0 < \theta_1 < \pi/g$, by Theorem 2.3, where g is the number of distinct principal curvatures of M_0. Then by (4.3),

$$(\text{grad } \tau)_p = -\tilde{\xi}_p,$$

where $\tilde{\xi}_p$ is a unit normal to the parallel hypersurface $\phi_t(M_0)$ at distance $t = \theta_1 - \tau(p)$ at the point $p = \phi_t(x)$, i.e.

$$\tilde{\xi}_p = -\sin(\theta_1 - \tau(p))x + \cos(\theta_1 - \tau(p))\xi_x.$$

We may assume for convenience that

$$\xi_x = \frac{\text{grad } \hat{V}(x)}{|\text{grad } \hat{V}(x)|} = \frac{\text{grad } \hat{V}(x)}{\tilde{g}},$$

for each x in $M_0 \cap U_0$.

We know that τ leads to a Cartan polynomial F, whose restriction to S^{n+1} will be called V. We now attempt to compare V and $\hat{V}$. To do this, consider

$$\tilde{\tau} = \frac{1}{\tilde{g}} \text{arc cos } \hat{V} \quad \text{on } U_0.$$

Then

$$\text{grad } \tilde{\tau} = \frac{1}{\tilde{g}} \left(\frac{-1}{(1-\hat{V}^2)^{1/2}}\right) \text{grad } \hat{V}. \tag{5.4}$$

Since $|\text{grad } \hat{V}| = \tilde{g}(1 - \hat{V}^2)^{1/2}$ by (5.3), the right-hand side of the above equation is a unit vector. Furthermore, it is normal to the level hypersurface of $\hat{V}$ at each point z of U_0. Since this hypersurface coincides with a parallel hypersurface of M_0 by Theorem 5.7, its normal coincides with the normal $\hat{\xi}$ defined earlier. Hence, if $p = \phi_t(x)$ for $x \in M_0$ as above, we have

$$\hat{\xi}_p = \frac{1}{\hat{g}} \frac{\operatorname{grad} \hat{V}\,(p)}{(1 - \hat{V}^2(p))^{1/2}}.$$

Thus τ and $\hat{\tau}$ have the same gradient, and therefore differ by a constant, i.e. $\hat{\tau} = \tau + a$, where $|a|$ is less than the maximum of $\{\pi/g, \pi/\hat{g}\}$.

Since grad τ = grad $\hat{\tau} = -\hat{\xi}$, we have from Remark 4.3 that

$$\Delta\hat{\tau} = \Delta\tau = \frac{1}{\sin g\tau}(n \cos g\tau - c/g), \tag{5.5}$$

where $c = (m_2 - m_1)g^2/2$, and m_1, m_2 have the usual meaning as multiplicities of principal curvatures of M_0. On the other hand, we compute from (5.3) and (5.4) that

$$\begin{aligned}
\Delta\,\hat{\tau} &= \operatorname{div} \operatorname{grad} \hat{\tau} \\
&= \frac{-1}{\hat{g}}\left(\langle \operatorname{grad}\,(1 - \hat{V}^2)^{1/2}, \operatorname{grad} \hat{V}\rangle + (1 - \hat{V}^2)^{-1/2}\Delta\hat{V}\right) \\
&= \frac{-1}{\hat{g}}\left((-1/2)(1 - \hat{V}^2)^{-3/2}(-2\hat{V})\,|\operatorname{grad} \hat{V}|^2 + (1 - \hat{V}^2)^{-1/2}\Delta\hat{V}\right) \\
&= -(1 - \hat{V}^2)^{-1/2}\,\hat{g}\,\hat{V} + (1 - \hat{V})^{-1/2}(n+\hat{g})\hat{V} - \frac{\hat{c}}{\hat{g}}\,(1 - \hat{V}^2)^{-1/2} \\
&= (1 - \hat{V}^2)^{-1/2}(n\,\hat{V} - \hat{c}/\hat{g}) \\
&= \frac{1}{\sin \hat{g}\,\hat{\tau}}\,(n \cos \hat{g}\,\hat{\tau} - \hat{c}/\hat{g}).
\end{aligned} \tag{5.6}$$

Here $\hat{g}$ and $\hat{c}$ are as given in (5.2) but have no geometric meaning, as yet.

We have shown that $\hat{\tau} = \tau + a$ for some constant a on U_0. Using this, and equating the two expressions (5.5) and (5.6) for $\Delta\hat{\tau}$, we get

$$\frac{n \cos g\tau - c/g}{\sin g\,\tau} = \frac{n \cos \tilde{g}(\tau + a) - \tilde{c}/\tilde{g}}{\sin \tilde{g}\,(\tau + a)}.$$

Since, along a suitable normal geodesic in U_0, a whole interval of values of τ are covered, the equation

$$(5.7) \qquad (n \cos g\tau - c/g) \sin \tilde{g}(\tau + a) = (n \cos \tilde{g}(\tau + a) - \tilde{c}/\tilde{g}) \sin g\tau$$

must hold as an identity in τ (each side being an analytic expression).

First note that the right side is zero at integral multiples of π/g, while since $n = (m_1 + m_2)g/2$, we have

$$n \cos g\tau - \frac{c}{g} = \pm \frac{(m_1 + m_2)}{2} g - \frac{(m_2 - m_1)}{2} g \neq 0,$$

when τ is an integral multiple of π/g. Thus $\sin g(\tau + a)$ must vanish at such points. Hence for any integer k,

$$\tilde{g}\,(\frac{k\pi}{g} + a) = \ell\pi,$$

for some integer ℓ. Setting $k = 0$ shows that a is an integral multiple of $\pi/\tilde{g}$, while $k = 1$ shows that π/g is an integral multiple of $\pi/\tilde{g}$. Thus, we may write

$$\tilde{g} = \alpha g$$

for some positive integer α. The restriction on $|a|$ reduces to $|a| < \pi/g$, since $g \leqslant \tilde{g}$.

If $\alpha = 1$, we must have $a = 0$, so that $\tau = \hat{\tau}$, and thus $V = \hat{V}$, $F = \hat{F}$, $c = \hat{c}$. Now assume that $\alpha > 1$. Substituting $a = \ell\pi/\tilde{g}$ for an integer ℓ into (5.7) yields

$$(n \cos g\tau - c/g) \sin(\alpha g\tau + \ell\pi) = (n \cos(\alpha g\tau + \ell\pi) - \tilde{c}/\tilde{g})(\sin g\tau).$$

Thus, either

$$(5.8) \qquad \frac{\sin \alpha g\tau}{\sin g\tau} = \frac{n \cos \alpha g\tau - \tilde{c}/\tilde{g}}{n \cos g\tau - c/g},$$

or

$$(5.9) \qquad \frac{\sin \alpha g\tau}{\sin g\tau} = \frac{n \cos \alpha g\tau + \tilde{c}/\tilde{g}}{n \cos g\tau - c/g},$$

depending on whether ℓ is even or odd. Set $\tau = \pi/\alpha g$. Then (5.8) yields

$$(5.10) \qquad -n - \tilde{c}/\tilde{g} = 0, \text{ i.e. } \tilde{c} = -\alpha gn, \text{ for } \ell \text{ even.}$$

Similarly (5.9) yields

$$(5.11) \qquad \tilde{c} = \alpha gn, \qquad \text{for } \ell \text{ odd.}$$

On the other hand, taking $\tau = 2\pi/\alpha g$ in (5.8) yields

$$(5.12) \qquad (\sin 2\pi/\alpha)\ (n - \tilde{c}/\tilde{g}) = 0, \qquad \text{for } \ell \text{ even,}$$

and (5.9) gives

$$(5.13) \qquad (\sin 2\pi/\alpha)\ (n + \tilde{c}/\tilde{g}) = 0, \qquad \text{for } \ell \text{ odd.}$$

But equations (5.12) and (5.13), respectively, contradict equations (5.10) and (5.11), unless $\sin(2\pi/\alpha) = 0$, i.e. $\alpha = 2$. Thus $\tilde{g} = 2g$. Then

$$|a| = |\ell\pi/2g| < \pi/g$$

implies that $\ell = 0$, 1, or -1. If $\ell = 0$, we have from (5.8) and (5.10)

$$(5.14) \qquad \frac{\sin 2g\tau}{\sin g\tau} = \frac{n\ (\cos 2g\tau + 1)}{n \cos g\tau - c/g}$$

which, using half-angle formulas, gives

$$2 \cos g\tau = \frac{2n \cos^2 g\tau}{n \cos g\tau - c/g}.$$

Thus $c = 0$, $\hat{c} = -2ng = -n\hat{g}$, $\tau = \hat{\tau}$, and

$$\hat{V} = \cos \hat{g}\hat{\tau} = \cos 2g\tau = 2 \cos^2 g\tau - 1 = 2V^2 - 1.$$

There is at most one homogeneous polynomial which extends $\hat{V}$, namely $2F^2 - r^{2g}$. So this must be $\hat{F}$. On the other hand, if $\ell = \pm 1$, we get from (5.9) and (5.11) that

$$\frac{\sin 2g\tau}{\sin g\tau} = \frac{n (\cos 2g\tau + 1)}{n \cos g\tau - c/g},$$

so that $c = 0$, $\hat{c} = 2ng = n\hat{g}$, $\hat{\tau} = \tau \pm \pi/2g$. Then,

$$\hat{V} = \cos \hat{g}\hat{\tau} = \cos (2g\tau \pm \pi) = - \cos 2g\tau$$

$$= - (2V^2-1),$$

and hence,

$$\hat{F} = -(2F^2 - r^{2g}),$$

as desired. In both of the last two cases, the conditions $c = 0$, $\hat{g} = 2g$, $\hat{c} = \pm \hat{g}n$ are satisfied.

6. THE GLOBAL STRUCTURE OF AN ISOPARAMETRIC FAMILY OF HYPERSURFACES.

Using the Cartan polynomials and the normal exponential map, Münzner [1] was able to obtain a very explicit global description of an isoparametric family of hypersurfaces in the sphere S^{n+1}. Specifically, he shows that each compact isoparametric hypersurface M divides S^{n+1} into two manifolds with boundary, each of which is a ball bundle over one of the two components M_+ and M_- of the focal set of M. By using standard theorems from algebraic topology, he is able to compute the Z_2-cohomology groups of M, M_+, M_-. He shows that the sum of the Z_2-Betti numbers $\beta(M) = 2g$, while

$\beta(M_+) = \beta(M_-) = g$, where g is the number of distinct principal curvatures of M. On the other hand, one can deduce that every non-degenerate spherical distance function L_p has $2g$ critical points on M, and g critical points on M_+ and M_-. Thus, these manifolds are all taut. Furthermore, it follows easily from Münzner's result that M is also totally focal, thereby proving the converse of Theorem 9.25 of Chapter 2. In this section, we prove Münzner's structure theorem and its consequences mentioned above.

Let F be a Cartan polynomial, V the restriction of F to S^{n+1}, and $M = M_0 = V^{-1}(0)$. At this point, we do not know that M must be connected. We do know by Corollary 5.12, that F is the Cartan polynomial of each connected component of M. For $x \in M$, $V(x) = 0$ and so from the construction of the Cartan polynomials, we know that $\tau(x) = \frac{\pi}{2g}$. Thus, the largest principal curvature of M is $\cot \theta_1 = \cot(\frac{\pi}{2g})$. Let $m_+ = m_1$ denote the multiplicity of this principal curvature, and let $f_+ : M \to S^{n+1}$ be the corresponding focal map,

$$f_+(x) = \phi(x, \frac{\pi}{2g}) = \cos \frac{\pi}{2g}\, x + \sin \frac{\pi}{2g}\, \xi(x).$$

By Theorem 2.3, $\theta_g = \pi - \frac{\pi}{2g}$, and so the first focal point of (M,x) in the opposite direction is

$$f_-(x) = \phi(x, \frac{-\pi}{2g}) = \cos(\frac{-\pi}{2g})\, x + \sin(\frac{-\pi}{2g})\, \xi(x).$$

Let $m_- = m_2$ denote the multiplicity of θ_g.

<u>Theorem 6.1</u>: <u>Let F be a Cartan polynomial and V its restriction to S^{n+1}.</u> <u>Let $M_t = V^{-1}(t)$, $-1 \leq t \leq 1$. Then each M_t is connected. Furthermore, M_1 and M_{-1} are the focal submanifolds $f_+(M_0)$ and $f_-(M_0)$, respectively.</u>

<u>Proof</u>: We let $M = V^{-1}(0)$ and define a map $d : M \times R \to S^{n+1}$ by

$$d(x,\tau) = \phi(x, \frac{\pi}{2g} - \tau) = \cos(\frac{\pi}{2g} - \tau)\, x + \sin(\frac{\pi}{2g} - \tau)\, \xi(x).$$

From the construction of the Cartan polynomials, we have on an open neighborhood of $M \times \{\frac{\pi}{2g}\}$,

$$V(d(x,\tau)) = \cos g\tau, \tag{6.1}$$

since $\theta_1 = \frac{\pi}{2g}$ on M. By analyticity, the relation (6.1) holds on all of $M \times R$. Note that $V = 1$ on the focal set $f_+(M) = d(M \times 0)$, while $V = -1$ on the focal set $f_-(M) = d(M \times \frac{\pi}{g})$. That is, $f_+(M) \subset M_1$ and $f_-(M) \subset M_{-1}$. We will show that $f_+(M) = M_1$ and $f_-(M) = M_{-1}$. The restriction of d to the set $\{x\} \times (0, \frac{\pi}{g})$ is an integral curve of the vector field $-\xi$. This integral curve is a geodesic in S^{n+1} which intersects M only at the point x by (6.1), and also intersects each M_t, $t \in (-1, 1)$, in precisely one point. Hence, $V^{-1}(-1,1)$ is the disjoint union of these integral curves of $-\xi$. The map d is clearly a local diffeomorphism on $M \times (0, \frac{\pi}{g})$, since its image contains no focal points. Since $V^{-1}(-1,1)$ is the disjoint union of the integral curves of $-\xi$, d is actually a diffeomorphism of $M \times (0, \frac{\pi}{g})$ onto $V^{-1}(-1,1)$. The image

$$d(M \times [0, \tfrac{\pi}{g}])$$

is clearly compact, and it contains the dense open set $V^{-1}(-1,1)$. Hence $d(M \times [0,\frac{\pi}{g}]) = S^{n+1}$. (The set $V^{-1}(-1,1)$ is dense, because if $V \equiv \pm 1$ on an open set in S^{n+1}, then $V \equiv \pm 1$ on all of S^{n+1} by the analyticity of V.)

Since $M_1 \subset d(M \times [0, \frac{\pi}{g}])$, the relation (6.1) implies that

$$M_1 = d(M \times 0) = \phi(M \times \tfrac{\pi}{2g}) = f_+(M).$$

Similarly, $M_{-1} = f_-(M)$. By Theorems 4.6 and 4.10 of Chapter 2, the focal submanifolds $f_+(M)$ and $f_-(M)$ have dimensions $n-m_+$ and $n-m_-$, respectively. Thus, they both have codimension greater than one, and so

$$S^{n+1} - (M_+ \cup M_-) = V^{-1}(-1,1)$$

is connected. Since $V^{-1}(-1,1)$ is diffeomorphic to $M \times (0,\frac{\pi}{g})$, we conclude that M is connected. Since each $M_t = d(M \times \{s\})$ for the appropriate choice of $s \in [0,\frac{\pi}{g}]$, each M_t is also connected.

Q.E.D.

We now prove the global structure theorem of Münzner [1]. $M = V^{-1}(0)$ is an isoparametric hypersurface as in the previous theorem.

Theorem 6.2: Let $k = \pm 1$, and let Z be a normal vector to the focal submanifold M_k in S^{n+1}. Let $\exp: NM_k \to S^{n+1}$ denote the normal exponential map for M_k. Then

(a) $V(\exp Z) = k \cos(g\,|Z|)$.

(b) Let $B_k = \{q \in S^{n+1} \mid k\,V(q) \geq 0\}$ and let $(B^{\perp}M_k, S^{\perp}M_k)$ be the bounded unit ball bundle in NM_k. Then

$$\psi_k:(B^{\perp}M_k, S^{\perp}M_k) \to (B_k, M)$$

where $M = V^{-1}(0)$ and $\psi_k(Z) = \exp(\frac{\pi}{2g} Z)$ is a diffeomorphism of manifolds with boundary.

Proof: Let $p \in M_k$ and Z be a unit normal to M_k at p, i.e. $Z \in S_p^{\perp}M_k$. Then, the focal map $f_k:M \to M_k$ maps an m_k-dimensional sphere onto p. As we showed by considering the map $h:f_k^{-1}(p) \to S_p^{\perp}M_k$ at the beginning on Section 2, there exists a point $x \in f_k^{-1}(p)$ such that Z is the tangent vector to the geodesic in S^{n+1} starting at x with initial direction $\xi(x)$. Thus, $\exp(\tau Z)$ and $d(x,\tau)$ traverse the same geodesic in S^{n+1} with unit speed as τ varies over R. From (6.1), we get that

$$V(\exp \tau Z) = \cos(g(\tau + \tau')) \text{ for some } \tau' \in [0,\frac{2\pi}{g}).$$

For $\tau = 0$, the left side equals k, and thus $\tau' = 0$ if $k = 1$, and $\tau' = \frac{\pi}{g}$ if $k = -1$. This proves (a). To prove (b), note that (a) implies that the family of all normal geodesics to M_k is the same as the family of all normal geodesics to M. Each point in $B_k - M_k$ lies on precisely one level hypersurface of V, and hence on precisely one normal geodesic to M_k. Thus, ψ_k is bijective and so it is a diffeomorphism, since it is clearly a local diffeomorphism.

Q.E.D.

Since the set of normal geodesics to each focal submanifold is the same as the set of normal geodesics to each of the parallel hypersurfaces, we immediately obtain:

Corollary 6.3: Let M_k, $k = \pm 1$, be a focal submanifold of an isoparametric hypersurface M. Then the focal set of M_k is the same as the focal set of M, i.e. it is $M_k \cup M_{-k}$.

We turn briefly to the question of tautness and derive some immediate consequences of Theorems 6.1 and 6.2. Let M be a compact submanifold of S^m. Then for $p \in S^m$, $x \in M$, the spherical distance function $L_p : M \to R$ has the form

$$L_p(x) = \cos^{-1} \langle p,x \rangle.$$

One has an index theorem for these spherical distance function analogous to Theorem 1.10 of Chapter 2 for Euclidean distance functions. The theorem states that L_p has a critical point at $x \in M$ if and only if p lies on a normal geodesic to M at x. The critical point of L_p is non-degenerate if and only if p is not a focal point of (M,x). The index of L_p at a non-degenerate critical point x is equal to the number of focal points (counting multiplicities) of (M,x) on the shortest geodesic segment from p to x.

Let M be a compact hypersurface in S^{n+1}, and let $\phi : M \times R \to S^{n+1}$ denote the normal exponential map. From the index theorem, we obtain immediately:

Proposition 6.4: Let M be a compact embedded hypersurface in S^{n+1}. If L_p

is non-degenerate on M, then the number of critical points of L_p on M is equal to the number of pre-images under ϕ of p in the set $M \times (-\pi, \pi]$.

Assume again that M is an isoparametric hypersurface in S^{n+1}. Note that Corollary 6.3 says that a distance function L_p is non-degenerate on M if and only if it is non-degenerate on each focal submanifold of M. From Theorems 6.1 and 6.2, we obtain:

Theorem 6.5: Let M be a compact isoparametric hypersurface in S^{n+1} with g distinct principal curvatures. Let M_+ and M_- denote the two focal submanifolds of M. Then,

(a) Every non-degenerate L_p has 2g critical points on M.

(b) Every non-degenerate L_p has g critical points on M_+ and g critical points on M_-.

Proof (a): Fix a choice of unit normal ξ to M and let $\cot \theta_1$ be the largest principal curvature of M. As in Theorem 6.1, define a map $d: M \times R \to S^{n+1}$ by

$$d(x, \tau) = \phi(x, \theta_1 - \tau) = \cos(\theta_1 - \tau)x + \sin(\theta_1 - \tau)\xi.$$

Let V be the restriction of the Cartan polynomial for M to S^{n+1}. Then, as in Theorem 6.1

$$V(d(x, \tau)) = \cos g\tau. \tag{6.2}$$

As shown in Theorem 6.1, the restriction of d to $M \times (0, \frac{\pi}{g})$ is a diffeomorphism onto $V^{-1}(-1,1) = S^{n+1} - (M_+ \cup M_-)$. The same is true if $(0, \frac{\pi}{g})$ is replaced by any interval $(j\pi/g, (j+1)\pi/g)$ for any integer j. Thus, as τ varies over $(-\pi, \pi]$, each p in $S^{n+1} - (M_+ \cup M_-)$ is covered 2g times by d, i.e. L_p has 2g critical points on M.

(b). We noted in Corollary 6.3, that L_p is non-degenerate on M_+ or M_- if and only if $p \in S^{n+1} - (M_+ \cup M_-)$. In that case, p lies on precisely one of the isoparametric hypersurfaces $M_t = V^{-1}(t)$, $-1 < t < 1$. There is exactly one great circle γ through p which is normal to each hypersurface in the family. γ is also the only great circle through p which meets M_+ or M_- orthogonally. The geodesic γ can be parametrized as

$$\gamma(\tau) = d(p,\tau).$$

As τ varies over $(-\pi,\pi]$, γ alternately intersects M_+ then M_- at values of $\tau = j\pi/g$, for j an integer, by equation (6.2). Thus, γ meets M_+ and M_- exactly g times each, i.e. L_p has g critical points on M_+ and g critical points on M_-.

Q.E.D.

After we have established that $\beta(M) = 2g$ and $\beta(M_+) = \beta(M_-) = g$, we will have immediately from Theorem 6.5 that M, M_+, M_- are taut.

Using the notation of the previous theorem, suppose p is a focal point of (M,x). Say $p \in M_+$ so that $V(p) = 1$. If p is also equal to $d(y,\tau)$ for some $y \neq x$ in M, then by equation (6.2), τ must be an even integral multiple of π/g. If p were in M_-, then τ would have to be an odd multiple of π/g. In any case, p would also be a focal point of (M,y). This shows that all of the critical points of L_p are degenerate. Hence, if M is an isoparametric hypersurface in S^{n+1}, then every L_p is either non-degenerate, or has only degenerate critical points, i.e. M is totally focal. We thus have the converse of Theorem 9.25 of Chapter 2.

Theorem 6.6: A compact isoparametric hypersurface in S^{n+1} is totally focal.

Remark 6.7: The focal submanifolds M_+ and M_- are not totally focal. For if $p \in M_+$, then p is a focal point of M_+, but L_p has a non-degenerate absolute minimum at p itself. The same holds for M_-, obviously.

In the summary of his first paper, Münzner makes some remarks concerning the global geometry of an isoparametric family which we paraphrase here. Consider the isoparametric family given by a Cartan polynomial V on S^{n+1}. Let $M = V^{-1}(0)$, $M_+ = V^{-1}(1)$, $M_- = V^{-1}(-1)$, $M_t = V^{-1}(t)$, $-1 < t < 1$ and

$\xi = \text{grad } V/|\text{grad } V|$ at points where grad $V \neq 0$. Let $p \in M_+$ and let Z be a unit normal to M_+ at p. We consider the great circle, $\gamma(\tau) = \exp(\tau Z)$, $0 \leq \tau \leq 2\pi$. The path begins at p and intersects M at $\tau = \pi/2g$. Note that the tangent vector to the curve at $\gamma(\pi/2g)$ is $\vec{\gamma}(\pi/2g) = -\xi$ since V is decreasing as one moves along γ at that point. The geodesic γ intersects M_- at $\gamma(\pi/g)$. As γ leaves that point, it begins travelling in the direction of grad V. Hence, when γ intersects M again at $\gamma(3\pi/2g)$, the tangent vector is $\vec{\gamma}(3\pi/2g) = \xi$. The geodesic continues on, intersecting M_+ at points $\gamma(j\pi/2g)$ for $j \equiv 0 \pmod 4$, intersecting M_- at $\gamma(j\pi/2g)$ for $j \equiv 2 \pmod 4$ and intersecting M at $\gamma(j\pi/2g)$ for $j \equiv 1 \pmod 2$. The tangent vector $\vec{\gamma}$ is alternately equal to ξ and $-\xi$ at consecutive points of intersection with M. Of course, similar statements could be made for any hypersurface M_t. The formula for the points of intersection of γ with M_t would just be slightly more complicated.

Part (a) of Theorem 6.2, shows that each M_t is a tube over each component of its focal set. In fact, M_t is realized as a sphere bundle over each focal submanifold in many different ways. As before, suppose that the principal curvatures of M_t are given as $\lambda_i = \cot \theta_i$, $1 \leq i \leq g$. Let $f_i : M_t \to S^{n+1}$ be the focal map,

$$f_i(x) = \cos \theta_i \; x + \sin \theta_i \; \xi(x).$$

As the discussion above shows, $f_1(M_t) = M_+$, and M_t is an m_+-dimensional sphere bundle over M_+ whose fibres are the leaves of the foliation T_1. Then, $f_2(M_t) = M_-$, and M_t is an m_--sphere bundle over M_- whose fibres are the leaves of T_2. Next, $f_3(M_t) = M_+$, $f_4(M_t) = M_-$, etc.

Here, we again must distinguish according to whether g is even or odd. Suppose first that g is even, and let $x \in M_t$. Then the focal point $f_j(x)$ and the antipodal focal point to $f_j(x)$ which comes from the same principal curvature λ_j are both on the same focal submanifold of M_t. Thus, both M_+ and M_- are invariant under the antipodal map of S^{n+1}. In this case of even g, only the foliations $T_1, T_3, \ldots, T_{g-1}$ can be realized as sphere bundles over M_+, while only $T_2, \ldots, T_g$ can be realized as sphere bundles over M_-. Note also that each foliation can be realized in two different ways as a sphere bundle over the appropriate focal submanifold. For example, for

$x \in M_t$, the leaf of the foliation T_1 through x is a sphere of radius θ_1 centered at the focal point $p = f_1(x)$ on M_1. This leaf is also a sphere of radius $\pi - \theta_1$ centered at the focal point antipodal to p, which is also on M_+.

If g is odd and $x \in M_t$, then the focal point $f_j(x)$ and its antipodal focal point lie on different components of the focal set of M_t. Thus, the submanifolds, M_+ and M_- are antipodal to one another, and each of the principal foliations can be realized in exactly one way as a sphere bundle over each of the two focal submanifolds.

The cohomology groups of an isoparametric hypersurface.

In his second paper, Münzner proves his major result: the number g of distinct principal curvatures of an isoparametric hypersurface M in S^{n+1} is 1, 2, 3, 4 or 6. This is a lengthy, delicate computation involving the cohomology rings of M and its focal submanifolds M_+ and M_-. The structure of these cohomology rings is determined by the topological situation described in Theorem 6.2(b), i.e. M divides S^{n+1} into two ball bundles over M_+ and M_-, respectively. Specifically,

Theorem (Münzner [1]): Let M be a connected compact hypersurface in S^{n+1} which divides S^{n+1} into two ball bundles over submanifolds M_+ and M_-. Then $\alpha = (1/2)\dim_R H^*(M,R)$ can only assume the values 1,2,3,4 and 6.

Here the ring R is Z if both M_+ and M_- are orientable and Z_2, otherwise. We omit the proof of this theorem. However, we will show that for an isoparametric hypersurface $\alpha = 2g$ which combined with the theorem above, yields the main results.

We now determine the cohomology modules over R of a compact isoparametric hypersurface M in S^{n+1}. Let V be the restriction of the Cartan polynomial for M to S^{n+1}. Since all of the parallel hypersurfaces $M_t = V^{-1}(t)$ are diffeomorphic, it suffices to consider the case $M = M_0$. Let

$$B_1 = \{x \in S^{n+1} \mid V(x) \geq 0\} \text{ and } B_{-1} = \{x \in S^{n+1} \mid V(x) \leq 0\}.$$

We know from Theorem 6.2, that each B_k is diffeomorphic to a ball bundle over the $(n-m_k)$-dimensional focal submanifold M_k. Of course, m_1 and m_{-1} are the two (possibly equal) multiplicities of the principal curvatures of M. If $\mu = m_1 + m_{-1}$ and g is the number of distinct principal curvatures of M, then $g = 2n/\mu$. Thus, an isoparametric hypersurface M satisfies the hypotheses of the following topological theorem of Münzner [1].

Theorem 6.8: *Let M be a compact connected hypersurface in S^{n+1} such that:*
(a) *S^{n+1} is divided into two manifolds (B_1, M) and (B_{-1}, M) with boundary along M.*
(b) *For $k = \pm 1$, B_k has the structure of a differentiable ball bundle over a compact manifold M_k of dimension $n-m_k$.*

Let the ring of coefficients $R = Z$ if M_1 and M_{-1} are both orientable and Z_2, otherwise. Let $\mu = m_1+m_{-1}$. Then $\alpha = 2n/\mu$ is an integer, and for $k = \pm 1$

$$\underline{H}^q(M_k) = \begin{cases} R \text{ for } q \equiv 0 \pmod{\mu},\ 0 \leq q < n \\ R \text{ for } q \equiv m_{-k} \pmod{\mu},\ 0 \leq q < n \\ 0 \text{ otherwise} \end{cases}$$

Further,

$$\underline{H}^q(M) = \begin{cases} R \text{ for } q = 0, n \\ \underline{H}^q(M_+) \oplus \underline{H}^q(M_-), \text{ for } 1 \leq q \leq n-1. \end{cases}$$

If M is an isoparametric hypersurface with g distinct principal curvatures, then $g = 2n/\mu = \alpha$. From the theorem, we see that $\alpha = \frac{1}{2}\,\beta(M;R)$, and thus $\alpha = \frac{1}{2}\,\beta(M;Z_2)$ for either choice of R. Thus, $\beta(M;Z_2) = 2g$, while $\beta(M_+) = \beta(M_-) = g$, and from Theorem 6.5, we get:

Theorem 6.9: *Let M be a compact connected isoparametric hypersurface in S^{n+1}, and let M_+ and M_- be the focal submanifolds of M. Then M, M_+ and M_- are taut.*

Remark 6.10: If $\Phi : S^{n+1} \to S^{n+1}$ is a conformal tranformation and M is an isoparametric hypersurface, then $\Phi(M)$ is taut by Theorem 1.4 of Chapter 2. $\Phi(M)$ is also a Dupin hypersurface (see Section 6 of Chapter 2), whose focal set is the union of submanifolds of codimension greater than one. The behavior of the focal submanifolds requires further study. It is likely that there are usually more than two focal submanifolds, and that these need not be taut. Consider, for example, the cyclides of Dupin in E^3 whose focal curves are not taut, in general. See Section 5 of Chapter 2. Note that Theorem 6.9 is a special case of the fact that a compact Dupin hypersurface is taut. (See Section 6 of Chapter 2 and Thorbergsson [1]).

Proof of Theorem 6.8: When convenient, we will abbreviate the subscripts +1 and −1 by + and −, respectively. By part (b) of the hypotheses, M_k is a deformation retract of B_k and also of the open ball bundle $B_k - M$, for $k = \pm 1$. Thus, we have for all q,

$$H^q(B_k) = H^q(M_k) = H^q(B_k - M), \quad k = \pm 1. \tag{6.3}$$

Suppose now that $1 \leq q \leq n-1$. Then $H^q(S^{n+1}) = H^{q+1}(S^{n+1}) = 0$. Hence, from the Mayer-Vietoris sequence for the exact triad (S^{n+1}, B_+, B_-), we get using (6.3) that

$$H^q(M) \cong H^q(B_+) \oplus H^q(B_-) = H^q(M_+) \oplus H^q(M_-). \tag{6.4}$$

Since $S^{n+1} - B_k = B_{-k} - M$, the Alexander duality theorem and (6.3) give for $1 \leq q \leq n-1$,

$$H^q(M_k) \cong H^q(B_k) \cong H_{n-q}(B_{-k} - M) = H_{n-q}(M_{-k}), \quad k = \pm 1. \tag{6.5}$$

Our convention on the coefficient ring R allows us to use Poincaré duality on the $(n-m_{-k})$-dimensional manifold M_{-k}. This and (6.5) give

$$H^q(M_k) \cong H_{n-q}(M_{-k}) \cong H^{q-m_{-k}}(M_{-k}), \quad k = \pm 1. \tag{6.6}$$

Two repetitions of the formula above lead to

$$(6.7)\qquad H^q(M_k) \cong H^{q-m_+-m_-}(M_k) \quad \text{for } 1 + m_{-k} \leq q \leq n-1.$$

Since M_k is connected, we have $H^0(M_k) = R$. Since M_{-k} has dimension $n-m_{-k}$, we have $H_q(M_{-k}) = 0$ for $q > n-m_{-k}$. Further, our convention on R gives $H_{n-m_{-k}}(M_{-k}) \cong R$. This and (6.5) yield

$$H^{m_{-k}}(M_k) \cong H_{n-m_{-k}}(M_{-k}) \cong R, \quad H^q(M_k) = 0 \text{ for } 0 < q < m_{-k}.$$

Using this and equation (6.6) or (6.7), one gets

$$H^q(M_k) = 0 \text{ for } m_{-k} < q < \mu, \quad H^\mu(M_k) \cong R.$$

Repeated use of formula (6.7) then gives the desired cohomology for M_k. Then, we get the cohomology of M immediately from (6.4) and the fact that M is connected.

Finally, we show that α is an integer. From the formulas for the cohomology of the M_k, we know that

$$\dim M_k \equiv 0 \pmod{\mu} \text{ or } \dim M_k \equiv m_{-k} \pmod{\mu}.$$

We consider two cases. First suppose that

$$\dim M_k \equiv 0 \pmod{\mu} \text{ for } k = 1 \text{ and } k = -1.$$

Then $n-m_+ = j\mu$ and $n-m_- = \ell\mu$, for some integers j, ℓ. Hence,

$$\mu = 2n - (n-m_+) - (n-m_-) = 2n - j\mu - \ell\mu = 2n - (j + \ell)\mu.$$

Thus, $\alpha = 2n/\mu = j + \ell + 1$. (Note that since $|j\mu - \ell\mu| = |m_- - m_+| < \mu$, we have $j = \ell$. Hence α is odd and $m_+ = m_-$.) In the other case, we can assume without loss of generality that $\dim M_+ \equiv m_- \pmod{\mu}$. But then

$$n = \dim M_+ + m_+ \equiv 0 \pmod{\mu},$$

and we see that $\alpha = 2n/\mu$ is, in fact, an even integer.

Q.E.D.

Applications to Dupin hypersurfaces

Suppose M is a compact embedded Dupin hypersurface in S^{n+1} with global field of unit normals ξ. Assume that M has g distinct principal curvature functions $\lambda_1,\ldots,\lambda_g$ given by

$$\lambda_i(x) = \cot \theta_i(x)$$

where $0 < \theta_1 <\ldots< \theta_g$ on M. Let $m_1,\ldots,m_g$ denote the respective constant multiplicities of $\lambda_1,\ldots,\lambda_g$. Let $f_+ : M \to S^{n+1}$ denote the focal map of M onto the first focal submanifold M_+ reached by going out a distance θ_1 from M in the direction ξ. Then M_+ is an immersed submanifold of dimension $n - m_1$. Similarly, let f_- be the focal map onto the first focal submanifold M_- in the direction $-\xi$. Then M_- is an immersed submanifold of dimension $n - m_{-1}$, where $m_{-1} = m_g$.

We now show that M divides S^{n+1} into a union of two ball bundles over M_+ and M_-. We first show that M_+ and M_- lie in different components of $S^{n+1} - M$. Let $p = f_+(x)$ for some x in M. Since the compact Dupin hypersurface M is taut, the spherical distance function L_p has an absolute minimum at x by Lemma 1.24 of Chapter 2. Therefore, the normal geodesic segment $[x,p]$ from x to p does not intersect M (except at x), and so p lies in the component W_+ of $S^{n+1} - M$ to which the normal field points. Hence, M_+ is contained in W_+, and similarly M_- is contained in the other component W_-. Since M is Dupin, the inverse image $f_+^{-1}(p)$ of any focal point p in M_+ consists of a discrete union of m_1-dimensional spheres (leaves of T_1). Each such leaf lies on the n-sphere in S^{n+1} centered at p of radius $r = \min L_p$. But since M has the STPP, this discrete collection must consist of only one leaf L. Thus, the immersion of the space of leaves M/T_1 onto M_+ given by Theorems 4.6 and 4.10 of Chapter 2 is injective, and M_+ is an embedded $(n-m_1)$-dimensional submanifold of S^{n+1}. For each x in the leaf L, the segment $[x,p]$ lies in the closure $\bar{W}_+$ of W_+. Further, if q is any point in $\bar{W}_+ - M_+$, then L_q has a non-degenerate minimum on M, which must be unique by tautness. Hence, q lies on exactly one of the segments $[x, f_+(x)]$ from a point x to the focal point $f_+(x)$ on M_+. Thus

$$\bar{W}_+ = \bigcup_{x \in M} [x, f_+(x)],$$

and the map $\pi: \bar{W}_+ \to M_+$ which takes the segment $[x, f_+(x)]$ to $f_+(x)$ is a ball bundle projection. The same proof shows that $\bar{W}_-$ is a ball bundle over the embedded $(n - m_{-1})$-dimensional focal submanifold M_-. Thus, the Z_2-homology of M, M_+ and M_- is given by Theorem 6.8. If we write the sum of the Z_2-Betti numbers of M as $\beta(M) = 2\alpha$, then α can only assume the values 1,2,3,4 or 6. We now show that α is always equal to the number g of distinct principal curvatures of a compact Dupin hypersurface and hence:

<u>Theorem 6.11</u>: <u>The number g of distinct principal curvatures of a compact embedded Dupin hypersurface M is 1,2,3,4 or 6.</u>

<u>Proof</u>: The normal exponential map $\phi: M \times R \to S^{n+1}$ satisfies the equation $\phi(x, t + 2\pi k) = \phi(x,t)$ for all integers k. Let Γ denote the set of critical points of ϕ. Because of the periodicity of ϕ, it suffices to consider ϕ on $M \times [-\pi,\pi]$, and identify $(x,-\pi)$ with (x,π). On this set, Γ consists of the 2g cross-sections

$$M \times \{\theta_i\}, \quad M \times \{\theta_i - \pi\}, \quad 1 \leq i \leq g.$$

Each of these cross-sections is mapped by ϕ onto a focal submanifold. The set $(M \times [-\pi,\pi]) - \Gamma$ consists of 2g components, each having 2 cross-sections for its boundary. The focal set $\phi(\Gamma)$ is the union of submanifolds of codimension greater than one, and thus $S^{n+1} - \phi(\Gamma)$ is connected. So any component of $\phi^{-1}(S^{n+1} - \phi(\Gamma))$ is a covering of $S^{n+1} - \phi(\Gamma)$ with covering map ϕ. Each component of $(M \times [-\pi,\pi]) - \Gamma$ contains at least one component of $\phi^{-1}(S^{n+1} - \phi(\Gamma))$. Thus, the number 2α of critical points of a non-degenerate L_q is at least as great as the number 2g of components of

$(M \times [-\pi,\pi]) - \Gamma$, i.e., $g \leq \alpha$.

We show that $g = \alpha$. If $g = 1$, then M is a sphere, while if $g = 2$, M must be a cyclide by Theorem 6.2 of Chapter 2. In either case, $g = \alpha$ as desired. Further, if $\alpha = 3$, then since $g \leq \alpha$, we must have $g = 3$, because $g = 1$ or 2 does not give the correct homology. Next suppose that $\alpha = 4$ and $g = 3$. By Theorem 6.8, we have $2n = \alpha(m_1 + m_{-1})$, so that $n = 2(m_1 + m_{-1})$ for $\alpha = 4$. If there are only 3 distinct principal curvature functions $\lambda_1 > \lambda_2 > \lambda_3$ with multiplicities ν_1, ν_2, ν_3, then since $m_1 = \nu_1$ and $m_{-1} = \nu_3$, we must have $\nu_2 = m_1 + m_{-1}$. If we examine the indices of non-degenerate critical points of distance functions with these values of ν_1, ν_2, ν_3, we find that there are no critical points of index $j = m_1 + m_{-1}$. This contradicts the fact that $\beta_j(M) = 1$. One obtains similar contradictions in the cases $\alpha = 6$, $g = 3$ and $\alpha = 6$, $g = 4$. Hence $g = \alpha$ in all cases, as desired.

Q.E.D.

These observations are due to Thorbergsson [1].

7. EXAMPLES OF ISOPARAMETRIC HYPERSURFACES.

We first give a detailed account of the cases with $g = 1$, 2 or 3 distinct principal curvatures. We then describe a particularly nice homogeneous example with four principal curvatures due to Cartan in dimension $n = 4$ and Nomizu for higher dimensions.

Example 7.1: The case $g = 1$.

It is well known that the only umbilic hypersurfaces in S^{n+1} are open subsets of great or small hyperspheres. Each such sphere is the intersection of S^{n+1} with a hyperplane in E^{n+2}. The spheres lying in hypersurfaces perpendicular to a given diameter of S^{n+1} make up an isoparametric family whose focal set consists of the two endpoints (poles) of the diameter.

Let p be a unit vector in E^{n+2} which we will use to determine a diameter of S^{n+1}. Let $F(z) = \langle p,z \rangle$, for $z \in E^{n+2}$, i.e. $F(z)$ is the linear height function ℓ_p. In Section 3, we found that

$$\text{grad}^E F = p, \quad |\text{grad}^E F|^2 = 1, \quad \Delta^E F = 0,$$

so that F satisfies the differential equations of Münzner with $g = 1$, $c = 0$. We consider the level sets

$$M_s = \{z \in S^{n+1} | \langle p,z\rangle = \cos s\}, \quad 0 \leqslant s \leqslant \pi.$$

Except for $M_0 = \{p\}$ and $M_\pi = \{-p\}$, each level set is a Euclidean n-sphere with radius $\sin s$ in the hyperplane situated $1 - \cos s$ units below the north pole p. From the point of view of the intrinsic geometry of S^{n+1}, M_s is the geodesic (distance) hypersphere $S(p,s)$ with center p and radius s. Of course, M_s may also be regarded as $S(-p,\pi-s)$. When $s = \pi/2$, the great sphere M_s is the unique minimal hypersurface in the isoparametric family.

Finally, the family M_s can also be realized as the set of orbits of a group action as follows. Let G_p be the subgroup of $SO(n + 2)$ which leaves the pole p fixed. Then G_p is a naturally embedded copy of $SO(n + 1)$, so that the orbit $G_p z$ has codimension 1 in S^{n+1} whenever $z \neq \pm p$. One easily shows that $G_p z$ is just the geodesic sphere through z with centers $\pm p$.

Example 7.2: The case $g = 2$.

Again, the standard examples of hypersurfaces in S^{n+1} with two constant principal curvatures are well-known. They may be regarded as the Cartesian product of hyperspheres in two complementary orthogonal Euclidean subspaces. Let

$$E^{n+2} = E^{j+1} \times E^{k+1}, \quad j+k = n.$$

For $z = (x,y) \in E^{n+2}$, set $F(z) = |x|^2 - |y|^2$. Then F satisfies

$$\text{grad}^E F = 2(x, -y), \quad |\text{grad}^E F|^2 = 4r^2, \quad \Delta^E F = 2(j-k),$$

so that the Münzner differential equations hold with $g = 2$, $c = 2(j-k)$. Let

$$M_s = \{z \in S^{n+1} | F(z) = \cos 2s\}, \; 0 \leqslant s \leqslant \tfrac{\pi}{2}.$$

Each M_s is the Cartesian product of a j-sphere of (Euclidean) radius cos s and a k-sphere of radius sin s, except for the two focal submanifolds

$$M_0 = \{x,y)|y = 0\} = S^j \times \{0\}, \quad M_{\pi/2} = \{(x,y)|x = 0\} = \{0\} \times S^k.$$

The M_s form a family of parallel hypersurfaces, all of which are tubes over each of the focal submanifolds. Since the focal submanifolds are totally geodesic, it is easy to compute the shape operator of M_s using Theorem 3.2 of Chapter 2. By considering M_s as a tube of radius s over the totally geodesic submanifold $S^j \times \{0\}$, we get that M_s has two constant principal curvatures,

$$\lambda = \cot(\frac{\pi}{2} - s) = \tan s, \quad \mu = -\cot s,$$

with respective multiplicities j, k. Of course, one can also obtain the principal curvatures by a direct calculation, using Theorem 5.1, or the general formula in Theorem 2.3.

The family M_s can also be realized as the set of orbits of a group action. Considering the decomposition of E^{n+2} above, the group $G = SO(j+1) \times SO(k+1)$ is naturally embedded as a subgroup of $SO(n+2)$. Take any point $z = (x,y)$ with x and y both non-zero. Then the isotropy subgroup of the G-action at z is isomorphic to $SO(j) \times SO(k)$, so that the orbit M of z has codimension 2 in E^{n+2} and hence codimension 1 in S^{n+1}. Considering the action of G on each factor, it is clear that the orbit consists of the product of standard spheres of radii $|x|$ and $|y|$, and hence it coincides with M_s where $\cos s = |x|$ and $\sin s = |y|$. Furthermore, the orbit of $(x,0)$ is the focal submanifold M_0 and the orbit of $(0,y)$ is $M_{\pi/2}$.

<u>Example 7.3</u>: <u>The case $g = 3$</u>.

In [2,3], Cartan determined all isoparametric hypersurfaces in spheres with three distinct principal curvatures. All the principal curvatures must have the same multiplicity $\nu = 1,2,4$ or 8 and the family can be defined by the polynomial

$$(7.1)\qquad x^3 - 3xy^2 + \frac{3}{2}x(X\bar{X} + Y\bar{Y} - 2Z\bar{Z}) + \frac{3\sqrt{3}}{2}y(X\bar{X} - Y\bar{Y})$$

$$+ \frac{3\sqrt{3}}{2}(XYZ + \overline{ZYX}).$$

In this formula, x and y are real parameters, while X, Y, Z are coordinates in the algebra F = R, C, Q (quaternions), O(Cayley numbers), respectively, for the cases $\nu = 1,2,4,8$. The sum $XYZ + \overline{ZYX}$ is twice the real part of the product XYZ. In the case $\nu = 8$, multiplication is not associative but the real part of XYZ is the same whether one interprets the product as (XY)Z or X(YZ). Thus, the corresponding isoparametric hypersurfaces are in S^4, S^7, S^{13} and S^{25}, respectively. The focal submanifolds obtained by setting $t = 0$ and $t = \pi/3$ are a pair of antipodal Veronese embeddings of FP^2. For the cases F = R, C and Q, we have the specific parameterization (which differs slightly from that given in Section 9 of Chapter 1) of the focal set M_0 where the polynomial is equal to 1

$$X = \sqrt{3}\,v\bar{w},\quad Y = \sqrt{3}\,w\bar{u},\quad Z = \sqrt{3}\,u\bar{v}$$

$$x = \frac{\sqrt{3}}{2}(|u|^2 - |v|^2),\quad y = |w|^2 - \frac{|u^2| + |v|^2}{2},$$

where u, v, w ε F and $|u|^2 + |v|^2 + |w|^2 = 1$. This map is invariant under the equivalence

$$(u, v, w) \rightarrow (u\lambda, v\lambda, w\lambda),\quad \lambda \varepsilon F,\quad |\lambda| = 1.$$

Thus, it is well-defined on FP^2, and it is easily shown to be an embedding (as in Section 9 of Chapter 1).

Finally, in the case F = R we give an alternate presentation of the family as a space of orbits. (An analogous construction can be made for the other algebras as well). As in Section 9 of Chapter 1, we consider E^9 as the space of 3×3 real matrices with inner product

$$\langle A, B\rangle = AB^t.$$

Let E^5 be the space of symmetric matrices with trace zero, and let S^4 be the unit sphere in E^5, i.e.

$$S^4 = \{A \mid A = A^t, \text{ trace } A = 0, |A| = 1\}.$$

Let SO(3) act on S^4 by conjugation. As in Section 9 of Chapter 1, this action is isometric and thus preserves S^4. For every $A \in S^4$, there exists a $U \in SO(3)$ such that UAU^t is diagonal. In fact, a direct calculation shows that every orbit of this action contains a representative B_t which is a diagonal matrix whose entries are

$$\sqrt{\frac{2}{3}}\ \{\cos(t - \tfrac{\pi}{3}),\ \cos(t + \tfrac{\pi}{3}),\ \cos\ (t + \pi)\}.$$

When all of the eigenvalues are distinct, the orbit is 3-dimensional. A typical example is the case

$$B_{\frac{\pi}{6}} = \text{diagonal } \{1/\sqrt{2},\ 0,\ -\ 1/\sqrt{2}\}.$$

The isotropy subgroup of $B_{\pi/6}$ under this action is the set of matrices in SO(3) which commute with $B_{\pi/6}$. One easily computes that this group is the group of diagonal matrices in SO(3) with entries ±1 along the diagonal. This group is isomorphic to the Klein 4-group $Z_2 \times Z_2$, and hence the orbit $M_{\pi/6}$ is diffeomorphic to $SO(3)/Z_2 \times Z_2$. Note also that $M_{\pi/6}$ is the minimal hypersurface in the family.

The focal submanifolds (lower dimensional orbits) occur when B_t has a repeated eigenvalue, i.e. when t is an integral multiple of $\pi/3$. For example, when $t = 0$, we have

$$B_0 = \text{diagonal } \{1/\sqrt{6},\ 1/\sqrt{6},\ -2/\sqrt{6}\}.$$

The isotropy subgroup for B_0 is the group $S(O(2) \times O(1))$ consisting of all matrices of the form

$$\left[\begin{array}{cc|c} \multicolumn{2}{c|}{\multirow{2}{*}{A}} & 0 \\ & & 0 \\ \hline 0 & 0 & \pm 1 \end{array}\right] \qquad A \in O(2)$$

having determinant 1. Thus, M_0 is diffeomorphic to $SO(3)/S(O(2) \times O(1))$ which is RP^2. In fact, M_0 is a Veronese surface. The other focal submanifold $M_{\pi/3}$ is the orbit of

$$B_{\pi/3} = \text{diagonal } \{2/\sqrt{6}, \ -1/\sqrt{6}, \ -1/\sqrt{6}\}.$$

$M_{\pi/3}$ is also a Veronese surface, antipodal to M_0.

Example 7.4: A homogeneous example with g = 4.

This example in full generality is due to Nomizu [1,2]. For dimension 4, it was given earlier by Cartan [4]. Let

$$(z,w) = \sum_{k=0}^{m} z_k \bar{w}_k$$

be the standard Hermitian inner product on C^{m+1}. The Euclidean metric on $C^{m+1} = E^{2m+2}$ is given by $\langle z,w \rangle = \text{Re}\,(z,w)$. If we write $C^{m+1} = E^{m+1} + iE^{m+1}$, then

$$\langle z,w \rangle = \langle x,u \rangle + \langle y,v \rangle$$

for $z = x + iy$, $w = u + iv$. Consider the function on C^{m+1} (we assume that $m \geq 2$; if $m = 1$ this construction reduces a product of two circles)

$$F(z) = \left|\sum_{k=0}^{m} z_k^2\right|^2 = (|x|^2 - |y|^2)^2 + 4\langle x,y\rangle^2) \text{ for } z = x + iy.$$

One computes that

$$|\text{grad}^E F|^2 = 16\, r^2 F, \quad \Delta^E F = 16\, r^2.$$

Then from Theorem 3.1, we see that the restriction V of F to the unit sphere S^{2m+1} in C^{m+1} satisfies

$$(7.2) \qquad |\text{grad } V|^2 = 16\, V(1-V), \quad \Delta V = 16 - V(16 + 8m).$$

Thus, V is an isoparametric function and by Corollary 5.9, the level hypersurfaces of V are isoparametric. From (7.2), we see that the focal submanifolds occur when V = 0 or 1. Setting V = 1 gives

$$\left|\sum_{k=0}^{m} z_k^2\right| = 1,$$

which is easily seen to imply that z lies in the set

$$M_0 = \{e^{i\theta}x \mid x \in S^m\},$$

where S^m is the unit sphere in the real space E^{m+1}. For $x \in S^m$, we have

$$T_x M_0 = T_x S^m + \text{span } \{ix\}.$$

Hence, the normal space,

$$T_x^{\perp} M_0 = \{iy \mid y \in S^m, \langle x,y\rangle = 0\}.$$

The normal geodesic to M_0 through x in the direction iy is

$$(7.3) \qquad \cos t\, x + \sin t\, iy.$$

At the point $e^{i\theta}x$ in M_0, one shows easily that

$$T^{\perp} M_0 = \{e^{i\theta}y \mid y \in S^m, \ \langle x,y\rangle = 0\}.$$

Hence, the normal geodesic to M_0 through the point $e^{i\theta}x$ in the direction $e^{i\theta}iy$ is

(7.4) $$\cos t\, e^{i\theta}x + \sin t\, e^{i\theta}iy = e^{i\theta}(\cos t\, x + \sin t\, iy).$$

Let $V_{m+1,2}$ denote the Stiefel manifold consisting of all orthonormal pairs of vectors (x,y) in E^{m+1}. From (7.3) and (7.4), we see that the tube of radius t over the focal submanifold M_0 is given by

$$M_t = \{e^{i\theta}(\cos t\, x + \sin t\, iy) \mid (x,y) \in V_{m+1,2}\}.$$

In fact, the map $f_t : S^1 \times V_{m+1,2}$ given by

(7.5) $$f_t(\theta, (x,y)) = e^{i\theta}(\cos t\, x + \sin t\, iy),$$

is an immersion which is a double covering of the tube M_t, since $f_t(0, (x,y)) = f_t(\pi, (-x,-y))$. Substituting (7.5) into the defining formula F shows that for $z \in M_t$,

$$V(z) = (\cos^2 t - \sin^2 t)^2 = \cos^2 2t.$$

Thus, the other focal submanifold occurs when $t = \frac{\pi}{4}$, and from (7.3) and (7.4), we see that it consists of all points of the form $e^{i\theta}(x + iy)/\sqrt{2}$ for (x,y) in $V_{m+1,2}$. On the other hand, setting $F = 0$, we get

$$\sum_{k=0}^{m} z_k^2 = 0, \text{ i.e. } |x| = |y| \text{ and } \langle x,y\rangle = 0.$$

Thus,

$$M_{\frac{\pi}{4}} = \{(x + iy)/\sqrt{2} \mid (x,y) \in V_{m+1,2}\},$$

is just an embedded image of the Stiefel manifold $V_{m+1,2}$, which has dimension $2m - 1$. Since adjacent focal points on a normal geodesic are at a distance $\pi/4$ apart, we know from Münzner's general result (Theorem 2.3) that M_t has 4 distinct principal curvatures

$$(7.6) \qquad \cot t,\ \cot(t + \tfrac{\pi}{4}),\ \cot(t + \tfrac{\pi}{2}),\ \cot(t + \tfrac{3\pi}{4}).$$

Since the focal submanifolds M_0 and $M_{\pi/4}$ have respective dimensions $m + 1$ and $2m - 1$, the principal curvatures in (7.6) must have respective multiplicities m-1, 1, m-1, 1. Of course, this information could have been computed directly from the immersion f_t, as was done by Nomizu. Note that F is not the Cartan polynomial for the family, since it does not satisfy the differential equations of Münzner in Theorem 4.1. On the other hand, as Takagi [1] noted, $\hat{F} = r^4 - 2F$ does satisfy the Münzner equations with $g = 4$ and $c = 8(m-2)$, i.e.,

$$|\text{grad } \hat{F}|^2 = 16\, r^6, \quad \Delta\hat{F} = 8(m - 2)r^2.$$

The hypersurface M_t is also the level set $\hat{F} = \cos 4t$, as expected for a Cartan polynomial. From (7.5), one sees that M_t admits a transitive group of isometries, isomorphic to $SO(2) \times SO(m + 1)$, and hence each M_t is an orbit hypersurface. This is the fifth example in Table 2 of the list of homogeneous isoparametric hypersurfaces of Takagi and Takahashi [1]. Later, Takagi [1] showed that if an isoparametric hypersurface M in S^{2m+1} has 4 distinct principal curvatures with multiplicities 1, m-1, 1, m-1, then M is congruent to an M_t of this example. In particular, such an M is homogeneous.

Finally, we note that each M_t, including each focal submanifold, is invariant under the S^1-action given by multiplication by $e^{i\theta}$. Hence, under the projection $\pi: S^{2m+1} \to CP^m$ of the Hopf fibration, each M_t projects to a submanifold of CP^m. The submanifold $\pi(M_0)$ is a naturally embedded totally geodesic RP^m in CP^m, while $\pi(M_{\pi/4})$ is just the complex quadric hypersurface Q^{m-1} (complex dimension) given by the equation in homogeneous coordinates

$$z_0^2 + \ldots + z_n^2 = 0.$$

The hypersurfaces M_t, $0 < t < \pi/4$, project to tubes over RP^m and Q^{m-1} which are real hypersurfces with 3 distinct constant principal curvatures in CP^m

(see Takagi [3], Cecil-Ryan [7], for more detail).

In the case of the focal submanifold M_0 the mapping $f_0:S^1 \times S^m \to M_0$ given by

$$f_0(\theta,x) = e^{i\theta}x$$

is a double covering of M_0, since the points (θ,x) and $(\theta+\pi, -x)$ are identified. Hence, M_0 can be considered as a quotient manifold with identifications given by f_0. The two spheres, $\{0\} \times S^m$ and $\{\pi\} \times S^m$ are attached via the antipodal map of S^m. Hence, M_0 is orientable if and only if the antipodal map on S^m preserves orientation of S^m. Thus, M_0 is orientable if m is odd and non-orientable if m is even. This illustrates Münzner's general discussion of the orientability of the focal submanifolds in Theorem C of [1, p. 59]. There he shows that in the case $g = 4$, either one of the focal manifolds M_+, M_- is orientable and the other is not, in which case, if M_+ is orientable then the multiplicity $m_+ = 1$, or both focal manifolds are orientable, in which case $m_+ + m_-$ is odd or $m_+ = m_-$ is even.

In the example at hand, $m_+ = 1$, $m_- = m-1$ and $m_+ + m_- = m$. Hence, if m is even, then Münzner's result implies that the focal submanifold $M_- = M_0$ corresponding to the principal curvature of multiplicity $m - 1$ is non-orientable, as we have shown above. On the other hand, when m is odd, then both focal submanifolds are orientable, which is consistent with Münzner's result, although not a consequence of it.

The reader is now referred to the papers mentioned in the introduction of this chapter, in particular, that of Ferus, Karcher and Münzner [1]. In that paper, they point out that isoparametric hypersurfaces are rigid, i.e. each isometry of such a hypersurface can be extended to an isometry of the sphere. Under such an isometry, parallel hypersurfaces are taken to parallel hypersurfaces, and hence the whole family is mapped isometrically onto itself. For $g \geq 4$, this fact follows from the classical rigidity theorem (see, for example, Kobayashi-Nomizu [1, vol. 3, p. 353]) since the rank of the second fundamental form of the hypersurface is at least three. For $g \leq 3$, the result follows from the classification of Cartan (Examples 7.1-7.3).

A surprising consequence of their construction is the existence of two examples with the same multiplicities (and hence the same dimension) which are not congruent by an isometry of the sphere. This can be seen by examining the second fundamental forms of the shape operators. On the other hand, if one takes the hypersurface from each of two such incongruent families at a certain fixed distance from the focal submanifold M_-, then the two hypersurfaces have the same principal curvatures and hence, by the Gauss equation, the same curvature tensor. That is, for any two points of the respective two hypersurfaces, there exists a linear isometry of the corresponding tangent spaces which transforms one curvature tensor into the other. Nevertheless, the two hypersurfaces are not intrinsically isometric, since any such isometry would extend to an isometry of the whole sphere taking one family to the other, and this does not exist.

References

Abresch, U.

[1] Notwendige Bedingungen für isoparametrische Hyperflächen in Sphären mit mehr als drei Hauptkrümmungen, Thesis, Univ. Bonn, 1982.

[2] Isoparametric hypersurfaces with four or six distinct principal curvatures, Math. Ann. 264(1983), 283-302.

Alexandrov, A.D.

[1] On a class of closed surfaces, Recueil Math. (Moscow) 4(1938), 69-77.

Banchoff, T.F.

[1] Tightly embedded 2-dimensional polyhedral manifolds, Amer. J. Math. 87(1965), 462-472.

[2] Critical points and curvature for embedded polyhedra, J. Differential Geometry 1 (1967), 245-256.

[3] The spherical two-piece property and tight surfaces in spheres, J. Differential Geometry 4 (1970), 193-205.

[4] The two-piece property and tight n-manifolds-with-boundary in E^n, Trans. Amer. Math. Soc. 161(1971), 259-267.

[5] High codimensional 0-tight maps on spheres, Proc. Amer. Math. Soc. 29(1971), 133-137.

[6] Tight polyhedral Klein bottles, projective planes and Moebius bands, Math. Ann. 207(1974), 233-243.

[7] Computer animation and the geometry of surfaces in 3- and 4-space, Proc. Internat. Congress of Math., Helsinki, 1978, 1005-1013.

[8] Non-rigidity theorems for tight polyhedra, Archiv der Math. 21(1970), 416-423.

Banchoff, T.F. and Kuiper, N.H.

[1] Geometrical class and degree for surfaces in three space, J. Differential Geometry 16(1981), 559-576.

Blaschke, W.

[1] Vorlesungen über Differentialgeometrie und geometrische Grundlagen von Einsteins Relativitätstheorie, Vol. 3, Springer, Berlin, 1929.

Blomstrom, C.A.

[1] Extrinsically symmetric and planar geodesic isometric immersions in pseudo-Riemannian space forms, Thesis, Brown University, 1984.

Bolton, J.

[1] Transnormal systems, Quart. J. Math. Oxford (2), 24(1973), 385-395.

[2] Tight immersions into manifolds without conjugate points, Quart. J. Math Oxford (2), 33(1982), 159-167.

Boothby, W.

[1] An introduction to differentiable manifolds and Riemannian geometry, Academic Press, New York, 1975.

Borel, A. and Hirzebruch, F.

[1] Characteristic classes and homogeneous spaces, I. Amer. J. Math. 80(1958), 458-538.

Borsuk, K.

[1] Sur la courbure totale des courbes, Ann. de la Soc. Math. Pol. 20(1947), 251-265.

Busemann, H.

[1] Convex surfaces, Wiley-Interscience, New York, 1958.

Cartan, E.

[1] Familles de surfaces isoparamétriques dans les espaces à courbure constante, Annali di Mat. 17(1938), 177-191.

[2] Sur des familles remarquables d'hypersurfaces isoparamétriques dans les espaces sphériques, Math. Z. 45(1939), 335-367.

[3] Sur quelques familles remarquables d'hypersurfaces, C.R. Congrès Math. Liège, 1939, 30-41.

[4] Sur des familles remarquables d'hypersurfaces isoparamétriques des espaces sphériques à 5 et à 9 dimensions, Revista Univ. Tucuman, Serie A, 1(1940), 5-22.

[5] Leçons sur la géométrie des espaces de Riemann, deuxième edition, Gauthiers-Villars, Paris, 1946.

Carter, S.

[1] The focal set of a hypersurface with abelian fundamental group, Topology 12(1973), 1-4.

Carter, S., Mansour, N.G., and West, A.

[1] Cylindrically taut immersions, Math. Ann. 261(1982), 133-139.

Carter S. and Robertson, S.A.

[1] Relations between a manifold and its focal set, Invent. Math. 3(1967), 300-307.

Carter, S. and West, A.

[1] Tight and taut immersions, Proc. London Math. Soc. 25(1972), 701-720.

[2] Totally focal embeddings, J. Differential Geometry 13(1978), 251-261.

[3] Total focal embeddings: special cases, J. Differential Geometry 16(1981), 685-697.

[4] A characterisation of isoparametric hypersurfaces in spheres, J. London Math. Soc. 26(1982), 183-192.

[5] Isoparametric systems and transnormality, 1983 preprint.

Cayley, A.

[1] On the cyclide, Quart. J. of Pure and Appl. Math. 12(1873), 148-165. See also Collected Papers, Vol. 9, 64-78.

Cecil, T.E.

[1] Geometric applications of critical point theory to submanifolds of complex projective space, Nagoya Math. J. 55(1974), 5-31.

[2] A characterization of metric spheres in hyperbolic space by Morse theory, Tohoku Math. J. 26(1974), 341-351.

[3] Taut immersions of non-compact surfaces into a Euclidean 3-space, J.

Differential Geometry 11(1976), 451-459.

Cecil, T.E. and Ryan, P.J.

[1] Focal sets of submanifolds, Pacific J. Math. 78(1978), 27-39.

[2] Focal sets, taut embeddings and the cyclides of Dupin, Math. Ann. 236(1978), 177-190.

[3] Distance functions and umbilic submanifolds of hyperbolic space, Nagoya Math. J. 74(1979), 67-75.

[4] Tight and taut immersions into hyperbolic space, J. London Math. Soc. 19(1979), 561-572.

[5] Conformal geometry and the cyclides of Dupin, Canadian J. Math. 32(1980), 767-782.

[6] Tight spherical embeddings, Proceedings 1979 Berlin Symposium in Global Differential Geometry, Lecture Notes in Mathematics No. 838, Springer-Verlag, Berlin, Heidelberg, New York, 1981, 94-104.

[7] Focal sets and real hypersurfaces in complex projective space, Trans. Amer. Math. Soc. 269(1982), 481-499.

[8] On the number of top-cycles of a tight surface in 3-space, to appear in J. London Math. Soc.

Chen, B.Y.

[1] Geometry of submanifolds, Marcel Dekker, New York, 1973.

Chen, C.S.

[1] On tight isometric immersions of codimension two, Amer. J. Math. 94(1972), 974-990.

[2] More on tight isometric immersions of codimension two, Proc. Amer. Math. Soc. 40(1973), 545-553.

[3] Tight embedding and projective transformation, Amer. J. Math., 101(1979), 1083-1102.

Chen, C.S. and Pohl, W.F.

[1] On classification of tight surfaces in a Euclidean 4-space, preprint.

Chern, S.S.

[1] La géométrie des sous-variétés d'un espace euclidien à plusieurs

dimensions, Enseignement Math. 40(1955), 26-46.

[2] Curves and surfaces in Euclidean space , Studies in global geometry and analysis, MAA Studies in Mathematics 4(1967), 16-56.

Chern, S.S. and Lashof, R.K.

[1] On the total curvature of immersed manifolds I, II, Amer. J. Math. 79(1957), 306-318, Mich. Math. J. 5(1958), 5-12.

Coghlan, L.

[1] Tight stable surfaces in E^3 and E^4, Thesis, Brown University, 1984.

Darboux, G.

[1] Lecons sur la théorie générale des surfaces, 2nd ed., Gauthier-Villars, Paris, 1941.

Dorfmeister, J. and Neher, E.

[1] Isoparametric triple systems of algebra type, Osaka J. Math. 20(1983), 145-175.

[2] An algebraic approach to isoparametric hypersurfaces I, II, Tohoku Math. J. 35 (1983), 187-224 and 225-247.

[3] Isoparametric triple systems of FKM-type I, Abh. Math. Sem. Hamburg 53(1983), 191-216.

[4] Isoparametric triple systems of FKM-type II, Manuscripta Math. 43(1983), 13-44.

[5] Isoparametric hypersurfaces, case g = 6, m=1, 1983 preprint.

Dupin, C.

[1] Applications de géométrie et de méchanique, Paris, 1822.

Eells, J. and Kuiper, N.H.

[1] Manifolds which are like projective planes, Publ. Math. I.H.E.S. 14(1962), 128-222.

Eilenberg, S. and Steenrod, N.

[1] Foundations of algebraic topology, Princeton U. Press, 1952.

Eisenhart, L.

[1] A treatise on the differential geometry of curves and surfaces, Ginn, Boston, 1909.

Fary, I.

[1] Sur la courbure totale d'une courbe gauche faisant un noeud, Bull. Soc. Math. France 77(1949), 128-138.

Fenchel, W.

[1] Über die Krümmung und Windung geschlossener Raumkurven, Math. Ann. 101(1929), 238-252.

Ferus, D.

[1] Über die absolute Totalkrümmung höher-dimensionaler Knoten, Math. Ann. 171(1967), 81-86.

[2] Totale Absolutkrümmung in Differentialgeometrie und-topologie, Lecture Notes in Mathematics no. 66, Springer-Verlag, Berlin, Heidelberg, New York, 1968.

[3] Symmetric submanifolds of Euclidean space, Math. Ann. 247(1980), 81-93.

[4] Notes on isoparametric hypersurfaces, Escola de geometria differencial, Univ. Estadual de Campinas, Brazil, 1980.

[5] The tightness of extrinsic symmetric submanifolds, Math. Z. 181(1982), 563-565.

Ferus, D., Karcher, H., Münzner, H.F.

[1] Cliffordalgebren und neue isoparametrische Hyperflächen, Math. Z. 177(1981), 479-502.

Fialkow, A.

[1] Hypersurfaces of a space of constant curvature, Ann. of Math. 39(1938), 762-785.

Fladt, K. and Baur, A.

[1] Analytische Geometrie spezieller Flächen und Raumkurven, Friedr. Vieweg and Sohn, Braunschweig, 1975.

Fox, R.H.

[1] On the total curvature of some tame knots, Ann. of Math. 52(1950), 258- 261.

Freudenthal, H.

[1] Zur ebenen Oktavengeometrie, Proc Akad. Amsterdam A 56; Indag. Math. 15 (1953), 195-200.

[2] Oktaven, Ausnahmegruppen und Oktavengeometrie, Math. Inst. Univ. Utrecht. Preprint. (1951).

Friedel, G.

[1] Les états mésomorphes de la matière, Ann. de Physique 18(1923), 273-474.

Goetz, A.

[1] Introduction to differential geometry, Addison-Wesley, Reading, Massachusetts, 1970.

Greenberg, M.

[1] Lectures on algebraic topology, Benjamin, New York, 1967.

Hahn, J.

[1] Isoparametrische Hyperflächen in pseudoriemannschen Räumen, Diplomarbeit, Bonn, 1983.

[2] Isoparametric hypersurfaces in the pseudo-Riemannian space forms, 1983 preprint.

Hebda, J.

[1] Manifolds admitting taut hyperspheres, Pacific J. Math. 97(1981), 119-124.

[2] Some new tight embeddings which cannot be made taut, 1983 preprint, St. Louis University.

Hempstead, B.W.

[1] Tight immersions in higher dimensions, Thesis, University of

Minnesota, 1970.

Hilbert, D. and Cohn-Vossen, S.

[1] Geometry and the imagination, Chelsea, New York, 1952.

Hirsch, M.

[1] Differential topology, Springer-Verlag, Berlin, Heidelberg, New York, 1976.

Hopf, H.

[1] Systeme symmetrischer Bilinearformen und euklidsche Modelle der projektiven Räume, Vierteljahrschrift der Naturforschenden Gesellschaft in Zürich 85(1940), 165-177.

Hsiang, W.Y. and Lawson, H.B.

[1] Minimal submanifolds of low cohomogeneity, J. Differential Geometry 5(1971), 1-38.

Hurwitz, A.

[1] Über die Komposition der quadratischen Formen, Math. Ann. 88(1923), 1-25.

Husemöller, D.

[1] Fibre bundles, Second Edition, Springer-Verlag, New York, 1975.

Ikawa, T.

[1] Totally real submanifolds on a flat Kähler space, Tensor 36(1982), 22-28.

[2] On critical point theory of totally real submanifolds, preprint.

[3] Morse functions and submanifolds of hyperbolic space, Rocky Mountain J. Math. 14(1984), 301-303.

Kelly, E.F.

[1] 0-tight equivariant imbeddings of symmetric spaces, Bull. Amer. Math. Soc. 77(1971), 580-583.

[2] Tight equivariant imbeddings of symmetric spaces, J. Differential

Geometry 7(1972), 535-548.

Klein, F.

[1] Vorlesungen über höhere Geometrie, third edition, Springer, Berlin, 1926.

Kobayashi, S.

[1] Imbeddings of homogeneous spaces with minimum total curvature, Tohoku Math. J. 19(1967), 63-70.

[2] Isometric imbeddings of compact symmetric spaces, Tohoku Math. J. 20(1968), 21-25.

Kobayashi, S. and Nagano, T.

[1] On filtered Lie algebras and geometric structures I, J. Math. Mech. 13(1964), 875-907.

Kobayashi, S. and Nomizu, K.

[1] Foundations of differential geometry, I, II, Wiley-Interscience, New York, 1963 and 1969.

Kon, M.

[1] Pseudo-Einstein real hypersurfaces in complex space forms, J. Differential Geometry 14(1979), 339-354.

Kühnel, W.

[1] Ein Productsatz für die zweiten Totalkrummüng, J. London Math. Soc. 14(1976), 357-363.

[2] Total curvature of manifolds with boundary in E^n, J. London Math. Soc. (2) 15(1977), 173-182.

[3] (n-2)-tightness and curvature of submanifolds with boundary, Internat. J. Math. and Math. Sci. 1(1978), 421-431.

[4] Total absolute curvature of polyhedral manifolds with boundary in E^n, Geom. Dedicata 8(1979), 1-12.

[5] Tight and 0-tight polyhedral embeddings of surfaces, Invent. Math. 58(1980), 161-177.

Kühnel, W. and Banchoff, T.F.
[1] The 9-vertex complex projective plane, Mathematical Intelligencer 5(1983), 11-22.

Kuiper, N.H.
[1] Immersions with minimal total absolute curvature, Coll. de géométrie diff., Centre Belge de Recherches Math., Bruxelles, 1958, 75-88.
[2] Sur les immersions a courbure totale minimale, Séminaire de Topologie et Géométrie Différentielle dirigé par Ch. Ehresmann, Faculté des Sciences., Paris, 1959, 1-5.
[3] La courbure d'indice k et les applications convexes, Séminaire de Topologie et Géométrie Différentielle dirigé par Ch. Ehresmann, Faculté des Sciences, Paris, 1960, 1-15.
[4] On surfaces in Euclidean three space, Bull. Soc. Math. Belg. 12(1960), 5-22.
[5] Convex immersions of closed surfaces in E^3, Comm. Math. Helv. 35(1961), 85-92.
[6] On convex maps, Nieuw Archief voor Wisk. 10(1962), 147-164.
[7] Der Satz von Gauss Bonnet für Abbildungen in E^n und damit verwendte Probleme, Jahr. der DMV 69(1967), 77-88.
[8] Minimal total absolute curvature for immersions, Invent. Math. 10(1970), 209-238.
[9] Morse relations for curvature and tightness, Proc. Liverpool Singularities Symp. II (Ed. by C.T.C. Wall) Springer Lecture Notes in Mathematics 209, 1971, 77-89.
[10] Tight topological embeddings of the Moebius band, J. Differential Geometry 6(1972), 271-283.
[11] Stable surfaces in Euclidean three space, Math. Scand. 36(1975), 83-96.
[12] Curvature measures for surfaces in E^n, Lobachevski Colloquium. Kazan, USSR (1976).
[13] Tight embeddings and maps. Submanifolds of geometrical class three in E^n, The Chern Symposium 1979 (Proc. Internat. Sympos., Berkeley, Calif., 1979) pp. 97-145, Springer-Verlag, Berlin, Heidelberg, New York, 1980.
[14] Taut sets in three space are very special, to appear in Topology.

[15] Polynomial equations for tight surfaces, Geom. Dedicata, 15(1983), 107-113.

[16] There is no tight continuous immersion of the Klein bottle into R^3, 1983 preprint.

[17] Geometry in total absolute curvature theory, to appear in 40-year Oberwolfach Anniversary Volume (1984).

Kuiper, N.H. and Meeks III, W.

[1] Total curvature for knotted surfaces, Invent. Math. 77(1984), 25-69.

Kuiper, N.H. and Pohl, W.F.

[1] Tight topological embeddings of the real projective plane in E^5, Invent. Math. 42(1977), 177-199.

Lancaster, G.

[1] Canonical metrics for certain conformally Euclidean spaces of dimension three and codimension one, Duke Math. J. 40(1973), 1-8.

Langevin, R. and Rosenberg, H.

[1] On curvature integrals and knots, Topology 15(1976), 405-416.

Lane, E.P.

[1] A treatise on projective differential geometry, University of Chicago Press, Chicago, 1942.

Lastufka, W.S.

[1] Tight topological immersions of surfaces in Euclidean space, J. Differential Geometry 16(1981), 373-400.

Lawson, H.B.

[1] Rigidity theorems in rank-1 symmetric spaces, J. Differential Geometry 4(1970), 349-357.

Levi-Civita, T.

[1] Famiglie di superficie isoparametrische nell'ordinario spacio euclideo, Atti Accad. naz. Lincei. Rend. Cl. Sci. Fis. Mat. Natur 26(1937), 355-362.

Lilienthal, R.

[1] Besondere Flächen, Encyklopädie der Math. Wissenschaften, Vol. III, 3, 269-354, B.G. Teubner, Leipzig, 1902-1927.

Liouville, J.

[1] Note au sujet de l'article précédent, J. de Math. Pure et Appl. (1) 12(1847), 265-290.

Little, J.A.

[1] On singularities of submanifolds of higher dimensional Euclidean spaces, Annali di Mat. pura ed appl. (IV) 83(1969), 261-336.

[2] Manifolds with planar geodesics, J. Differential Geometry 11(1976), 265-285.

Little, J.A. and Pohl, W.F.

[1] On tight immersions of maximal codimension, Invent. Math. 13(1971), 179-204.

Maeda, Y.

[1] On real hypersurfaces of a complex projective space, J. Math. Soc. Japan 28(1976), 529-540.

Magid, M.

[1] Lorentzian isoparametric hypersurfaces, 1983 preprint.

Massey, W.

[1] Algebraic topology: an introduction, Harcourt, Brace and World, New York, 1967.

Maunder, C.R.F.

[1] Algebraic topology, Van Nostrand Reinhold, London, 1970.

Maxwell, J.C.

[1] On the cyclide, Quarterly J. of Pure and Applied Math. 34 (1867). See also Collected Works, Vol. 2, p 144-159.

Meeks III, W.

[1] The topological uniqueness of minimal surfaces in three dimensional Euclidean space, Topology 20(1981), 389-410.

Milnor, J.W.

[1] On the total curvature of knots, Ann. of Math. 52(1950), 248-257.

[2] On the total curvature of closed spaces curves, Math. Scand. 1(1953), 289-296.

[3] Morse Theory, Ann. Math. Stud. 51, Princeton U. Press, 1963.

[4] Topology from the differentiable viewpoint, U. of Virginia Press, Charlottesville, Va., 1965.

[5] On manifolds homeomorphic to the 7-sphere, Ann. of Math. 64(1956), 399-405.

Miyaoka, R.

[1] Minimal hypersurfaces in the space form with three principal curvatures, Math. Z. 170(1980), 137-151.

[2] Complete hypersurfaces in the space form with three principal curvatures, Math. Z. 179(1982), 345-354.

[3] Compact Dupin hypersurfaces with three principal curvatures, 1983 preprint.

Morse, M.

[1] The existence of polar non-degenerate functions on differentiable manifolds, Ann. of Math. 71(1960), 352-383.

Morse, M. and Cairns, S.

[1] Critical point theory in global analysis and differential topology, Academic Press, New York, 1969.

Morton, H.R.

[1] A criterion for an embedded surface in R^3 to be unknotted, Springer Lecture Notes in Mathematics No. 722, 1977, 93-98.

Münzner, H.F.

[1] Isoparametric Hyperflächen in Sphären, I and II, Math. Ann. 251(1980), 57-71 and 256(1981), 215-232.

Nagano, T.

[1] Transformation groups on compact symmetric spaces, Trans. Amer. Math. Soc. 118(1965), 428-453.

Nirenberg, L.

[1] Rigidity of a class of closed surfaces, Non-linear problems, Ed. R.E. Langer, Proc. Symposium Math. Research Center, U. Wisconsin, Madison 1962, 177-193.

Nomizu, K.

[1] Some results in E. Cartan's theory of isoparametric families of hypersurfaces, Bull. Amer. Math. Soc. 79(1973), 1184-1188.

[2] Elie Cartan's work on isoparametric families of hypersurfaces, Proc. Symposia in Pure Math., Amer. Math. Soc. 27(Part I), (1975), 191-200.

[3] Characteristic roots and vectors of a differentiable family of symmetric matrices, Lin. and Multilin. Alg. 2(1973), 159-162.

[4] On isoparametric hypersurfaces in the Lorentzian space forms, Japan. J. Math. 7(1981), 217-226.

Nomizu, K. and Rodriguez, L.L.

[1] Umbilical submanifolds and Morse functions, Nagoya Math. J. 48(1972), 197-201.

Nomizu, K. and Smyth, B.

[1] Differential geometry of complex hypersurfaces II, J. Math. Soc. Japan 20(1968), 498-521.

O'Neill, B.

[1] The fundamental equations of a submersion, Mich. Math. J. 13(1966), 459-469.

Otsuki, T.

[1] Minimal hypersurfaces in a Riemannian manifold of constant curvature,

Amer. J. Math. 92(1970), 145-173.

[2] Minimal hypersurfaces with three principal curvature fields in S^{n+1}, Kodai Math. J. 1(1978), 1-29.

Ozawa, T.

[1] On tight PL-manifolds, 1982 preprint.

[2] Relations between tightness and k-tightness for PL-manifolds in Euclidean spaces, 1982 preprint.

[3] Curvatures of PL-complexes in R^N and tightness, 1982 preprint.

[4] Products of tight continuous functions, Geom. Dedicata 14(1983), 209-213.

Ozeki, H. and Takeuchi, M.

[1] On some types of isoparametric hypersurfaces in spheres I and II, Tohoku Math. J. 27(1975), 515-559 and 28(1976), 7-55.

Palais, R.

[1] A global formulation of the Lie theory of transformation groups, Mem. Amer. Math. Soc. 22(1957).

Pinkall, U.

[1] Dupin'sche Hyperflächen, Dissertation, Univ. Freiburg, 1981.

[2] Dupin'sche Hyperflächen in E^4, 1983 preprint.

[3] Hypersurfaces of Dupin, 1983 preprint.

[4] Curvature properties of taut submanifolds, 1983 preprint.

Reckziegel, H.

[1] Krümmungsflächen von isometrischen Immersionen in Räume konstanter Krümmung, Math. Ann. 223(1976), 169-181.

[2] On the eigenvalues of the shape operator of an isometric immersion into a space of constant curvature, Math. Ann. 243(1979), 71-82.

[3] Completeness of curvature surfaces of an isometric immersion, J. Differential Geometry 14(1979), 7-20.

Robertson, S.A.

[1] Generalized constant width for manifolds, Mich. Math. J. 11(1964), 97-105.

[2] Smooth curves of constant width and transnormality, Bull. London Math. Soc. 16(1984), 264-274.

Rodriguez, L.L.

[1] The two-piece property and convexity for surfaces with boundary, J. Differential Geometry 11(1976), 235-250.

[2] The two-piece property and relative tightness for surfaces with boundary, Thesis, Brown University, 1973.

Ryan, P.J.

[1] Homogeneity and some curvature conditions for hypersurfaces, Tohoku Math. J. 21(1969), 363-388.

[2] Hypersurfaces with parallel Ricci tensor, Osaka J. Math. 8(1971), 251-259.

Salmon, G.

[1] A treatise on the analytic geometry of three dimensions, Vol. 1, Chelsea, New York, 1958.

Samelson, H.

[1] Orientability of hypersurfaces in R^n, Proc. Amer. Math. Soc. 22(1969), 301-302.

Schouten, J.A.

[1] Über die konforme Abbildung n-dimensionaler Mannigfaltigkeiter mit quadratischer Maßbestimmung auf eine Mannigfaltigkeit mit euklidischer Maßbestimmung, Math. Z. 11(1921), 58-88.

Segre, B.

[1] Famiglie di ipersuperficie isoparametrische negli spazi euclidei ad un qualunque numero di demensioni, Atti Accad. naz Lincie Rend. Cl. Sci. Fis. Mat. Natur 27(1938), 203-207.

Segre, C.

[1] Su un classe di superficie degl'iperspazi legata colle equazioni

lineari alle derivate parziale di 2° ordine, Torino Atti 42(1907), 1047.

[2] Le linee principali di una superficie di S_5 e une proprietà caratteristica della superficie di Veronese, Rom. Acc. L. Rend. (5) 30(1921), 200 and 227.

Smyth, B.

[1] Differential geometry of complex hypersurfaces, Ann. of Math. (2) 85(1967), 246-266.

Stoker, J.

[1] Differential geometry, Wiley, New York, 1969.

Tai, S.S.

[1] Minimum embeddings of compact symmetric spaces of rank one, J. Differential Geometry 2(1968), 55-66.

Takagi, R.

[1] A class of hypersurfaces with constant principal curvatures in a sphere, J. Differential Geometry 11(1976), 225-233.

[2] On homogeneous real hypersurfaces in a complex projective space, Osaka J. Math. 10(1973), 495-506.

[3] Real hypersurfaces in a complex projective space with constant principal curvatures I and II, J. Math. Soc. Japan 27(1975), 43-53, and 27(1975), 507-516.

Takagi, R. and Takahashi, T.

[1] On the principal curvatures of homogeneous hypersurfaces in a sphere, Differential Geometry in honor of K. Yano, Kinokuniya, Tokyo, 1972, 469-481.

Takeuchi, M. and Kobayashi, S.

[1] Minimal imbeddings of R-spaces, J. Differential Geometry 2(1968), 203-215.

Terng, C.L.

[1] Isoparametric submanifolds and their Coxeter groups, 1984 preprint.

Teufel, E.

[1] Totale Krümmung und totale Absolutkrümmung in der sphärischen Differentialgeometrie und Differentialtopologie, Thesis, Univ. Stuttgart, 1979.

[2] Eine differentialtopologische Berechnung der Totalen Krümmung und Totalen Absolutkrümmung in der Sphärischen Differentialgeometrie, Manuscripta Math. 31(1980), 119-147.

[3] Anwendungen der differentialtopologischen Berechnung der Totalen Krümmung und Totalen Absolutkrümmung in der sphärischen Differentialgeometrie, Manuscripta Math. 32(1980), 239-262.

[4] Differential topology and the computation of total absolute curvature, Math. Ann. 258(1982), 471-480.

Thom, R.

[1] Les singularites des applications differentiables, Ann. Inst. Fourier (Grenoble) 6 (1955-56), 43-87.

Thomas, T.Y.

[1] On closed spaces of constant mean curvature, Amer. J. Math. 58(1936), 702-704.

[2] Extract from a letter of E. Cartan concerning my note: On closed spaces of constant mean curvature, Amer. J. Math. 59(1937), 793-794.

Thorbergsson, G.

[1] Dupin hypersurfaces, Bull. London Math. Soc. 15(1983), 493-498.

[2] Highly connected taut submanifolds, Math. Ann. 265(1983), 399-405.

[3] Tight immersions of highly connected manifolds, 1984 preprint.

Valentine, F.

[1] Convex sets, Robt. E. Krieger, Huntington, N.Y. 1976.

Voss, K.

[1] Eine Verallgemeinerung der Dupinschen Zykliden, Tagungsbericht

41/1981, Geometrie, Mathematisches Forschungsinstitut, Oberwolfach, 1981.

van der Waerden, B.L.

[1] Modern algebra, Ungar, New York, 1949.

Weinstein, A.

[1] Distance spheres in complex projective spaces, Proc. Amer. Math. Soc. 39(1973), 649-650.

Willmore, T.J.

[1] Tight immersions and total absolute curvature, Bull. London Math. Soc. 3(1971), 129-151.

[2] Total curvature in Riemannian geometry, Ellis Horwood Limited, West Sussex, England, 1982.

Wilson, J.P.

[1] The total absolute curvature of an immersed manifold, J. London Math. Soc. 40(1966), 362-366.

[2] Some minimal imbeddings of homogeneous spaces, J. London Math. Soc. 1(1969), 335-340.

Yamaguchi, S. and Ikawa, T.

[1] On the index theorem of Sasakian submanifolds in an odd-dimensional sphere, TRU-Math. 18(1982), 61-71.

Summary of notation

The following is a list of notations which are used frequently in the book. and whose meaning is usually assumed to be known.

1. R and C denote the real and complex number fields, respectively.

E^m:	vector space of m-tuples of real numbers $(x_1,\ldots,x_m)$
C^m:	vector space of m-tuples of complex numbers $(z_1,\ldots,z_m)$
$\langle x,y\rangle$:	standard inner product $\sum_{i=1}^{m} x_i y_i$ on E^m
$\lvert x\rvert$:	length of the vector $x = \langle x,x\rangle^{1/2}$
S^m:	unit sphere in E^{m+1}
H^m:	m-dimensional hyperbolic space of constant sectional curvature −1.
D^m:	m-dimensional unit disk in E^m
RP^m:	m-dimensional real projective space
CP^m:	m-dimensional complex projective space
QP^m:	m-dimensional quaternionic projective space
GL(n;R):	general linear group
O(n):	orthogonal group for the standard inner product on E^n
SO(n):	special orthogonal group
grad F:	gradient vector field of a function $F:E^m \to R$
ΔF:	Laplacian of $F:E^m \to R$
$\mathrm{grad}^S F$:	gradient of the restriction of F to the unit sphere S^{m-1}
$\Delta^S F$:	Laplacian of the restriction of F to S^{m-1}
Z:	the ring of integers
Z_2:	the field of integers modulo 2

T^2: 2-dimensional torus

K^2: 2-dimensional Klein bottle

2. M is smooth n-dimensional manifold and $\phi: M \to R$ a smooth function.

$M_r(\phi)$: $\{x \in M \mid \phi(x) \leqslant r\}$

$M_r^-(\phi)$: $\{x \in M \mid \phi < r\}$

$M_r^+(\phi)$: $\{x \in M \mid \phi(x) > r\}$

$H_k(M;F)$: k-th homology module with coefficients in the field F

$\beta_k(M;F)$: $\dim_F H_k(M;F)$, the k-th F-Betti number of M

$\beta(M;F)$: $\sum_{k=0}^{n} \beta_k(M;F)$, sum of the F-Betti numbers

$\beta_k(\phi,r,F)$: $\dim_F(H_k(M_r(\phi);F))$

$\mu_k(\phi,r)$: number of critical points of ϕ on $M_r(\phi)$ of index k

$\mu_k(\phi)$: number of critical points of ϕ on M of index k

$\mu(\phi)$: number of critical points of ϕ on a compact manifold M,

$$\mu(\phi) = \sum_{k=0}^{n} \mu_k(\phi)$$

$\gamma(M)$: Morse number of M = min $\mu(\phi)$, ϕ is a Morse function on M

$\chi(M)$: Euler characteristic of M

[X,Y]: Lie bracket of two vector fields X and Y on M

3. For a manifold M immersed in a Riemannian manifold $(\tilde{M},g)$.

f: the immersion $f: M \to \tilde{M}$

$T_x M$: tangent space to M at x

$T_x^\perp M$: normal space to M at x

TM: tangent bundle of M

NM: normal bundle of M

BM: bundle of unit normals to M

$\tilde{\nabla}_X$: covariant differentiation with respect to a vector X for Levi-Civita connection $\tilde{\nabla}$ of $\tilde{M}$

∇_X: covariant derivative for induced metric of M

D_X: covariant derivative in Euclidean space

ξ: unit normal vector (field) on M

$\nabla^\perp$: normal connection

$\alpha(X,Y)$: second fundamental form defined as a mapping

$$T_xM \times T_xM \to T_x^\perp M$$

A_ξ: shape operator determined by normal vector ξ, i.e. the endomorphism of T_xM satisfying

$$g(A_\xi X, Y) = g(\alpha(X,Y), \xi)$$

A: shape operator for a hypersurface (determined by unit normal field ξ)

h: mean curvature of a hypersurface $= \frac{1}{n}$ trace A, n = dim M

K: Gaussian curvature of a surface

$\lambda(x)$: principal curvature of a hypersurface M at a point x

$T_\lambda(x)$: eigenspace for a principal curvature λ at x

T_λ: principal foliation for a principal curvature λ of constant multiplicity

M/T_λ: space of leaves of a principal foliation

f_λ: focal map corresponding to a principal curvature function λ

$f_\lambda(M)$: image of f_λ; sheet of the focal set of M determined by λ

4. For an immersion $f:M \to E^m$.

$\tau(M;f)$: total absolute curvature of f

ℓ_p: linear height function $\ell_p:E^m \to R$, $\ell_p(q) = \langle p,q\rangle$; restriction $\ell_p:M \to R$, $\ell_p(x) = \langle p,f(x)\rangle$

$\mu(p)$: number of critical points of ℓ_p on M

$\Omega(p)$: top-set of $\ell_p = \{x \in M | \ell_p(x) = \max_{y \in M} \ell_p(y)\}$

$\Omega(p,q)$: top^2-set $= \{x \in \Omega(p) | \ell_q(x) = \max_{y \in \Omega(p)} \ell_q(y)\}$

L_p: distance function $L_p : E^m \to R$, $L_p(q) = |p-q|^2$;

restriction $L_p : M \to R$, $L_p(x) = |p-f(x)|^2$

$Hf(M)$: convex hull of $f(M)$

$\partial Hf(M)$: boundary of convex hall of $f(M)$

ν: Gauss map, $\nu : BM \to S^{m-1}$

Index